©1996

GRAD LEVEL

Particle Physics at the New Millennium

Springer
*New York
Berlin
Heidelberg
Barcelona
Budapest
Hong Kong
London
Milan
Paris
Santa Clara
Singapore
Tokyo*

Byron P. Roe

Particle Physics at the New Millennium

With 134 Figures
Including Nine in Color

 Springer

Byron P. Roe
Randall Lab, Dept. of Physics
University of Michigan
Ann Arbor, MI 49109

On the cover: A historic photograph of a neutrino interaction involving the weak neutral current $\nu N \to \nu X$, seen in the large heavy-liquid bubble chamber Gargamelle at CERN. See also page 182.

Library of Congress Cataloging-in-Publication Data
Roe, Byron P.
  Particle physics at the new millennium/Byron P. Roe.
    p. cm.
  Includes bibliographical references and index.
  ISBN 0-387-94615-2 (hardcover)
  1. Particles (Nuclear physics) 2. Nuclear physics. I. Title.
QC793.2.R64 1996
539.7′2—dc20   95-44879

Printed on acid-free paper.

© 1996 Springer-Verlag New York, Inc.
All rights reserved. This work may not be translated or copied in whole or in part without the written permission of the publisher (Springer-Verlag New York, Inc., 175 Fifth Avenue, New York, NY 10010, USA), except for brief excerpts in connection with reviews or scholarly analysis. Use in connection with any form of information storage and retrieval, electronic adaptation, computer software, or by similar or dissimilar methodology now known or hereafter developed is forbidden.
The use of general descriptive names, trade names, trademarks, etc., in this publication, even if the former are not especially identified, is not to be taken as a sign that such names, as understood by the Trade Marks and Merchandise Marks Act, may accordingly be used freely by anyone.

Production managed by Natalie Johnson; manufacturing supervised by Jacqui Ashri.
Photocomposed using the author's LaTeX files.
Printed and bound by R.R. Donnelley and Sons, Harrisonburg, VA.
Printed in the United States of America.

9 8 7 6 5 4 3 2 1

ISBN 0-387-94615-2 Springer-Verlag New York Berlin Heidelberg   SPIN 10518233

Preface

P.385 = BIBLIOGRAPHY

The twentieth century has been a golden age of physics. The discoveries of relativity and of quantum mechanics have been two philosophical revolutions, each of which has fundamentally changed humanity's view of the world. Layer after layer has been peeled off the structure of matter, from molecules to atoms to nuclei to neutrons and protons to quarks. The weak and strong interactions have been discovered, and the weak interactions have been found to be deeply intertwined with electromagnetism. The weak, electromagnetic, and strong interactions have been found to obey a profound symmetry, which only becomes apparent at very high energies. The Standard Model has been developed, which essentially explains all present data. Particle physics at the edge of the millennium stands on a very high plateau.

Nonetheless, the Standard Model is known to be only a way station; it cannot be complete. Some physicists believe that the great quest is almost over and that early in the millennium the fundamental laws of nature will be completely laid bare. Others suspect that there will be still more revolutions, comparable to the development of quantum mechanics and relativity, ahead of us.

This book is intended to provide physics graduate students with an overview of the field at this splendid time. For experimental physicists, this overview may be sufficient. Theorists may desire to supplement this volume with a more theoretically inclined text such as the one by T.D. Lee.[1] The text by R.N. Cahn and G. Goldhaber[2] is recommended as an excellent reprint volume, containing copies of many historically important publications in this field.

There are, basically, two ways of approaching the subject, which can be called the "top down" and "bottom up" methods. In the "top down" method, one starts from the basic postulates of present theory and derives the experimental phenomena. In the treatment of electromagnetism, this corresponds to starting with Maxwell's equations and deriving the many electromagnetic phenomena we see.

For particle physics, the author strongly prefers the "bottom up" approach. In this approach, the basic postulates are constructed piece by piece, and at each postulate the evidence driving one to that postulate is given.

The present Standard Model is very beautiful and it might be right, if certainly not complete. However, as we have seen many times in physics, a beautiful present day theory may not survive, and a new theory then will be built using fragments from the old one. In this situation it is essential for a working physicist to have a clear idea of what evidence pushes what conclusions.

The author has tried to introduce mainly the formalism needed to test hypotheses and look at the data. The student should have had a course in quantum mechanics and should be familiar with special relativity and with matrix manipulations. A knowledge of elementary group theory would be useful, but is not required. No other special knowledge is assumed.

Feynman diagrams are introduced to be used, for much of the book, only in a qualitative (but still powerful!) way. However, in Chapter 12, the set of rules needed to write down a matrix element from Feynman diagrams and a Lagrangian is described and illustrated with several examples. Dirac notation and relativistic quantum methods are introduced. This book in no way substitutes for a course in relativistic quantum field theory. However, a minimum number of concepts are introduced where needed to allow contact with data.

For a one semester course, chapters 12–17 and perhaps chapters 2–3 could be omitted, or left for individual reading. If desired, sections 6.5, 6.6, 7.3, 7.13, 8.4, 9.5, 10.2, 10.3, and 11.7–11.14 could also be left out without affecting the rest of the course.

The author wishes to thank Gail Hanson who taught from a preliminary version of the text and provided the author with extremely valuable feedback, including corrections, additional material, and exercises. In addition, she read the final manuscript for errors.

The author also wishes to thank M. Veltman, R. Akhoury, J. Wells, and W. Loinaz for their very great help in elucidating the extent to which the form of the Standard Model is fixed by the need to avoid bad high energy behavior. He wishes to thank Y.-P. Yao for his help on a number of theoretical questions, for his critical reading of Chapter 11 and his valuable suggestions for improvements.

The author also wishes to thank B. Demkowski and R.C. Ball for their enormous help in preparing the text.

Contents

	Preface		v
1.	**Preliminaries**		1
	1.1	Introduction	1
	1.2	The Fundamental Particles and Forces, Fermions, Bosons	1
		1.2 a. The Particles	2
		1.2 b. The Interactions	3
		1.2 c. Comments on Particles and Forces	4
		1.2 d. Some Common, Strongly Interacting Particles	5
	1.3	Relativity	6
	1.4	Conventions and Units	9
	1.5	Mandelstam Variables–s, t, u	10
	1.6	Feynman Diagrams	11
	1.7	Exercises	16
2.	**Electromagnetic Interactions Used for Detection of Charged Particles**		19
	2.1	Rutherford Scattering	19
	2.2	Multiple Scattering	25
	2.3	Bremsstrahlung	27
	2.4	Conversion of γ-Rays to e^+e^- Pairs	32
	2.5	Direct Pair Production	33
	2.6	Cherenkov Radiation	33
	2.7	Exercises	35

viii   Contents

3. Particle Accelerators and Detectors 37
   3.1   Particle Accelerators 37
   3.2   Present High-Energy Accelerators 38
         3.2 a. U.S. Accelerators 38
         3.2 b. Accelerators in Western Europe 41
         3.2 c. Accelerators Elsewhere 43
   3.3   A Quick Survey of Some Major Types of Detectors ... 44
         3.3 a. Detectors Measuring Ionization 45
         3.3 b. Other Types of Particle Detectors 48
         3.3 c. Calorimeters 51
   3.4   Exercises 54

4. Invariance, Symmetries, and Conserved Quantities ... 57
   4.1   Preliminary Discussion 57
         4.1 a. Invariance Formalisms 57
         4.1 b. Conserved Electromagnetic Current 58
   4.2   Translation Invariance 59
   4.3   Rotation Invariance 60
   4.4   Time Invariance 61
   4.5   Parity 61
   4.6   Time Reversal Invariance 63
         4.6 a. Detailed Balance 63
   4.7   Illustration of Uses of Invariance Principles: Pions 64
         4.7 a. Pion Spin 65
         4.7 b. Pion Intrinsic Parity 67
   4.8   Charge Conservation 70
   4.9   Gauge Invariance 71
   4.10 Charge Conjugation 73
   4.11 Illustration of Charge Conjugation: Particle–Anti-Particle Pairs 75
   4.12 P, C, and T. Who Is Invariant Under What? 76
   4.13 Isotopic Spin Invariance 77

		4.13 a. The Two-Nucleon System and Multinucleons	78
	4.14	Illustration of I-Spin Considerations: Pions and the Pion Nucleon System	79
	4.15	Further Use of I-Spin: Nucleon–Anti-Nucleon Pairs	82
	4.16	Electromagnetism and Isotopic Spin	83
	4.17	G-Parity	84
	4.18	Strangeness	85
		4.18 a. K^0 and $\overline{K}^0$ Decays	87
		4.18 b. K^0–$\overline{K}^0$ Oscillations	89
		4.18 c. CP Violation	90
	4.19	Exercises	93
5.	**Hadron–Hadron Scattering**		97
	5.1	Wave Optics Discussion of Hadron Scattering	97
	5.2	Breit–Wigner Resonance Formula	100
	5.3	S-Matrix, Phase Space, and Dalitz Plots	103
	5.4	Exercises	108
6.	**The Quark Model**		110
	6.1	Introduction	110
	6.2	Baryons	113
	6.3	Mesons	120
	6.4	Simple Model Relations	123
	(6.5)	Mass Relations	125
	(6.6)	Baryon Magnetic Moments	128
	6.7	Heavy Quarks	129
	6.8	Color	138
	6.9	Free Quarks	143
	6.10	Exercises	144
7.	**Weak Interactions**		147
	7.1	Brief Outline of Dirac Theory	147
		7.1 a. Second Quantization	151

	7.2	Nuclear β-Decay 153
		7.2 a. Parity Considerations 153
		7.2 b. Nuclear β-Decay Summary 155
	7.3	Kurie Plots; Neutrino Mass Limits 159
	7.4	Zero Mass Neutrinos 161
	7.5	Evaluating Matrix Elements; Traces 163
	7.6	$\pi \to l + \bar{\nu}$ Decay 165
	7.7	Further Weak Interaction Results; Several Neutrinos; Current–Current? 168
	7.8	Muon Decay 171
	7.9	Conserved Vector Current (CVC) 174
	7.10	Strangeness-Changing Weak Decays 178
		7.10 a. Cabibbo Hypothesis 178
		7.10 b. Non-Leptonic Weak Interactions 179
	7.11	Neutral Currents 180
	7.12	Generalized Cabibbo Theory: The Cabibbo–Kobayashi–Maskawa Matrix 184
	7.13	Penguin Diagrams; Applications to $\Delta I = 1/2$ and to CP Violation 187
	7.14	Intermediate Bosons 188
	7.15	Neutrino Oscillations 189
	7.16	Exercises 192
8.	**Elastic and Inelastic Scattering** 196	
	8.1	Electron–Muon Elastic Scattering 197
	8.2	Muon Pair Production in e^+e^- Annihilation 199
	8.3	Electron–Proton Elastic Scattering 200
	8.4	Geometrical Interpretation of Form Factors 203
	8.5	Electron–Proton Inelastic Scattering 204
	8.6	Inelastic Charged Current ν–p Scattering 206
	8.7	Form Factors in the Quark–Parton Picture 208
	8.8	Exercises 225

9. The Strong Interaction: Quantum Chromodynamics and Gluons 228

9.1 Evidence for Gluons 228

9.2 Color Forces 229

9.3 Are Quarks Fractionally Charged? 233

9.4 Asymptotic Freedom; $\alpha_s(Q^2)$ 234

9.5 Q^2 Evolution of Structure Functions 242

9.6 Exercise 245

10. The Standard Model 246

10.1 $SU(2) \otimes U(1)$ Weinberg–Salam Model; Neutral Currents 246

10.2 Weak Neutral Current Neutrino Interactions 251

10.3 Triangle Anomalies 252

10.4 Exercises 254

11. Spontaneous Symmetry Breaking: The Higgs Mechanism 255

11.1 Gauge Theory in QED for the Dirac Equation 256

11.2 Non-Abelian Gauge Transformations 257

11.3 Spontaneous Symmetry Breaking 258

11.4 Combining Gauge Invariance and Spontaneous Symmetry Breaking 260

11.5 The Standard Model: Gauge Theory and Symmetry Breaking Applied to the Weinberg–Salam Model 262

11.6 The Standard Model Lagrangian 265

11.7 Introduction to Study of Unitarity Restrictions 268

11.8 Preliminary Definitions and Assumptions 269

11.9 High-Energy Behavior 269

11.10 Further Feynman Rules and Other Needed Formalities 270

11.11 Neutrino Electron Scattering 272

11.12 $\nu\bar{\nu} \to W^+W^-$ 274

11.13 The GIM Mechanism 278

11.14 Comments on These Results 278

11.15	How Constant are the Coupling Constants?	280
11.16	Exercises	280

12. Extensions of the Standard Model: Grand Unification .. 281

12.1	Introduction	281
12.2	The Group $SU(5)$	283
12.3	Further Unification	288
12.4	Exercises	291

13. Physics at the Z 292

13.1	Exercises	313

14. High-Energy Processes at Low Q^2 314

14.1	Pomeranchuk Theorem; Froissart Bound; Regge Analysis	315
14.2	Phenomenology of High-Energy Collisions	319
14.3	Total Cross Sections and Elastic Scattering	323
14.4	Multiplicity Distributions	327
14.5	Exercises	327

15. Heavy Quark Effective Field Theory 328

15.1	Mass Relations	331
15.2	Form Factors for Weak Decay	332
15.3	Heavy Baryon Semileptonic Decays	337
15.4	Pseudoscalar and Vector Meson Decay Constants	337
15.5	Comments on Corrections to Zeroth Order	339
15.6	Determining CKM Matrix Elements	339
15.7	Exercises	340

16. Monopoles 341

16.1	Monopoles in Electromagnetic Theory	341
16.2	Monopoles in Extended Gauge Theories	343
16.3	t'Hooft Polyakov Monopoles	345
16.4	Other Topological Defects; Cosmic Strings	345
16.5	Catalysis of Proton Decay by Monopoles	346

	16.6	Experimental and Observational Limits on Monopole Flux 347
	16.7	Exercises 349

17. Present Status of Particle Physics and Outlook 350

	17.1	Present Results That Might Hint at Further Phenomena 350
	17.2	Some Theoretical Questions 351
	17.3	Present Experimental Challenges 352
	17.4	Final Comment 357

Appendices

A. **Review of Lagrangians and Perturbation Theory; the Heisenberg and Interaction Pictures** 358

	A.1	Review of Classical Lagrangian and Hamiltonian Formalism 358
	A.2	Perturbation Theory; the Heisenberg and Interaction Pictures 359

B. **Proof of the Noether Theorem** 363

C. **Clebsch–Gordan Coefficients** 365

D. **Generators for $SU(3)$** 370

E. **Feynman Rules and Calculation of Matrix Elements** 372

F. $W^+ + W^- \to W^+ + W^-$ **and a General Theorem for Restrictions to the Standard Model Lagrangian** 377

	F.1	$W^+ + W^- \to W^+ + W^-$ Scattering 377
	F.2	The General Theorem 381

References 385

Index 396

1
Preliminaries

1.1 INTRODUCTION

The section on the Fundamental Particles and Forces in this chapter is meant as an overall introduction to the present "zoo" of particles and forces. Most of the course will be spent working out the intricate relations just dimly indicated in that section. The sections on relativity, conventions, and Mandelstam variables are mainly to define notation.

It is well past the scope of this text to explicate Feynman diagrams in detail, but even used in a qualitative way, they are a very powerful tool. In later sections of the text additional meaning is given to elementary Feynman diagrams so that simple matrix elements can be calculated.

1.2 THE FUNDAMENTAL PARTICLES AND FORCES, FERMIONS, BOSONS

In the late nineteenth century the atomic theory of matter was not yet established. Many still believed matter to be continuous with no smallest molecules of a given kind of matter. The great German physicist, Boltzmann, committed suicide at least partly in despair of convincing his colleagues that the atomic theory was valid. However, very shortly thereafter, the atomic theory was firmly established, with Einstein's paper on Brownian motion serving as a final crucial link in the logic chain. The only forces known at the time were electromagnetism and gravity.

In the first quarter of the twentieth century, the atom was found to be composed of electrons and a very small positively charged nucleus containing most of the mass of the atom. The weak interactions were added to the list of forces.

In the second quarter of the twentieth century, the nucleus was found to consist of neutrons and protons, held together by newly postulated strong forces.

In the third quarter of the twentieth century, the protons and neutrons were found to be made up of still smaller entities, which are called quarks, and the forces were found to be unified to a degree not suspected hitherto.

1. Preliminaries

Table 1.1 Fundamental Fermions

Generation	Flavor	Charge	Spin	Colors	mass
1	u	+2/3	1/2	Yes (3)	"few MeV"
	d	-1/3	1/2	Yes (3)	"few MeV" $m_d > m_u$
	e^-	-1	1/2	No	0.511 MeV
	ν_e	0	1/2	No	≤ 7 eV
2	c (charm)	+2/3	1/2	Yes (3)	≈ 0.9 GeV
	s (strange)	-1/3	1/2	Yes (3)	≈ 0.1 GeV
	μ^-	-1	1/2	No	0.1057 GeV
	ν_μ	0	1/2	No	≤ 0.27 MeV
3	t (top)	+2/3	1/2	Yes (3)	≈ 180 GeV
	b (bottom)	-1/3	1/2	Yes (3)	≈ 5 GeV
	τ^-	-1	1/2	No	1.776 GeV
	ν_τ	0	1/2	No	≤ 31 MeV

Will there be still smaller divisions and more unity? In this text, the present state of knowledge of particle physics will be treated and some of the new directions that studies of matter are pursuing will be outlined.

1.2 a. The Particles

The fundamental fermion particles are grouped into three generations. The particles in a higher generation differ from those in a lower generation only in mass. The interactions of the corresponding particles in a higher generation are identical to those in a lower generation (except for "mixing" parameters). Charges are measured in units of the magnitude of the charge on the electron. Strongly interacting particles (known as "quarks") have a quantum number called color and come in three colors. The different kinds of particles are known as different "flavors." The fundamental fermions are listed in Table 1.1.

Each particle has an associated anti-particle, which is identical in mass with the particle, but which has opposite charge and color. The observed nuclear particles are made up of sums of quarks. For example, the proton is $p = uud$, the neutron is $n = udd$, and the pion is $\pi^+ = u\bar{d}$, etc. (Note that the $\bar{d}$ is the anti-particle of the d.) No one has ever seen a quark! It is thought likely that they cannot exist as free particles, but only as confined objects with other quarks.

There are conservation laws associated with these particles. For the quarks, the sum of quarks minus anti-quarks is conserved. The e, μ, τ, and the neutrinos are collectively known as leptons. For the leptons, the leptons minus anti-leptons are separately conserved within each generation.

1.2 b. The Interactions

Each interaction has one or more particles (bosons) associated with it that "carry" the force. Define a quantity called cross section, σ, to measure the strength of an interaction. Suppose, classically, there is a random particle beam of unit area striking a black disk of area equal to σ. The fraction of the beam that is absorbed is equal to $\sigma/1 = \sigma$. The effective area of a particle that interacts with an incoming beam is called its cross section. An interaction that is more probable ("stronger") will make the cross section larger. For decays, a more probable interaction will make the particle decay more quickly, i.e., have a shorter lifetime.

Strong Interaction: Gluons of spin one carry the force, typical cross sections are $\sigma \simeq 10^{-26}$ cm^2, and typical lifetimes for particles decaying by this force are 10^{-23} s. There are eight gluons, differing from each other only by a quantum number called color. They interact directly among themselves (i.e., the gluons have the strong charge). These are strong interactions.

Electromagnetic Interaction: Photons of spin one carry the force, typical cross sections are $\sigma \simeq 10^{-29}$ cm^2 and typical lifetimes for particles decaying by this force are 10^{-20} s. There is one kind of photon. Photons do not directly interact among themselves (i.e., the photons are neutral; they have zero electric charge).

Weak Interaction: W^+, W^- ($\simeq 80$ GeV) and Z^0 ($\simeq 91$ GeV), all of spin one, carry the force. These bosons cause β decays. Typical cross sections are $\sigma \simeq 10^{-40}$ cm^2 and typical lifetimes for particles decaying by this force are 10^{-8} s. The W^+, W^-, and Z do interact among themselves. It is now known that the electromagnetic and weak forces are closely related, really different manifestations of the same force.

Gravity: Gravitons of spin two carry the force (= 0 mass)

Other: There are other postulated forces such as Superweak Forces, which are proposed to account for CP violation, but none are confirmed. Some of these proposed forces will be discussed later in the text.

The u and d quarks have strong, electromagnetic, weak, and gravitational interactions. The e has only electromagnetic, weak, and gravitational. The ν_e has neither strong nor electromagnetic interactions. The same is true for the other generations. Note that neutrinos are believed to have a gravitational interaction, independent of whether their rest mass is or is not zero. In fact, any particle carrying energy is believed to interact with the gravitational field. Photons, whose rest mass, if any, is exceedingly small, have been observed to be gravitationally deflected when passing near the sun. This has been measured during solar eclipses by observing a distortion of star positions near the sun.

1.2 c. Comments on Particles and Forces

The lowest mass charged lepton is the electron. For neutrinos, all present experimental evidence is consistent with all neutrinos having zero rest mass; however, there are presently intensive searches underway for neutrino masses.

The quarks can be considered to have a quantity called baryon number, B, associated with them. A quark has $B = 1/3$, and an anti-quark has $B = -1/3$. Mesons are made of quark and anti-quark combinations with $B = 0$. Thus the $\pi^+ = u\bar{d}$, $K^+ = u\bar{s}$, $D^+ = c\bar{d}$, $B_s^0 = s\bar{b}$, etc.

Baryons are composed of three quarks and have $B = 1$. For example, a proton $(p) = uud$, $\Lambda = sud$, $\Lambda_c^+ = cud$, etc. In practice, besides these "valence quarks," both mesons and baryons can have a sea of quark–anti-quark pairs and gluons. This will be discussed further in Chapters 6 and 8.

Quantum electrodynamics is the quantum study of electromagnetism and concerns interactions involving a real or virtual photon. It was the first relativistic field theory to be developed and has served as a prototype for the other field theories. During this course it will be seen that, in fact, the weak and electromagnetic interactions are connected, and a single theory of electroweak interactions describes both.

The strong interactions provide the nuclear glue to hold the quarks together to make hadrons, the strongly interacting particles, the mesons and baryons. Consider the valence quarks for a proton, uud. The proton has spin one-half and, using Fermi statistics, must be represented by an antisymmetric state, $\psi = \psi_{\text{spatial}} \times \psi_{\text{spin}}$. The spatial ground state is expected to be symmetric in the three particles. This seems highly likely kinematically, since otherwise there are nodes and higher curvature for the wave functions, which leads to higher energies. One can build such a state from uud, since one u can be spin up and the other spin down, making them distinguishable.

However, a Δ^{++} particle is composed of uuu, and one cannot get a state anti-symmetric in all three particles using just the spin functions. This can be resolved by introducing a new quantum number, color, which can have one of three values, say red, yellow, or blue. Then $\Delta^{++} = u_R u_Y u_B$, and the three u's are distinguishable particles.

Color is a "hidden" quantum number; only colorless particles, i.e., color singlets, are allowed as real particles. Virtual particles may be colored. This solves another problem. Without color it is not clear why there are no particles consisting of qq or $\bar{q}\bar{q}$, or even just q. However, these cannot be colorless and are forbidden by color. Anti-particles have anti-color, $\bar{R}, \bar{Y}, \bar{B}$, and hence the mesons $\bar{q}q$ are allowed. $R\bar{R}, Y\bar{Y}, B\bar{B}$ = white; $RYB, \overline{RYB}$ = white.

The gluons have to be bi-colored, $R\bar{B}$, etc. There are eight gluons, since $3 \times 3 = 9$ and one is subtracted for the symmetric colorless combination. The interactions of quarks and gluons conserve color. These concepts will be made sharper later in the text.

1.2 d.   Some Common, Strongly Interacting Particles

FERMIONS: The strongly interacting fermions are called baryons. They have odd half-integer spin and obey Fermi–Dirac statistics. (A wave function involving two identical fermions goes to minus itself under the interchange of the two fermions.) Below, we list a few of the low-lying states. All of those listed have spin 1/2. Later in the text other particles with higher spins will be listed. In fact, the number of known states today runs into the hundreds.

Proton = p = uud (Nucleus of the hydrogen atom)

Neutron = n = udd

Lambda = Λ = sud

Sigma plus = Σ^+ = suu

Sigma naught = Σ^o = sud (Note: Λ and Σ^o are both different states of the same three quarks, sud.)

Sigma minus = Σ^- = sdd

Cascade naught = Ξ^o = ssu

Cascade minus = Ξ^- = ssd

Omega minus = Ω^- = sss

Lambda-c = Λ_c^+ = cud

BOSONS: These particles have integer spins and obey Bose–Einstein statistics. (A wave function involving two identical bosons goes to plus itself under interchange of the two bosons.) Again we list below a few of the low-lying states and will introduce more later in the text. All those listed below have spin zero:

pi-plus $= \pi^+ = u\bar{d}$
pi-naught $= \pi^o =$ combination of $\bar{u}u$ and $\bar{d}d$.
pi-minus $= \pi^- = \bar{u}d$
k-plus $= K^+ = u\bar{s}$
k-naught $= K^o = d\bar{s}$
k-naught-bar $= \overline{K}^o = \bar{d}s$
k-minus $= K^- = \bar{u}s$
D-plus $= D^+ = c\bar{d}$
D-zero $= D^o = c\bar{u}$
D-zero-bar $= \overline{D}^o = \bar{c}u$
D-minus $= D^- = \bar{c}d$
D-sub $s = D_s^+ = c\bar{s}$

1.3 Relativity

This section is intended as a very brief review of the fundamental points needed for this course and is intended mainly to establish notation. When a particle moves at close to the speed of light, the energy and momentum relations are different from those at low speeds:

$$\vec{p} = \gamma m \vec{v}, \qquad (1.1)$$

$$E = \gamma m c^2, \qquad (1.2)$$

where

$$\gamma = \frac{1}{\sqrt{1-\beta^2}}, \qquad (1.3)$$

$\beta = v/c$; $c =$ velocity of light; $m =$ mass of the particle at rest. From the above formulas, it is easy to show that

$$E^2 = (\vec{p}c)^2 + (mc^2)^2,$$
$$\beta = \frac{pc}{E}; \quad \gamma = \frac{E}{mc^2}. \qquad (1.4)$$

Note that if $v = 0$, $E = mc^2$. Mass can be considered "frozen energy."

If $m = 0$, v is always $c(!)$, i.e., $E = pc$ (for $m = 0$) and γm is indefinite ($\frac{0}{0}$). This actually happens! The photon has $m = 0$.

Next consider "four-vectors." Two types of four-vectors are distinguished depending on whether the index is a superscript, p^α, (contravariant vector), or a subscript, p_α, (covariant vector). For the contravariant vector momentum let

$$p^\alpha = (E/c, p_x, p_y, p_z). \tag{1.5}$$

The index α is taken from 0-3, and the quantity with index zero is E/c.

The metric tensor, $g^{\alpha\beta}$ allows one to change between the two types of vectors:

$$x_\alpha g^{\alpha\beta} = x^\beta. \tag{1.6}$$

Here, a summation convention is implied. The repeated index α is to be summed over and the process is known as contraction of indices. The metric tensor used in this text is

$$g^{\alpha\beta} = g_{\alpha\beta} = \delta_{\alpha\beta}(-1 + 2\delta_{\alpha 0}). \tag{1.7}$$

Thus, there are only the diagonal components (1,-1,-1,-1) and the covariant momentum vector is

$$p_\alpha = (E/c, -p_x, -p_y, -p_z). \tag{1.8}$$

For derivatives, a subtlety should be noted.[3] The derivative of a scalar with respect to a contravariant vector yields a covariant vector and vice versa. ($\partial F/\partial x^\alpha = y_\alpha$; $\partial F/\partial x_\alpha = y^\alpha$.)

The scalar product $q \cdot k$ is then defined as

$$q^\alpha k_\alpha = q^\alpha k^\beta g_{\beta\alpha} = (E_q E_k/c^2 - p_{qx}p_{kx} - p_{qy}p_{ky} - p_{qz}p_{kz}) \tag{1.9}.$$

For a single particle $p^2 = (E/c)^2 - \vec{p}^2 = m^2 =$ constant. This quantity is also constant for a system of particles. For such a system, "m" is called the effective mass and is the energy in the center-of-mass system where $\vec{p}$ is 0.

How is contact made with non-relativistic (slow speed) physics? Let $\beta = v/c \ll 1$. This implies that

$$\gamma = 1/\sqrt{1 - \beta^2} \approx 1/(1 - \beta^2/2) \approx 1 + \beta^2/2;$$

$$E = \gamma mc^2 \approx (1 + \beta^2/2)mc^2 = mc^2 + mv^2/2.$$

Thus for $v \ll c$, E equals the energy of the rest mass plus the non-relativistic kinetic energy.

1. Preliminaries

How does one transform from one coordinate system to another moving with respect to it? Let the new (primed) system be moving with a constant velocity $\vec{v}$ with respect to the original (unprimed) system. Suppose $\vec{v}$ is in the $+x$ direction and let β, γ refer to this velocity, i.e., the velocity of the coordinate system. Consider a particle of mass m with momentum $\vec{p}$ and energy E in the original system. Then the transformation equations (Lorentz transformation) are:

$$p'_y = p_y; \quad p'_z = p_z;$$
$$p'_x = \gamma(p_x - \beta E/c); \quad (1.10)$$
$$E' = \gamma(E - \beta p_x c).$$

A scalar product of two four-vectors is invariant under a Lorentz transform.

Position and time form a four-vector, just as momentum and energy do. The similar transformations applied to the contravariant four-vector (ct, x, y, z) predict that the lifetime (τ) of an unstable moving particle is $\tau = \gamma \tau_o$, where τ_o is the lifetime of the particle as seen in the rest frame of the particle.

Consider some examples of these transforms. Suppose a particle of mass M decays into two particles of masses m_1 and m_2. To reconstruct M^2 from the decay products, one can use conservation of energy and momentum:

$$\vec{p}_M = \vec{p}_1 + \vec{p}_2; \quad E_M = E_1 + E_2;$$

$$M^2 c^2 = (E_1 + E_2)^2/c^2 - (\vec{p}_1 + \vec{p}_2)^2 = \text{(sum of four-momenta)}^2.$$

A commonly used quantity is the rapidity, y,

$$y = \frac{1}{2} \ln \frac{E + p_z}{E - p_z},$$

where z is the direction of motion.

Why is this a useful quantity? Consider a Lorentz transform along the z axis:

$$p'_z = \gamma(p_z - \beta E); \quad E' = \gamma(E - \beta p_z),$$

$$y' = \frac{1}{2} \ln \frac{E' + p'_z}{E' - p'_z} = \frac{1}{2} \ln \frac{E + \beta p_z + p_z - \beta E}{E - \beta p_z - p_z + \beta E}$$

$$= \frac{1}{2}\ln\frac{(1-\beta)(E+p_z)}{(1+\beta)(E-p_z)} = \frac{1}{2}\ln\frac{1-\beta}{1+\beta} + y = \tanh^{-1}\beta + y.$$

Thus the rapidity distribution is invariant and the Lorentz transformation corresponds to a translation.

1.4 CONVENTIONS AND UNITS

Some fundamental constants needed are

$$\hbar = \frac{h}{2\pi} = 1.055 \times 10^{-34} \text{ J-s} = \text{Planck's Constant} \quad [\frac{mL^2}{T}],$$

$$c = 2.998 \times 10^8 \text{ m/s} = \text{speed of light in vacuo} \quad [\frac{L}{T}].$$

Often in particle physics one takes units such that $c = 1$ and $\hbar = 1$. These are called "natural units." One gets back to normal units by dimensional analysis. Mass, momentum, and energy are then all in units of GeV. Length ($\hbar/mc$) and time ($\hbar/mc^2$) are in units of GeV^{-1}. Some relations that are useful for this purpose are

$$1 \text{ eV} = 1.6 \times 10^{-19} \text{ joules}; \; 1 \text{ MeV} = 10^6 \text{ eV}; \; 1 \text{ GeV} = 10^9 \text{eV};$$

$$\hbar = 6.582 \times 10^{-22} \text{ MeV-s};$$

$$\hbar c = 0.197 \text{ GeV-fermi, where 1 fermi} = 1 \text{ fm} = 10^{-13} \text{ cm};$$

$$1 \text{ barn} = 10^{-24} \text{ cm} = 100 \text{ fm}^2 \tag{1.11}$$

$$\alpha \equiv e^2/(4\pi\epsilon_o \hbar c)(\text{mks units}) \approx 1/137;$$

One often uses e in Heaviside – Lorentz units $= e_{mks}/\sqrt{\epsilon_o}$.

Magnetic moments are expressed in terms of magnetons. For the electron, the Bohr magneton is $\mu_B = e\hbar/(2m_e) = 5.788 \times 10^{-11}$ MeV/T. For the nucleon, the nuclear magneton is $\mu_N = e\hbar/(2m_p) = 3.1525 \times 10^{-14}$ MeV/T. In terms of these quantities, a magnetic moment can be written as $geJ/(2m) = g\mu J/\hbar$, where J is the angular momentum quantum number and g is a dimensionless constant. Note that the quantity $J/\hbar$ is dimensionless.

10   1. Preliminaries

The "fine structure constant," α, defined above, is the Coulomb repulsive energy between two electrons one natural unit of length apart, divided by the rest mass energy of the electron. It is dimensionless:

$$\alpha = \frac{e^2}{4\pi\epsilon_o(\hbar/mc)} \times \frac{1}{mc^2} = \frac{e^2}{4\pi\epsilon_o\hbar c}.$$

If $\hbar = c = 1$, and Heaviside units are used, then $e = \sqrt{4\pi\alpha}$.

1.5 Mandelstam Variables–s, t, u

A simple set of standard kinematic variables are defined here. Consider the reaction $A + B \to C + D$. Let p_A, p_B, p_C, and p_D be the four momenta of A, B, C, and D, respectively. Define the Mandelstam variables in natural units by

$$s = (p_A + p_B)^2 = E_{cm}^2,$$
$$t = (p_A - p_C)^2, \quad (1.12)$$
$$u = (p_A - p_D)^2.$$

t is known as the momentum transfer squared. It can be shown (see Exercise 1.1) that

$$s + t + u = m_A^2 + m_B^2 + m_C^2 + m_D^2. \quad (1.13)$$

In the laboratory if the target has mass m_T and the incoming particle beam (mass m_B) is a high-energy beam,

$$s = (E + m_T)^2 - \vec{p}^{\,2} = m_B^2 + m_T^2 + 2Em_T \approx 2Em_T. \quad (1.14)$$

Colliding-beam experiments involve the collision of two beams of particles going in opposite directions. Assume for now that the particles in the two beams have equal mass and energy. Examine colliding-beam versus fixed-target experiments. For the colliding-beam experiment, the laboratory system is the center-of-mass system. At Fermilab, protons and antiprotons, each of 900 GeV, collide. There, $E_{cm} = \sqrt{s} = 2 \times 900$ GeV $= 1800$ GeV $= 1.8$ TeV.

To achieve the same $\sqrt{s}$ for a fixed-target experiment using the preceding equation involves $\sqrt{s} = 1800 = \sqrt{2mE}$ or $E = 1700$ TeV, which is almost two thousand times the energy of the particles needed for the colliding-beam facility.

Next, consider the process $p + p \to p + p + \bar{p} + p$. What is the minimum incoming proton energy needed onto a fixed target for the interaction to occur (threshold energy)? s is the square of the sum of the two incoming particle momenta and hence is equal to the square of the sum of the four-momenta of all of the final-state particles by conservation of energy and momentum. Since s is an invariant, one can use any convenient coordinate system. Choose the center of mass. Then the minimum value for s will occur when all final-state particles are at rest, giving $s_{\min} = (4m_p)^2$. Since s is invariant, this will be the minimum value in any system. Consider now the laboratory system, $p_A = (E, \vec{p})$; $p_B = (m_p, 0)$ in natural units:

$$s = (p_A + p_B)^2 = (p_A^2 + p_b^2 + 2 p_A \cdot p_B) = 2m_p^2 + 2(E m_p - p \times 0)$$

$$= s_{\min} = 16 m_p^2.$$

Thus $2 E m_p = 14 m_p^2$ or $E = 7 m_p$.

1.6 Feynman Diagrams

In Appendix A, a perturbation expansion for non-relativistic quantum mechanics is developed. Relativity introduces further complications, but fortunately a powerful tool is available. Feynman diagrams provide a very powerful graphical method of picturing the interactions in perturbation theory. Each graph is really a precise mathematical notation and the matrix element corresponding to each graph can be written down from the graph according to very specific rules. Thus, the graphs provide pictures representing probability amplitudes.

The general mathematical theory of Feynman diagrams is the subject of full advanced quantum mechanics texts and is well past the scope of this text. However, even qualitative use of these diagrams provides the reader with powerful tools, and if a very few of the mathematical correspondences are added, they can greatly enhance one's intuition.

Consider the diagram shown in Figure 1.1. The solid straight lines are fermion lines, an electron for the top line and a muon for the bottom line. The wiggly line is a photon line. The horizontal axis is imagined to be the time axis, positive to the right. Generally, lines correspond to particles propagating through space–time and dots correspond to vertices, places where interactions occur. The present diagram can be viewed as an electron emitting one photon which is then absorbed by the muon (or vice versa), i.e., $e + \mu \to e + \mu$. This corresponds to the lowest order of perturbation theory. It is equivalent to the first term in the expansion of U_I in Appendix A with ϵH_{int} interacting only once. However, unlike

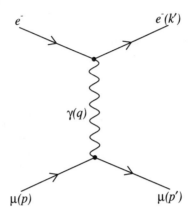

Figure 1.1 A first-order Feynman diagram for $e - \mu$ scattering.

that example, the Feynman methodology includes spins, relativity, Fermi-Dirac and Bose statistics, and is equivalent to a full field theory. Every legal diagram is a term in the perturbation series and vice versa. Each part of the figure has mathematical meaning. For most of this text, the detailed correspondences will not be used, but some features are very useful in obtaining an intuitive picture of the processes.

We can view this diagram as having two vertices, i.e., places where a particle appears or vanishes. From each vertex the matrix element gets a factor of "e," corresponding to the coupling constant, the strength of the interaction.

For each internal line, the wiggly photon line here, the matrix element gets a factor of $-ig_{\mu\nu}/(q^2 - m^2)$, where q is the momentum and m is the mass of the exchanged particle ($m = 0$ for the present case since a photon is being exchanged). This factor is called the "propagator." There are additional parts of the propagator that depend on the spin of the exchanged particle. They will be omitted for now and are discussed separately when necessary.

In this picture, the solid lines have arrows indicating direction. A major conceptual stroke was Feynman's realization that an anti-particle could be viewed in these diagrams as a particle going backward in time. In Dirac theory anti-particles come from negative-energy solutions to the Dirac equation. Feynman viewed these as positive-energy particles moving backward in time and with charge reversed from the positive-energy solution. Thus, in Figure 1.2, the first diagram refers to the process $e + K \to \pi + \nu$ and the second to $K \to \pi + e + \nu$. The intermediate particle is a W boson in Figure 1.2. Both diagrams correspond to the same matrix element as

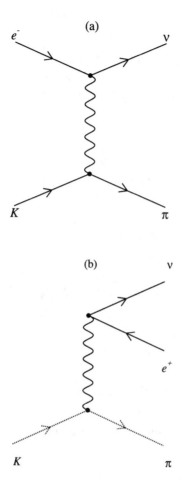

Figure 1.2 Two processes with the same Feynman diagram.

both diagrams have the same topology! Thus, the diagrams indicate profound relationships between quite different physical processes. An electron going backwards is seen as a positron. The only difference is that in the matrix element, one uses the negative of the laboratory momentum for the positron, as will be discussed later in the text.

Consider again Figure 1.1, but rotate it counter clockwise by 90°, i.e., view it sideways. Instead of $e + \mu \to e + \mu$, the diagram represents $e^+ + e^- \to \mu^+ + \mu^-$. The first interaction is known as a "t-channel process," since the propagator term is proportional to $1/(k - k')^2$, the inverse of the Mandelstam variable, t. The second interaction is known as an "s-channel process," since the propagator term is proportional to the inverse of the Mandelstam variable s.

At each vertex, four-momentum is conserved. However, the photon is an intermediate particle in this process, existing only a short time and not detectable. The effective mass of the photon, the square of its four-momentum, does not have to be equal to the physical mass of zero. Indeed in the t-channel process, the effective mass squared is forced to be negative, the momentum transfer being "spacelike," while in the s-channel process, the effective mass squared is positive, corresponding to a "timelike" momentum transfer.

The preceding facts are the major ones needed for the present. For future reference, a description of the mathematical terms corresponding to each part of Figure 1.1 is given to illustrate how the detailed prescription works:

1. $-ie$ comes from each electromagnetic vertex, where $\alpha = e^2/4\pi$ in Heaviside–Lorentz units. (Remember $\hbar = c = 1$.)

2. The top lines stand for the four-vector particle current $j^\nu_{(e)}$. [It will be seen in a later chapter that $j^\nu_{(e)} = \bar{u}(k')\gamma^\nu u(k)$, where the u's are four-component column "spinors" to give the spin components, and γ^ν is a 4×4 matrix.] Here the ν is a spacetime index from 0 to 3. All of this refers to the electrons.

3. The bottom lines stand for the current $j^\nu_{(\mu)}$ $[= \bar{u}(p')\gamma^\nu u(p)]$. All of this refers to the muon.

4. The wiggly vertical line is the "propagator," $-ig_{\mu\nu}/q^2$, since m is 0. $q = k - k'$. Note $q^2 = t_{\text{Mandelstam}}$.

These prescriptions can be combined to get the total amplitude for the reaction

$$\text{Amplitude} = -iM_{\beta\alpha} = (-ie)j^\mu_{(e)}\left(\frac{-ig_{\mu\nu}}{q^2}\right)(-ie)j^\nu_{(\mu)}. \tag{1.15}$$

The repeated indices ν and μ mean that the ν and μ are summed from 0 to 3. There is conservation of four-momentum at each vertex.

Sometimes diagrams can have lines that form closed loops. See Figure 1.3. In this methodology the internal four-momentum is integrated over all closed loops.

Consider the second-order diagrams shown in Figure 1.3, which correspond to some possible diagrams with two photons. They are equivalent to the second term in the perturbation series for U_I with two ϵH_{int} terms. Besides the straightforward first term there are five others. (The "crossed" diagram similar to Figure 1.3a, but with the photons crossing over and connected to the opposite bottom vertices, has been omitted.)

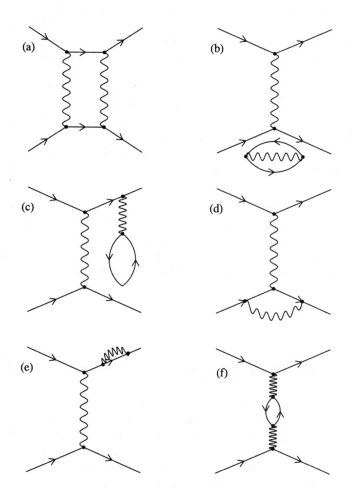

Figure 1.3 Some second-order Feynman diagrams for $e - \mu$ scattering.

These extra terms are called renormalization terms. Figure 1.3b is called a vacuum polarization diagram, Figure 1.3c and 1.3e are called wave function renormalization diagrams, Figure 1.3d is called a vertex renormalization diagram, and Figure 1.3f is called a propagator renormalization diagram. However, for example, the vertex renormalization diagram can be viewed as already included in the first-order term if by "vertex" is meant the vertex with all of the second-order, third-order, ... terms already included; when e is measured using a vertex, what is measured is the already renormalized vertex. This is an important point because these renormalization terms formally give logarithmic infinities when integrated over the momenta of the closed loops, i.e., $\propto \ln p_{max}$. In spite of the fact that agreement can be obtained between this theory and experiment to better than ten decimal places, which is the most accurate physical theory ever devised, there is something fundamentally wrong with it. The renormalization procedure allows one to sweep the problems under the rug. To get finite physical results, one must have the "bare" electron mass infinite and the "bare" electric charge infinite. The renormalized quantities are finite. Is there a natural p_{max} cutoff? Is space "grainy"? Should there be a cutoff at $p = E/c$ of the whole universe? Perhaps a whole new formulation is needed. Later in the course much more serious divergence problems—quadratic rather than logarithmic—will be found. However, these logarithmic singularities are still singularities and await a good explanation by future physicists.

The Feynman diagrams introduced in this section provide a pictorial and beautiful way of picturing and organizing even complex interactions. These diagrams will be used throughout the text, at first intuitively rather than quantitatively. However, in Chapter 11 and in Appendix E, the procedure for obtaining matrix elements using this procedure will be described and illustrated.

1.7 Exercises

1.1 Show, using the Mandelstam variables, the following relation, given in the text: $s + t + u = m_A^2 + m_B^2 + m_C^2 + m_D^2$.

1.2 The next several problems involve Lorentz transforms. Let the coordinates of the laboratory system be x, t and those of a system based on a particle be x', t'. Prove the formula given in the text that the lifetime of a particle as seen in the laboratory system is longer than as seen in the particle system. To do this recall that one is measuring an interval of time $\Delta t'$ at a fixed x' (not a fixed x) and asking for the Δt.

1.3 Derive the formula for the relativistic Lorentz–Fitzgerald contraction, i.e., that a moving meterstick appears shorter in the laboratory than in a frame centered on the meterstick (or particle). To do this recall that one is trying to measure Δx in the laboratory for the two ends of the meterstick, with Δt zero in the laboratory.

1.4 A charged pion with energy 1 GeV in the laboratory decays into a muon and a muon neutrino. Find the maximum possible angle in the laboratory between the pion and the muon. Hint: Work with the Lorentz transform. Find the maximum $\tan \theta_{lab}$ using the center-of-mass to laboratory frame relations.

1.5 In the laboratory frame of reference, particle 1 is at rest with total relativistic energy E_1, and particle 2 is moving to the right with total relativistic energy E_2 and momentum p_2.

a) Use the relativistic momentum–energy Lorentz transformation equations to show that the frame in which the total momentum of the system is zero is moving to the right with velocity

$$v = c\frac{cp_2}{E_1 + E_2}$$

relative to the laboratory frame. This is the center-of-momentum frame usually called the "center-of-mass" frame.

b) Now let the two particles have the same rest mass m, and let the total relativistic energy of the system in the laboratory frame be E_{lab}. Evaluate E_{cm}, the total relativistic energy of the system in the center-of-mass frame, and show that

$$E_{cm} = \sqrt{2mc^2 E_{lab}}.$$

1.6 The π^0 lifetime has been determined by studying the decay from rest of the K^+ meson in the mode $K^+ \to \pi^0 + \pi^+$. The average distance traveled by the π^0 in a block of photographic emulsion before it decays in the easily observable mode $\pi^0 \to e^+ + e^- + \gamma$ is measured, and from the calculated velocity of flight of the π^0, its lifetime is obtained. Given that the lifetime is 0.8×10^{-16} s, predict the mean distance traveled by a π^0 before it decays.

1.7 A K^0 meson decays *isotropically* in its rest frame into $\pi^+\pi^-$. The mass of the K^0 is 0.498 GeV, and the mass of the π^- is 0.140 GeV. Assume that the K^0 has laboratory momentum p_K. Express your results in terms of the pion momentum in the kaon rest frame. Calculate the laboratory distributions of

a) The pion energy E.

b) The pion transverse momentum component, $p_\perp$, with respect to the kaon direction.

1.8 Consider the interaction $\pi^- p \to \pi^0 n$. The momentum of the incident pion beam is 100 GeV. The photons from the decay are detected 10 m downstream from the target. The scattering angle is small; the mass difference of the charged and neutral pion can be neglected as can the proton neutron mass difference.

a) What is the laboratory momentum and angle of the produced π^0 for a four-momentum transfer squared (between the p and the n) of $q^2 = 0.2$ GeV^2?

b) What spatial resolution is required of the photon detector to resolve the two photons?

18   1. Preliminaries

1.9 Find the deBroglie wavelength for an electron with $p = m_e c$; for a proton with $p = m_p c$.

1.10 Find the effective mass (i.e., the center of mass energy) of a system consisting of a proton at rest in the laboratory and a charged pion with energy 10 GeV in the laboratory.

1.11 The classical radius of the electron is the radius given by setting the approximate electric potential energy (e^2/r) equal to the mass energy $(m_e c^2)$. Find the value of this radius in centimeters. Suppose a uniform ball of charge had the total charge of the electron and a radius equal to the classical radius of the electron. What would the potential energy of this ball of charge be? Note that these kinds of considerations set a scale for the size of electromagnetic contributions to the mass of a charged particle.

1.12 Suppose a charged particle travels in an arc of a circle within a constant magnetic field. If one draws the straight line chord between the two ends of the arc, then the sagitta is defined as the maximum distance between the chord and the arc (i.e., the length of the line drawn between the center points of the chord and arc). Assuming that the radius, R, of the arc is much greater than the length, L, of the chord, find a relation between the sagitta, s, and R, L. This relation is of great practical use when measuring the momentum of a charged particle.

1.13 List the various possible B mesons composed of a b-quark and some anti-quark. For each meson list the charge of the meson.

1.14 For each of the following processes, state the interaction involved:

a) $n \rightarrow p + e^- + \bar{\nu}$.

b) $\Delta^{++} \rightarrow p + \pi^+$.

c) $\pi^0 \rightarrow e^+ + e^- + \gamma$.

d) $\Sigma^+ \rightarrow n + \pi^+$.

e) $\mu^- \rightarrow e^- + \bar{\nu}_e + \nu_\mu$.

1.15 Draw the Feynman diagrams for second-order scattering (four vertices) of two quarks by the exchange of gluons. Gluons have the strong charge, so that there can be graphs with three-gluon vertices, i.e., one gluon turning into two gluons. In this way gluons differ from photons.

2
Electromagnetic Interactions Used for Detection of Charged Particles

In this and the next chapter a brief examination of the interactions and tools physicists need to perform experiments will be given. In this chapter interactions used for experimental detection of particles will be outlined and in the following chapter particle accelerators will be discussed as well as detection tools that use the interactions given in the present chapter. These discussions will provide some insight into how the experimental results discussed throughout the text can be obtained.

2.1 RUTHERFORD SCATTERING

Rutherford scattering is just electromagnetic elastic scattering as shown in Figure 2.1. The classical cross section for scattering through an angle θ in the laboratory by means of this process is (cgs units)

$$\frac{d\sigma}{d\Omega} = \frac{1}{4}\left(\frac{Zze^2}{pv}\right)^2 \frac{1}{\sin^4\theta/2}, \qquad (2.1)$$

where $d\Omega = \sin\theta\, d\theta\, d\phi$. Here ϕ is the azimuthal angle. Ze and ze are the charges of the target and incoming particles, respectively; p and v are the momentum and velocity of the incoming particle. Note that the propagator equals $1/(it)$ where $t \approx -2p^2(1-\cos\theta) = -4p^2\sin^2\theta/2$. The probability involves the amplitude squared and hence $(1/t)^2$. Thus the propagator provides the $1/(\sin^4\theta/2)$ factor in the cross section. This factor is characteristic of the exchange of a zero-mass particle, and thus a $1/r^2$ force.

Physicists wish to dig deeper than just having a formula. They wish to understand in as intuitive a way as possible the physical processes leading to the result. Once the result is given, they then wish to find as many of the physical consequences of the result as possible. The discussion in the present situation is a fine example of these parts of the physicists' craft.

To gain further insight into the present process, consider a classical physics derivation of the Rutherford formula. Let b be the "impact parameter," i.e., the initial distance of a particle in an atom from the incoming

20   2. Electromagnetic Interactions Used for Detection

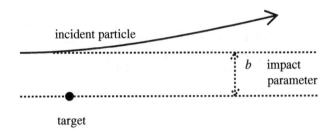

Figure 2.1 Impact parameter for an incident particle scattering on a target.

particle in the direction perpendicular to the direction of motion of the incoming particle. (See Figure 2.1.) Assume small angle scattering.

As the particle goes by, the perpendicular impulse transmitted to the incident particle is $p_\perp = \int F_\perp dt = \int zeE_\perp dt$. However, $\int dt$ can be written as $(1/v) \int dx$, since $dx = vdt$. This latter integral can, in turn, be related to the integral over the gaussian cylinder of radius b, i.e., $\int E_\perp dA = \int E_\perp 2\pi b dx$ and $\int E_\perp dA = $ flux $= 4\pi Ze$ (Gauss's law). Furthermore,

$$\int E_\perp dx = \frac{4\pi Ze}{2\pi b} = 2Ze/b.$$

Hence:

$$\text{Impulse} = p_\perp = \frac{ze}{v} \frac{2Ze}{b} = \frac{2Zze^2}{bv}, \tag{2.2}$$

$$\theta \approx \frac{p_\perp}{p_\|} \approx \frac{2Zze^2}{bvp}. \tag{2.3}$$

The cross section for this happening at impact parameter b within db is just the area of the ring of radius b and width db. Thus, $d\sigma = |2\pi b db|$. From Equation 2.3 for θ one obtains:

$$|db| = \left|\frac{2Zze^2 d\theta}{vp\theta^2}\right|.$$

2.1 Rutherford Scattering

The solid angle integrated over ϕ is $d\Omega = 2\pi \sin\theta d\theta \approx 2\pi\theta d\theta$. Hence

$$\frac{d\sigma}{d\Omega} = \frac{2\pi b db}{2\pi\theta d\theta} = \frac{2Zze^2 b}{\theta^3 vp},$$

$$\frac{d\sigma}{d\Omega} = \frac{4Z^2 z^2 e^4}{(pv)^2 \theta^4} = \frac{1}{4}\frac{Z^2 z^2 e^4}{(pv)^2 (\theta/2)^4}.$$

(2.4)

This is essentially the Rutherford formula, and the derivation has introduced the useful concept of the impact parameter.

So far the discussion has concerned collisions with "particles in an atom." The meaning of this phrase is collisions with the atomic electrons and with the electric field of the nucleus. Interesting parameters are the angle through which the incident particle is scattered and the amount of energy that it loses. Which kind of scattering is more important for each of these processes, that on nuclei or that on atomic electrons? Suppose the incoming particle has $z = 1$.

Consider first the scattering angle. For collisions with a nucleus there is a factor Z^2 that tends to make the nuclear scattering more important. Even though there are only $1/Z$ as many collisions, for collisions leading to scattering through a given (small) angle, the scattering with the nucleus wins over scattering with an electron by a factor of Z. (Note, however, that for high q^2 scattering, i.e., $|q^2| \gtrsim 50 - 100$ MeV^2, the nucleus does not scatter coherently, and this factor vanishes.)

For a given energy loss, the story is different; it is easier to transfer energy to a lighter particle. Note that $m_e/m_{nucleus} \sim m_e/(2Zm_p)$, when it is noted that the atomic number, A, is of the order of $2Z$ for many nuclei. The energy lost is just the energy given to the target and $p_{target} \approx p_\perp = 2Ze^2/(bv)$, where the Z of the target is one for collisions with electrons. Most of the energy loss turns out to come from collisions in which the target is given energy small compared to its rest mass energy. Thus, $E_{target} \approx p^2/2m_{target} = 4Z^2 e^4/b^2 v^2 2m_{target} \approx Ze^4/b^2 v^2 m_{proton}$ for collisions with nuclei and $2e^4/b^2 v^2 m_{electron}$ for collisions with electrons. Hence it is seen that $E_{target} \propto 1/m_{target}$. For a given impact parameter, the nucleus has a factor Z in E_{target}, but this is cancelled because there are Z times as many electrons as nuclei. This leaves the ratio of masses. For energy loss, the collisions with electrons dominate by a factor of the order of $2m_p/m_e$.

Consider energy losses further. Again use a classical picture. Let $\Phi(E, E')dE'dx$ be the probability that a particle of energy E has a collision in dx (g/cm^2) giving up an amount of energy in a range dE' of E'. Note that dx (g/cm^2) $= dx$ (cm) $\times \rho$ where ρ is the density of the material. The number of collisions with impact parameter b is equal to the number of electrons with impact parameter between b and $b + db$ in $dx = 2\pi b db dx N_\circ Z/A$, where $N_\circ$ is Avogadro's number.

Realizing that $E' = E_{target}$, the relations above for b imply (assuming E_{target} to be non-relativistic):

$$b^2 = \frac{2e^4}{E'v^2 m_e},$$

$$|2bdb| = \left|\frac{2e^4 dE'}{(E')^2 v^2 m_e}\right|.$$

Let

$$C = \pi N_o \frac{Z}{A} \frac{e^4}{m_e^2} = 0.150 \frac{Z}{A}. \qquad (2.5)$$

Then

$$\Phi(E, E') dE' dx = C \frac{2m_e}{v^2} \frac{dE'}{(E')^2} dx. \qquad (2.6)$$

Here dx is in g/cm^2. This was a non-relativistic derivation; however, this formula is essentially correct for relativistic E', but does have corrections depending on the spin of the particle.

Note the proportionality to $1/(E')^2$. Because of that factor most of the ionization loss [proportional to $\int E' dE'/(E')^2$] comes from the low-energy collisions. In fact, the energy loss can be written as

$$-\frac{dE}{dx} dx = \int E' \Phi(E, E') dE' dx = C \frac{2m_e}{v^2} \ln \frac{E'_{max}}{E'_{min}} dx; \qquad (2.7)$$

b_{max} is related to E'_{min} and b_{min} to E'_{max}.

As v approaches c, the incident particle electric field is contracted in the direction of motion. $\vec{E}_\perp$ increases by γ. However, the integral of the force pulse remains constant, since the duration of the pulse decreases. The approximate duration of the force pulse is $\tau = b/v$ classically, and becomes $b/(v\gamma)$ relativistically. If $1/\tau$ is far less than the vibration frequencies, ν_i, in the atom, then the impulse is averaged over many electron revolutions. Over part of a revolution it is speeding the electron up and over part it is slowing it down. The net result is a cancellation and the electron will absorb no net energy. One requires $1/\nu > \tau = b/(v\gamma)$, or $b_{max} \approx v\gamma/\nu \approx h v \gamma/E_{ioniz}$.

If b_{max} becomes much greater than the distance between atoms, there is dielectric shielding, which causes dE/dx to flatten at high energies and not increase logarithmically. This is known as the "density effect."

Next consider b_{min}. Classically, $v_{electron} < 2v_{incident} = 2v$ or $E_e < (1/2)m(2v)^2$. This would give $b_{min(cl)} \approx e^2/(mv^2)$. However, it turns out that a more severe limit is given by the requirement that the Coulomb field of the incident particle not vary much over the Compton wavelength of the electron. The wavefunction can be considered as spread over this distance and values of b less than this are not meaningful. Take the incident particle to be at rest. Then $v_{electron} = -v$. This gives $p_{electron} = mv\gamma$ and the Compton wavelength $= \hbar/(mv\gamma)$. Hence, $b_{min} \approx \hbar/(mv\gamma)$.

Note that

$$\ln(E'_{max}/E'_{min}) \approx \ln((p_\perp^2)_{max}/(p_\perp^2)_{min}) \approx \ln(b_{max}^2/b_{min}^2)$$

$$\approx 2\ln((hv\gamma/E_{ioniz})(mv\gamma/\hbar)) = 2\ln(2\pi mv^2\gamma^2/E_{ioniz}).$$

(2.8)

This agrees with the following results of a more exact calculation for dE/dx (still ignoring spin and density effects):

$$-\frac{dE}{dx} = C\frac{4m_e z^2}{v^2}\left[\ln\left(\frac{2m_e\beta^2}{I(1-\beta^2)}\right) - \beta^2\right].$$

(2.9)

Here I is the effective ionization potential of the atom and z is the charge of the incident particle (assumed 1 above) in units of e. $I = 16Z^{0.9}$ eV for $Z > 1$ to an accuracy of about 10%. Note that dE/dx is a function of the γ (or β) of the particle only. Thus, $(dE/m)/(dx/m) = d\gamma/(dx/m)$ is a function of γ only. It is a universal function for the electromagnetic ionization energy loss for all particles. (In the next section it will be noted, however, that electrons, because of their low mass, have a large additional energy-loss mechanism.) A good summary of energy-loss calculations is given in the review article of S.P Ahlen.[4]

Figure 2.2 shows a plot of the energy loss versus E/m. Note that at a given low momentum, a proton, because of its low velocity, can have a much higher rate of energy loss than a lower mass particle. The minimum in the energy-loss curve varies with Z from a value of 4.1 MeV/g/cm² in hydrogen to 1.1 MeV/g/cm² in uranium with the most typical values between 1.1 and 2 MeV/g/cm². The minimum occurs for $\beta \approx 0.96$ ($\gamma \approx 3.6$). At high energies, the energy loss becomes approximately constant for solids at a value about 10–15% above the minimum value and for gases at up to 50% above the minimum value.

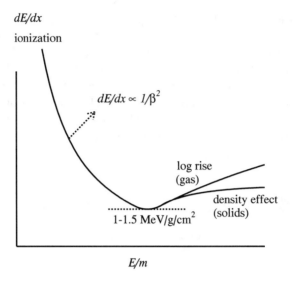

Figure 2.2 Coulomb energy loss versus E/m for a charged particle heavier than an electron.

When E' is sufficiently large in a collision that the individual struck electron can be seen in a particle detector, this particle is called a "delta ray" or "knock-on electron." Kinematically the maximum energy E' that can be given to an electron is

$$E'_{max} \approx \frac{2p_{inc}^2 m_e}{m_{inc}^2 + m_e^2 + 2E_{inc}m_e} = \frac{2m_e}{\frac{m_{inc}^2 + m_e^2}{p_{inc}^2} + \frac{2E_{inc}m_e}{p_{inc}^2}} \longrightarrow p_{inc}. \quad (2.10)$$

E'_{max} approaches p_{inc} only at very high energy. For incident protons the second term in the denominator equals the first term in the denominator only at $E_{inc} = 850$ GeV.

Finally consider straggling, i.e., the variations in energy loss. The variance of dE/dx is given by $\langle (E - \langle E \rangle)^2 \rangle = \langle E^2 \rangle - \langle E \rangle^2$. In the preceding equation, $\langle E \rangle$ was examined. Consider $\langle E^2 \rangle$:

$$\int (E')^2 \Phi(E, E') dE' dx = C \frac{2m_e}{v^2} (E'_{max} - E'_{min}) dx. \quad (2.11)$$

E'_{max} dominates this term and dominates the straggling. Thus, although most of the energy loss is from the many low-energy transfer collisions, most of the straggling is from the few high-energy transfer collisions. (This high-energy transfer tail is called the "Landau tail.") Because the straggling is due to a few collisions, it remains significant even at high energies.

In order to measure $\langle dE/dx \rangle$ for particle identification, one can sample dE/dx in many layers of gas for a single particle. The Landau tail can be removed by taking a "truncated mean," e.g., the mean of, say, the lowest 70% of the measurements. This allows a determination of the Lorentz factor, γ, which, along with a measurement of momentum, gives the particle mass. This method works best at low energies, where $dE/dx \propto 1/\beta^2$. (A similar method based on another kind of energy loss, transition radiation, discussed in the next chapter, works at high energies where $dE/dx \propto \ln \gamma$.)

In this section the most basic electromagnetic scattering process, Rutherford scattering, has been examined. It has been seen that, for a particle traveling through matter, collisions with electrons are the most important for energy loss, while the collisions with nuclei are the most important for angular deflections. The energy loss per centimeter and its variance, as well as the angular effects of Coulomb scattering, are the essential beginnings needed to design particle detectors and experiments. In the next section, the effects of a great many successive scatterings will be treated.

2.2 MULTIPLE SCATTERING

Consider singly charged particles with $m > m_e$. As particles travel through matter they tend to make a great many very small angle collisions as shown in the previous section. Therefore, the net angle scattered and the net displacement can be treated in a statistical manner. The three-dimensional angle is not distributed in a normal distribution because of the phase space factor $\theta d\theta$. For a net small angle deflection, the distribution of the scattering angle projected onto a plane containing the initial trajectory of the particle does have an approximately normal distribution:

$$P(\theta_y)d\theta_y = \frac{1}{\sqrt{2\pi \langle \theta_y^2 \rangle}} e^{\frac{-\theta_y^2}{2\langle \theta_y^2 \rangle}} d\theta_y, \qquad (2.12)$$

where

$$(\theta_y)_{rms} \equiv \langle \theta_y^2 \rangle^{1/2} = \frac{13.6 \text{MeV}}{p v} Z_{inc} \sqrt{\frac{x}{X_\circ}} \left(1 + .038 \ln \frac{x}{X_\circ}\right), \qquad (2.13)$$

$$X_\circ = \frac{716.4 A \text{ g/cm}^2}{Z(Z+1) \ln(287/\sqrt{Z})}, \qquad (2.14)$$

x is the length along the initial trajectory direction. Z_{inc} is the Z of the incident particle. Z and A refer to the material through which the particle

2. Electromagnetic Interactions Used for Detection

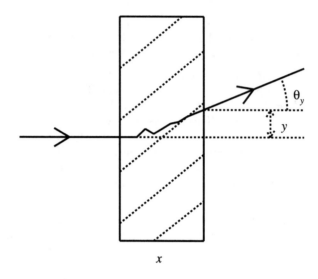

Figure 2.3 Multiple scattering quantities.

travels. X_o is called the radiation length of the material for electrons. The preceding formula for X_o is accurate to a few percent. The expected projected (y) displacement also has an approximately normal distribution. It and the correlation between y and θ are given by

$$(y)_{rms} = \langle y^2 \rangle^{1/2} = \frac{x}{\sqrt{3}} (\theta_y)_{rms}; \quad \langle y\theta \rangle = \frac{x}{2} (\theta_y)^2_{rms}. \qquad (2.15)$$

See Figure 2.3.

Thus, multiple scattering gives a distribution with rms angle θ proportional to the square root of the distance traveled. The distribution is a normal distribution and the tail, therefore, falls much faster than that of the distribution for single Rutherford scattering considered in the previous section. At sufficiently large angles, single scattering will dominate over multiple scattering.

In a typical experiment, an initial beam of particles will be defined. It will scatter on a target and the final directions of particles will be examined. Single and multiple Coulomb scattering set limits on how well the initial beam can be focused and the final directions of the particles can be measured.

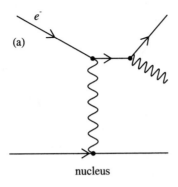

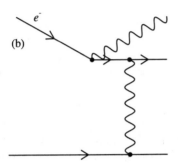

Figure 2.4 Bremsstrahlung Feynman diagrams.

2.3 BREMSSTRAHLUNG

When a charged particle travels through a medium, it can interact with the nuclei and electrons in the medium and emit a γ-ray (photon). As will be seen, this is the major source of energy loss for high-energy electrons, but is almost negligible for heavier particles until very high energies.

Bremsstrahlung means "braking radiation," and this is radiation induced by the deceleration of the electron. The modern way to calculate this is to write down the Feynman diagrams shown in Figure 2.4 and calculate the matrix elements. However, an intuitive semiclassical treatment can help provide a physical intuition for the process.

The electric field of one of the particles, either the electron or the nucleus, will be regarded as a set of virtual photons, which then scatter off of the field of the other particle. This procedure is known as the Weizsäcker–Williams method. Used with care it can provide almost the entire standard bremsstrahlung formula. For the present, a number of simplifying approximations will be made, which considerably simplify the mathematics, but still allow the essential physical features to be seen.

Consider a charged particle traveling at very near the speed of light past a stationary observer. The observer sees the electric field in the direction of motion squashed into a smaller distance than at rest by the Lorentz–Fitzgerald contraction, the contraction factor being $1/\gamma$. In the transverse direction, relativistic transformations of the electric field increase the field by a factor of γ. If the impact parameter (see Figure 2.1) is b, then the time of maximum field at a distance b is $\Delta t \approx 2b/(\gamma c)$. This corresponds to the field, in the rest system of the traveling particle, when the particle is a distance b in the longitudinal direction past the minimum impact point.

The field energy within db of b is $U(b)db = 2\pi bdbc\Delta t \times S(b)$. Here $S(b) = [(\vec{E})^2 + (\vec{H})^2]/(8\pi)$ is the energy density and the first terms are the volume factor. Here $\vec{E} \approx \vec{H} \approx Ze\gamma/b^2$, so $S(b) = Z^2e^2\gamma^2/(4\pi b^4)$. Hence, $U(b) \approx \gamma Z^2 e^2 db/b^2$.

In order to find the frequency spectrum of this energy, one must, in principle, Fourier analyze the pulse. However, a simple approximation will be made here. Note that if $\nu \gg 1/\Delta t$, the frequency spectrum will drop to zero since Δt is the characteristic time of the pulse. The frequency spectrum will be approximated by a simple square wave, constant up to $\nu_m = 1/\Delta t = c\gamma/b$, and zero for larger frequencies.

Introducing a frequency dependence into the energy function, one can then write $U(b,\nu)hd\nu = U(b)hd\nu/(h\nu_m)$. Translating this into a photon spectrum where $N(b, E', \gamma)$ represents the number of photons within db of b with energy within dE' of E' for a particle with energy E, one has $E'N(b, E', E) = U(b, \nu(E'))$. Substituting, one obtains

$$N(b, E', E) = \frac{Z^2 e^2 \gamma}{hb^2(c\gamma/b)E'} = \frac{e^2}{\hbar c} \times \frac{Z^2}{2\pi bE'} \approx \frac{\alpha Z^2}{2\pi bE'}. \qquad (2.16)$$

Integrating over b one obtains $N(E', E) = [\alpha Z^2/(2\pi)] \ln(b_{max}/b_{min}) dE'/E'$ for the spectrum of photons accompanying the particle. A more careful derivation differs from this by a factor of four, yielding

$$N(E', E) = \frac{2}{\pi} \alpha Z^2 \ln\left(\frac{b_{max}}{b_{min}}\right) \frac{dE'}{E'}. \qquad (2.17)$$

What are the appropriate maximum and minimum values for b? This depends on the process. It was found that $\nu_m = c\gamma/b$ and hence $b \leq hc\gamma/E'$, for a photon of energy E'. In Rutherford scattering it was found that if b became very large, an atomic screening limit occurred, and a similar result is found here. A lower limit for b can be found by insisting that $h\nu_m$ not be greater than the total energy of the particle, γMc^2. Hence, $h\nu_m = hc\gamma/b \leq \gamma Mc^2$ or $b_{min} \geq h/(Mc)$, approximately the Compton wavelength of the particle. In the absence of screening, $b_{max}/b_{min} \approx (hc\gamma)/E' \times Mc/h = E/E'$.

2.3 Bremsstrahlung

Next apply this to bremsstrahlung. Consider the Feynman diagram in Figure 2.4a. In principle, either the electron or nucleus could be taken as the incident particle depending on the reference frame used. The Weizsäcker–Williams method evaluated in the laboratory considers the field of the incident electron as a set of virtual photons, which are then Compton scattered by the nucleus. Call this process one. The same method used in a frame in which the electron is at rest considers the field of the nucleus as the set of virtual photons. Call this process two.

Photon scattering is proportional to $1/M^2$, where M is the mass of the scatterer. Semiclassically, one can see this because, for the scattering, one needs to accelerate the particle and this is harder for the heavier particle. Hence (except for very low energies), process two dominates and, for process two, one transforms into the system in which the electron is at rest. Here it is appropriate to use classical Thomson scattering, since the low-energy photons here will be Doppler shifted up to be high-energy photons in the laboratory, and high-energy photons in this system will be low energy in the laboratory. The classical dipole radiation formula is

$$\left\langle \frac{dE}{dt} \right\rangle = \frac{2}{3}\frac{e^2 \langle a^2 \rangle}{c^3}, \qquad (2.18)$$

where $\vec{a} = e\vec{E}/m$. Above, it was noted that the energy/(area-s) is $(\vec{E})^2 c/(4\pi)$. The energy/(area-s) $\times \sigma_T =$ the energy scattered per second. Thus,

$$\frac{(\vec{E})^2 c}{4\pi}\sigma_T = \frac{2}{3}\frac{e^2}{c^3}\frac{e^2(\vec{E})^2}{M^2}.$$

This implies that the Thomson scattering cross section is given by

$$\sigma_T = \frac{8\pi}{3}\left(\frac{e^2}{m_e c^2}\right)^2 \left(\frac{m_e}{M}\right)^2 = \frac{8\pi}{3}r_e^2 \left(\frac{m_e}{M}\right)^2 = 6.6 \times 10^{-25}\left(\frac{m_e}{M}\right)^2 \text{cm}^2. \qquad (2.19)$$

Thomson scattering is independent of frequency and contains the expected $1/M^2$ dependence on the mass of the scattering particle.

One can now transform to the laboratory system, where the incident particle has energy E and put the pieces of this derivation together. Let the probability of a bremsstrahlung within dE' of E' in a distance dx (g/cm^2) be $\Phi(\gamma, E')dE'dx$. Note that from the Lorentz transformation, with the starred system representing the electron rest system, $\nu = \gamma(1 + \beta\cos\theta^*\nu^*) \approx \gamma\nu^*$. Thus if $\Phi =$ number of particles/cm$^2 \times$ number of photons per particle × cross section for scattering, one has

$$\Phi(E, E') = \left(\frac{N_o}{A}dx\right) N(\gamma, E')\sigma_T dE'$$

$$= \frac{N_o}{A}\frac{2}{\pi}Z^2\alpha^2 \ln\left(\frac{b_{max}}{b_{min}}\right) \frac{8\pi}{3}r_e^2 \left(\frac{m_e}{M}\right)^2 dx\frac{dE'}{E'}.$$

The factor $(2/\pi)(8\pi/3) = 16/3$ here becomes a factor of four in a more careful approach. If the incoming particle is not an electron, the factor of $(m_e/M)^2$ greatly reduces the probabilities. The characteristic dE'/E' dependence appearing here is responsible for many of the qualitites, peculiar to bremsstrahlung, that will be examined in this section.

In the electron rest system, the angular distribution of the radiation is broad, but in the laboratory one has $\tan\theta = \sin\theta^*/\gamma(\beta + \cos\theta^*) \approx 1/\gamma$. A good estimate for the characteristic laboratory angle turns out to be $(mc^2/E)\ln(E/mc^2)$.

The general bremsstrahlung formula can be written as

$$\Phi(E, E')dE'dx \approx F \times \frac{dE'}{E'}\frac{dx}{X_o}, \qquad (2.20)$$

where X_o is the radiation length for electrons defined in the previous section. Note that the $Z(Z+1)$ factor there takes account of the radiation from interactions with the nuclei (Z^2) and with the atomic electrons (Z). F is a slowly varying function of E'/E, which is about $4/3$ for small E'/E and close to 1 for E'/E near 1. For heavier incident particles the same formula applies but m_e is replaced by the mass of the particle. m_e enters X_o through r_e (the classical radius of the electron, e^2/m_ec^2), and r_e enters squared. r_e^2 is part of the constant term, 716.4 g/cm^2, in Equation 2.14. Hence, the radiation length is extremely long for heavier particles. For the simplest approximation, the function F is given by

$$F = \frac{E - E'}{E}\left[\frac{T}{T + 1/18}\left(\frac{E - E'}{E} + \frac{E}{E - E'} - \frac{2}{3}\right) + \frac{4}{9}\frac{1}{4T + 2/9}\right],$$

where $T = \ln(183Z^{-1/3})$.

Modifications to this formula, which reduce the lowest-energy part of the bremsstrahlung, occur in dense materials for very-high electron energies (LPM effect).[5] This effect occurs because, after interaction with one atom, the intermediate electron state can live long enough, from uncertainty principle considerations, to be scattered by a second atom in the

2.3 Bremsstrahlung

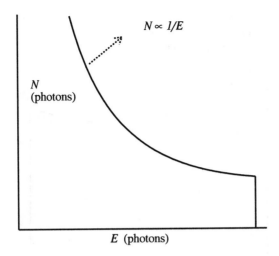

Figure 2.5 Bremsstrahlung: N(photons) versus E(photons).

material, thus destroying the coherence of the state. For electron energies in the TeV range, this effect starts to become significant in describing their energy losses. For electrons of 100 GeV or lower, only the very-low-energy part of the spectrum is strongly affected. Very roughly, the effect is large for $E'/E < 4 \times 10^{-5} E(\text{GeV})/X_0(\text{cm})$.

Note that in one radiation length an electron loses all but $1/e$ of its energy by bremsstrahlung on the average, i.e., $\langle E \rangle = E_o e^{-x/X_o}$.

For low-energy electrons, the ionization loss becomes larger than the radiation loss. The critical energy E_c is the energy when the average values of the two losses are equal. It is given approximately by $E_c = 610/(Z + 1.24)$ MeV for solids and liquids and by $E_c = 710/(Z+0.92)$ MeV for gases.

Straggling is very high for the bremsstrahlung process because there is a lot of energy in a few high-energy photons. Bremsstrahlung goes as $1/E'$ and knock-ons go as $1/(E')^2$. This means that for bremsstrahlung $\langle (E')^2 \rangle \propto (E'_{max})^2$, and one cannot define a definite range for the particles. This is a crucial factor in the design of experiments and detectors.

In Figure 2.5 a plot of the cross sections for emitting a photon in traversing dx versus energy is shown for Pb, and in Figure 2.6 $E \times$ number of photons is shown for Pb, i.e., the energy emitted due to photons of a given energy, versus energy.

Bremsstrahlung, considered in this section, provides the major energy-loss mechanism for high-energy electrons traversing matter. It is a mechanism with a very large variance.

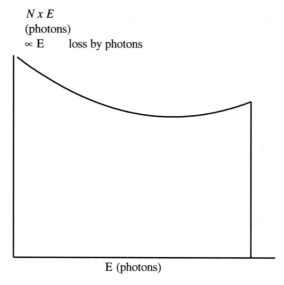

Figure 2.6 Bremsstrahlung: $E \times N(\text{photons})$ versus $E(\text{photons})$.

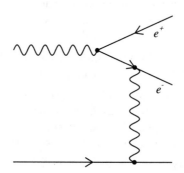

Figure 2.7 γ-ray pair conversion Feynman diagram.

2.4 Conversion of γ-Rays to e^+e^- Pairs

In pair production γ-rays traveling through matter, as they pass near nuclei, convert to e^+e^- pairs. The relevant Feynman diagram is shown in Figure 2.7. Note that by switching the directions of the top e^+ line and the initial photon line, the diagram in Figure 2.7 becomes the same as the diagram for bremsstrahlung in Figure 2.4. The diagrams are topologically the same. It seems reasonable that the considerations are similar to those for bremsstrahlung, and the concept of radiation length also enters here.

This process dominates for high-energy γ-rays. If a beam of high-energy γ-rays initially has intensity $I_\circ$, then, after a distance x, the beam will have

intensity
$$I = I_o e^{-x/(9/7X_o)}, \qquad (2.21)$$

where X_o is the radiation length and $9/7X_o$ is called the conversion length.

At energies below about 10 MeV photoelectric absorption, which goes as $1/E^2$, and Compton scattering, which goes as $1/E$, become important sources of loss and they become dominant below about 2 MeV. For high energies, the pair process, considered here, dominates.

At high energies, an electron in matter emits bremsstrahlung photons. These photons convert into e^+e^- pairs which then, in turn, emit bremsstrahlung radiation. Thus, an electromagnetic shower is built up. It will be noted in the next chapter, that this electromagnetic shower becomes the basis of detectors specific to high-energy electrons and photons.

2.5 DIRECT PAIR PRODUCTION

It has been seen that when a charged particle passes near a nucleus, the particle can radiate a γ-ray. The particle can also directly produce an e^+e^- pair. The Feynman diagrams for this latter process are shown in Figure 2.8. The formula for the cross section is quite complicated but it is approximately $\propto f(\gamma)/m_e^2$, where γ refers to the incident particle. At extremely high energies and when integrated over the pair energy, $f(\gamma) \propto \ln^3 \gamma$. Since the direct pair cross section rises with energy, eventually the energy loss due to this mechanism overtakes that due to ionization loss.

By 400–500 GeV, for muons, the direct pair energy loss is almost the same magnitude as the ionization loss. The variance is larger than for ionization, but much smaller than for bremsstrahlung. The range of a particle remains a meaningful concept, unlike the case with electron bremsstrahlung.

2.6 CHERENKOV RADIATION

When the velocity of a charged particle in a medium is faster than the velocity of light in the medium, then the particle radiates light. This phenomenon is known as Cherenkov radiation. The condition for this to occur is $v_{\text{particle}} > c/n$ ($= v$ of light in the medium), where n is the index of refraction of the medium. This is a phenomenon very similar to the emission of bow waves by ships when they are traveling through water faster than the velocity of surface waves in the water.

From the diagram shown in Figure 2.9 it is seen that coherent light is emitted with an angle to the incident direction of the particle of

2. Electromagnetic Interactions Used for Detection

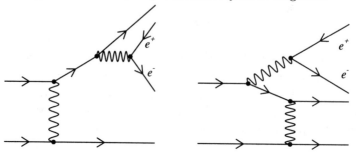

Figure 2.8 Direct pair production Feynman diagrams.

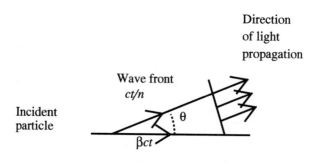

Figure 2.9 Cherenkov radiation.

$$\cos\theta = \frac{ct/n}{\beta ct} = \frac{1}{\beta n}.$$

Cherenkov radiation is a negligible source of energy loss, but the light emitted is very useful for particle detection. The rate of energy loss is

$$\frac{dE}{dx} = \frac{4\pi^2 z^2 e^2}{c^2} \int \left(1 - \frac{1}{\beta^2 n^2}\right) \nu d\nu, \qquad (2.22)$$

where z refers to the charge of the incident particle, ν is the frequency of the light, and cgs units are used, where $e = 4.8 \times 10^{-10}, c = 3 \times 10^{10}$, and E is in ergs ($= 1/1.6 \times 10^{-12}$ eV). This can be rewritten as

$$\frac{dE(\text{eV})}{dx} = 6.32 \times 10^{-27} z^2 \int \left(1 - \frac{1}{\beta^2 n^2}\right) \nu d\nu. \qquad (2.23)$$

The preceding formula allows the number of photons in a given frequency region to be found, since the energy of a given frequency photon is known

($E = h\nu$). Because of the factor ν, in the integral, blue light (higher frequency) is enhanced in the radiation. For water ($n = 1.33$), if $z = 1$, $\beta = 1$, then $dE/dx \approx 500$ eV/cm for visible light.

Cherenkov radiation provides a convenient signal for particle detectors, using detectors sensitive to light in approximately the visible region.

2.7 Exercises

2.1 Consider Rutherford scattering for large incoming particle energy and small scattering angle. Assume both the incoming particle and target are singly charged. Show that in this approximation, $p_\perp^2 = -q^2 = -t$. Find $p_\perp$ corresponding to impact parameter b of 10^{-13} cm. This sets an approximate scale for the regime in which proton form factor effects might start to enter. Hints: $E' \approx E$ for this case, since θ is small. For $p_\perp$, use $p_\perp c$, which has the units of energy.

2.2 Calculate b_{max} in iron for $\beta\gamma = 10$. Use the formula given in the text to find the mean ionization potential of iron for this calculation. Calculate the mean spacing of atoms in iron using the density of iron ($\rho = 7.87$). Note that b_{max} is far larger than the atomic spacing. Thus, by this energy, one would expect coherent effects, i.e., dielectric constant effects to be important. This tends to shield the material from the field of the particle, and this density effect then flattens the logarithmic relativistic rise of the energy loss.

2.3 Derive the formula given in the text for the maximum laboratory energy of a knock-on electron for a given incident E, m. For incident π^+ mesons of 1, 10, 20, and 200 GeV/c, find E'_{max}. Find E'_{max} for protons of the same momenta.

2.4 Derive the formulas for y_{rms} and $\langle y\theta \rangle$ given in the text. One does not need the specific form of the scattering given in Equation 2.12. One does need to assume that the scattering probability as a function of x is $p(\theta, x)d\theta dx \propto p(\theta)d\theta dx$, i.e., is independent of x. Also assume all angles are small, i.e., $\sin\theta \approx \tan\theta \approx \theta$.

2.5 Calculate the multiple scattering ($\sqrt{\langle\theta^2\rangle}$) and rms displacement ($\sqrt{\langle y^2\rangle}$) for a muon of 1 GeV momentum passing through 20 cm of Fe.

2.6 A 10 GeV parallel beam of positively charged muons passes through a 0.5 cm wall of Al followed by a 5 m path in nitrogen at STP. Initially the beam has an approximately normal distribution in the transverse directions with $\sigma = 1.0$ mm. What is the σ of the beam after passing through the Al and N_2, i.e., by how much has the multiple scattering increased the size of the beam? For Al, $X_0 = 8.9$ cm; for N_2 at STP, $X_0 = 4.7 \times 10^4$ cm.

2.7 Since multiple scattering falls exponentially and Rutherford scattering falls as a power of angle, eventually multiple scattering must have a single scattering tail. Consider integral probabilities and assume the angles are small so $\sin\theta \approx \theta$.

a) Integrate the Rutherford cross section over ϕ from 0 to 2π and θ from θ_1 to ∞. (In fact, the θ integration should only go to π. However, since the scattering angle is assumed small, the effect of extending the integration to infinity is negligible and makes the calculation simpler.) This is the cross section for scattering through an angle at least as large as θ_1.

b) Find the probability of Rutherford scattering through an angle at least as large as θ_1 in going through a thickness x of material of density ρ, atomic number Z, and atomic weight A.

c) For multiple scattering, integrate the probability for three-dimensional scattering from θ_1 to ∞. Note that this involves the product of the probabilities for each projection, $P(\theta_x)P(\theta_y)$. This product is integrable over θ, although the individual projected probabilities are not.

d) Equate the probabilities from parts b and c. Show that each of these depend only on the product $p\theta_1/x$. Hence, the probability at crossover is independent of thickness and momentum.

e) Find approximately the probability at crossover for a relativistic particle in iron. It is easiest numerically to try different numbers to see where the two probabilities are equal. For iron, $X_0 = 1.76$ cm; $\rho = 7.87$ g/cm^3; $Z = 26$; and $A = 55.85$.

2.8 From the formula given in the text for the radiation length, calculate the radiation and conversion lengths for Al, Fe, and Pb in g/cm^2 and in cm.

2.9 It is indicated in the text, from the formula given for Cherenkov radiation, that one can show for a highly relativistic particle traveling through water, dE/dx is about 500 eV/cm for visible light. Calculate this from the formula and estimate the number of photons per cm to which this corresponds. What is the Cherenkov angle for this case? Visible light corresponds to wavelengths between about 4000 to 7000 Å.

3
Particle Accelerators and Detectors

3.1 Particle Accelerators

Particle physics has continually pushed to higher and higher energies. This push is driven by simple considerations. Because of the deBroglie relation, $\lambda = h/p$, the search for structures of smaller and smaller sizes inevitably leads to higher and higher energies. hc is 1.24 GeV-fm, where one fermi is 10^{-13} cm. To examine distances on the scale of 10^{-16} cm requires energies of the order of 1000 GeV (1 TeV).

Modern particle accelerators are either circular machines or linear machines. In circular machines the path of particles is bent into a circle by magnetic fields while the particles pass repeatedly through regions of accelerating electric fields (which are at rf frequencies). The particles travel in bunches along the circumference with the bunches distributed such that the regions of rf electric field have the proper phase to accelerate the particles. Circular machines have the large advantage that each accelerating region is used many times and only moderate acceleration per turn is required. Most machines today are several stage accelerators, really composites of several accelerators, each moving the particles up through a particular energy region. The circumference of the final stage of modern day machines is becoming very large (tens of miles).

Electrons have a severe problem at high energy in circular machines owing to their low mass. When a charged particle is bent in a magnetic field, there is an acceleration (the centripetal acceleration). When a charged particle is accelerated, it radiates electromagnetic radiation. For particles bent in a magnetic field this is called "synchrotron radiation." It becomes so severe for electrons that it is virtually impractical to make circular machines for electrons with energy in excess of about 100 GeV, and even at lower energies linear machines are sometimes preferred.

The machines listed below are all circular machines unless specifically indicated as linear. As an aside, a useful formula for the radius of the circle a charged particle travels in a magnetic field is $pc = 0.3BR$ where pc is the particle momentum in GeV, B is the magnetic field in tesla (w/m^2), and R is the radius in meters. This also means that for small angles of bend $\Delta p_\perp \simeq 0.3 \int B dl$.

38   3. Particle Accelerators and Detectors

Besides being divided by circular and linear, accelerators are also divided by fixed target and colliding beam. In a fixed target accelerator only one beam of particles is accelerated and it then strikes a fixed target. It is possible to get very high interaction rates this way. Typically beams of 10^{13} particles or more can be obtained and beams are accelerated a few times per minute. A large fixed target accelerator can serve many different experiments at once by splitting the primary beam into several parts and then using several secondary beams from each part of the primary beam. In a colliding beam accelerator two beams of particles are accelerated and the beams are collided with one another. One can obtain much higher center-of-mass energies this way, but the interaction rates in the past have been much smaller. The next generation of colliding proton beam machines is expected to have interactions at the rate of 100–1000 MHz, so this limitation will not be present. The number of experiments in a colliding beam facility is much smaller than in a fixed target facility, since the number of regions in which the beams intersect is small. However, the groups performing each experiment tend to be very large and the detectors are very elaborate.

For colliding beam machines, the crucial intensity parameter is the luminosity L defined such that the event rate for a process with cross section σ is given by $L\sigma$. The relation among L, the current in each beam, and the size of each beam is examined in Exercise 3.4.

Although both kinds of accelerators are active and planned, much of current interest centers on the colliding beam accelerators. It is exceedingly interesting to see what nature will say at the highest available energies.

3.2 PRESENT HIGH-ENERGY ACCELERATORS

High-energy accelerators are very expensive and few enough in number that essentially all of them can be listed here.

3.2 a.   U.S. Accelerators

The present center of U.S. particle physics research is Fermilab located in Batavia, Illinois, near Chicago. (See Figure 3.1 in the color insert.) Fermilab has a fixed target proton accelerator (about 3 miles in circumference) at near 1 TeV (1000 GeV) and a colliding beam facility with proton and anti-proton beams, each near 1 TeV colliding. There are two intersection regions for the colliding beam facility and two experiments, CDF (Central Detector Facility) and D0. Fermilab is now considering programs to double the energy of the machine and/or greatly increase the intensity. The discovery of the Υ resonance, which signaled the presence of the b quark, occurred in a fixed target experiment at Fermilab, and the discovery of the t quark occurred at the Fermilab colliding beam facility.

SLAC (Stanford Linear Accelerator Center), Stanford, California, has concentrated on electron-positron physics. (See Figure 3.2 and Figure 3.3, the latter in the color insert.) SLAC has a fixed target 50 GeV electron linear accelerator (2 miles long); a low-energy e^+e^- colliding beam facility (SPEAR), which goes up to about 5 GeV in the center-of-mass; a medium-energy e^+e^- collider, PEP, going up to about 36 GeV in the center-of-mass; and a high-energy colliding beam e^+e^- facility, SLC (SLAC Linear Collider), going up to near 100 GeV in the center-of-mass. The low-energy and high-energy colliders have only one intersection region used for particle physics. The medium-energy collider, PEP, had several, but it is now closed down. The initial discovery of the second generation c quark occurred at SPEAR as did the discovery of the τ lepton, which was the first indication of a third generation of particles. SLC participated in acquiring the evidence limiting the number of generations of fundamental particles to three.

The low- and medium-energy colliders are circular machines fed by the linear accelerator. The high-energy machine, SLC, is a novel accelerator, which may well be a prototype for the future. It consists of two beams accelerated by the linac, which are bent and intersect once. It requires very small beams, of the order of a few microns in diameter, but avoids the very large radii needed in multiturn circular machines to guard against synchrotron radiation. The technique appears promising. A luminosity of $3.5 \times 10^{29}/\text{cm}^2/\text{s}$ has been obtained at SLC. Future electron accelerators may use colliding beams from two linacs. SLAC is discussing a 0.3–0.5 TeV center of mass e^+e^- facility of this type [NLC (Next Linear Collider)].

SLAC is presently building a medium-energy circular e^+e^- collider, PEPII, of high intensity and with asymmetric energy beams to study CP violation in b quark physics (b-factory). The beams will be 9 and 3.5 GeV. The design luminosity is $10^{33}/\text{cm}^2/\text{s}$, and the planned completion date is 1999.

BNL (Brookhaven National Laboratory), Long Island, New York, has a 30 GeV proton accelerator with high intensity beams used for lower-energy and decay experiments. The Ω^- particle, which helped in the understanding of strong interaction symmetries, was discovered at BNL as was the evidence that the muon and electron neutrinos are distinct. The maximum beam intensity is 6×10^{13} protons per pulse at 2 s intervals. They are now also accelerating heavy ions in the machine.

BNL is building a larger relativistic heavy ion collider (RHIC), which will be 100 GeV/amu per beam for Au and 250 GeV for protons. The design luminosity is $1.2 \times 10^{27}/\text{cm}^2/\text{s}$ for Au–Au collisions and $1.5 \times 10^{31}/\text{cm}^2/\text{s}$ for pp collisions. Completion is planned for mid-1999. (See Figures 3.4–3.5.)

CESR (Cornell Electron Synchrotron), Ithaca, New York, is a 16 GeV center-of-mass e^+e^- colliding beam machine. It happens that the region it covers is a region of considerable interest for the study of the b quark and,

40   3. Particle Accelerators and Detectors

Figure 3.2 Aerial view of SLAC. The two mile long linear accelerator can be seen as can the end stations. *(Photograph courtesy of Stanford Linear Accelerator Center.)*

although the energy is low, it has proven to be a very useful machine. There is now one experimental set up, CLEO. CESR has plans to increase their luminosity to become a b-factory. The 1995 luminosity is $6 \times 10^{32}/\text{cm}^2/\text{s}$.

Los Alamos (New Mexico) has a fixed target, low-energy, high-intensity "pion factory," LAMPF, which has been active in low-energy experiments. LAMPF accelerates protons to a kinetic energy of 800 MeV. It accelerates 5×10^{13} protons/pulse and has 120 pulses/second.

Finally, mention should be made of the now abandoned SSC (Superconducting Supercollider) project. This was to have been an enormous colliding beam accelerator with a ring about 53 miles in circumference. Two beams of protons with a total energy of 40 TeV in the center of mass were to have collided. The $11 billion project was started on a site near Dallas, Texas, and, in an action tragic for particle physics, abandoned in 1993 by

3.2 Present High-Energy Accelerators

Figure 3.4 Aerial view of Brookhaven National Laboratory. The large ring in the upper left is for the RHIC heavy ion accelerator. The smaller ring below that one is the AGS accelerator. *(Photograph courtesy of Brookhaven National Laboratory.)*

the U.S. Government after an expenditure, including closing costs, of about $3 billion.

3.2 b.   Accelerators in Western Europe

The major European accelerator center is CERN (Orginisation Européenne pour la Recherche Nucléaire). This laboratory is located near Geneva, Switzerland, on the Swiss–French border, and is supported by a consortium of European countries. It has a large fixed target accelerator (SPS, 400 GeV protons), a large proton–anti-proton colliding beam facility with 800 GeV in the center of mass, a low-energy anti-proton storage ring (LEAR), and a high-energy e^+e^- collider (LEP). The $\bar{p}p$ collider has a luminosity of $6 \times 10^{30}/\text{cm}^2/\text{s}$. It has two large experimental areas and some parasitic

Figure 3.5 Magnets in the Brookhaven alternating gradient synchrotron (AGS). *(Photograph courtesy of Brookhaven National Laboratory.)*

experimental areas, but is not now active. The discoveries of the neutral weak currents and the weak gauge bosons, the Z and W, were made at the CERN hadron machines.

The intensity obtainable in the anti-proton storage ring LEAR depends on the extraction mode and the anti-proton energy. It can deliver 3×10^6 protons/s for a 1 h period or 10^9 protons in shots of 10^7. The beam momentum can be as low as 60 MeV/c or as high as 1.95 GeV/c.

The e^+e^- collider, LEP, 27 km in circumference, has four experimental areas. It has been operating near the Z^0 peak, about 91 GeV. In 1996 it will go to about 190 GeV, past the W-pair threshold, after an improvement program. The four experiments are known as L3, DELPHI, OPAL, and ALEPH. Some of these LEP experiments are as elaborate as entire medium size laboratories. The L3 experiment, which is the most elaborate of these, cost close to $200 million. LEP started operations in 1989, obtaining about 100K–150K Z^0 per experiment in the first year of operation. By 1996 each experiment is expected to have about 10 million Z^0 events.(See Figure 3.6 in the color insert and Figure 3.7.) The bulk of the evidence limiting the number of generations of fundamental particles to three and precision tests of the Standard Model were obtained at LEP. Starting in 1996, LEP2 will examine WW interactions, search for Higgs and other possible particles, and further examine the Standard Model.

CERN is also building additional facilities. They are building a large hadron collider (LHC) using the LEP tunnel. It will accelerate two beams of protons or of heavy ions. Present plans call for completion of a first phase in 2004 with about 9 TeV available in the center of mass at a luminosity of about $10^{33}/cm^2/s$ for protons. A second phase is then planned to be completed in 2008, which will have 14 TeV available in the center of mass at a luminosity of $10^{34}/cm^2/s$ for protons.

CERN has discussed a large linear e^+e^- collider (CLIC) with about 1 TeV in the center of mass in the farther future.

DESY (Deutsche Elektronen Synchrotron), Hamburg, Germany, has built a large e–p colliding beam facility (HERA), which collides a beam of 30 GeV electrons with a beam of 820 GeV protons. This facility started running in 1992. The design luminosity is $1.5 \times 10^{31}/cm^2/s$. Two experiments, Zeus and H1, are under way. HERA experiments are examining weak and electromagnetic scattering of electrons on protons in unique kinematic regions. They are also discussing a linear e^+e^- collider for the future.

3.2 c. Accelerators Elsewhere

IHEP at Serpukhov, Russia, has a 70 GeV fixed target proton machine. They are building a large new machine, UNK. It will initially be a fixed target proton machine of 600 GeV. At approximately two year intervals it had been scheduled to go to 3 TeV fixed target, then 6 TeV center-of-mass pp colliding beam. However, the schedule is in considerable flux because of the political and economic instability in the region. A 1 TeV center-of-mass linear e^+e^- colliding beam facility had been planned next. (It is being designed at Novosibirsk, Siberia, but is to be installed at Serpukhov.) Finally, an ep colliding beam facility using the facilities above is hoped to be built eventually.

KEK at Tsukuba, Japan, has a 64 GeV center-of-mass e^+e^- colliding beam machine (TRISTAN) now in operation with two experiments (VENUS and AMY). The luminosity is $3.7 \times 10^{31}/cm^2/s$. Japan has also been discussing plans for a linear e^+e^- collider, and is building an e^+e^- colliding beam b-factory. This latter machine, KEKB, is planned for 1998 completion. The e^+e^- beams will be asymmetric (3.5×8) GeV and the design luminosity is $2 \times 10^{33}/cm^2/s$ which will be extendable to $10^{34}/cm^2/s$.

44   3. Particle Accelerators and Detectors

Figure 3.7 View of the LEP tunnel. *(Photograph courtesy of the CERN Photographic Service.)*

Beijing, China, has a low-energy (4.4 GeV in the center of mass) e^+e^- colliding beam machine (BEBC) in operation. The luminosity is $10^{31}/\text{cm}^2/\text{s}$.

This has been a brief, but essentially complete, survey of the large present day particle accelerators. These devices require very large efforts, and it is likely that most future giant accelerators will be true international efforts.

3.3   A QUICK SURVEY OF SOME MAJOR TYPES OF DETECTORS

In this section some of the principal types of particle detectors are described briefly. Good surveys in some detail are given in references 6–8.

3.3 a. Detectors Measuring Ionization

The simplest chamber of this type is the ionization chamber. Here, as a particle passes through a gas, it makes a string of ions and free electrons. An electric field is set up between two plates (see Figure 3.8a). The electrons drift quickly to the positive plate, the ions more slowly to the negative plate. The drift velocities of the electrons vary greatly depending on the gas and the electric field, but tend to be in the range of a few centimeters per microsecond. The average drift velocity depends on a balance between the acceleration due to the electric field and collisions with molecules. For a constant electric field the drift velocity is a constant. The drift velocity depends on mass as $1/\sqrt{M}$; ions drift much more slowly than electrons. A major worry in ionization chambers is that if the electrons become attached to atoms to make negative ions, the signal rise time deteriorates badly. The signals due to a single track are very low with an ionization chamber. One usually cannot see individual particles, and a time-integrated signal is used.

It is natural to try to find a detector with larger signals. An early attempt used a thin wire a few mils in diameter (1 mil = 0.001 inch = 25.4 μm) as a positive electrode. The field outside the wire depends on the distance from the wire center as $1/r$; as electrons drift near the wire, they eventually have sufficient energy to ionize atoms (see Figure 3.8b). These new electrons are then accelerated in the field and make additional ions. If the voltage difference is sufficient, a spark can occur. This was the basis of the Geiger–Müller counter. A "quencher" gas is often used to absorb photons from the cascading process that otherwise might travel far down the gas and start new discharges.

If a very high field is pulsed between parallel plates, a spark can be made to occur along the path of the particle. This was the principle of an early electronic chamber allowing for track visualization, the spark chamber. Position accuracies of the order of 1 mm are obtainable and time resolutions are about 1 μs. A typical dead time, the time after particles traverse the chamber before the device can be made sensitive again, is about 0.1 s. Typical pulsed electric fields required are in the 10–20 kV/cm range.

In a proportional wire chamber (PWC) the positive electrode is again a thin wire; however, the field used is lower. It is then possible to control the cascading process so that it does not run away and just give a spark. This allows much faster recovery and the signals are proportional to the amount of ionization. Amplifications up to about 10^5 are possible. A typical signal for a minimum ionizing particle traversing 1 cm of gas, with 5×10^4 amplification and an effective capacitance of 20 pF, is about 20 mV. If a wire of high resistance is used (e.g., a stainless steel wire), then one can read out both sides of the wire and use the ratio of the pulse heights to measure where on the wire the discharge occurred. Accuracies to about 1 % of the length of the wire are possible.

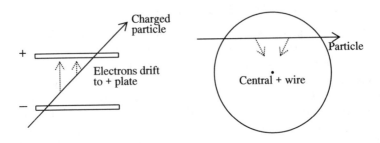

(a) Ionization chamber.   (b) Proportional chamber.

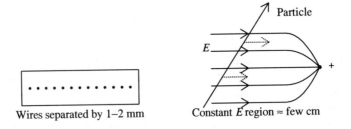

(c) Multiwire proportional chamber.   (d) Drift chamber.

Figure 3.8 Various detectors measuring ionization directly.

A multi-wire proportional chamber (MWPC) is a proportional counter with many wires spaced typically a few millimeters apart (see Figure 3.8c). Spatial inaccuracies are usually ≥ 300 μm. Rise times of the signals in PWC's and MWPC's are typically of the order of 0.1 μs and resolution times of better than a nanosecond (1 ns = 10^{-9} s) can be achieved. Rise times are determined by the size of the region over which amplification occurs, the size of the region over which electrons are collected, and the capacitance of the electronic circuit used. With small distances and a fast gas, rise times of 10 ns or better are possible. Dead times are typically 200 ns. The resolution, if only a single wire is hit, is $\sigma = $ anode spacing$/\sqrt{12}$.

Drift chambers are a different modification of a PWC. Here the idea is to have a region where the electric field is kept very constant before the field starts to rise near the wire (see Figure 3.8d). The drift velocity for electrons is then constant for most of the electron path. For constant drift velocity, the time at which a signal arrives at the wire measures the distance from the wire at which the particle passed through the detector. Drift distances range up to 10 cm or so. A typical drift chamber will have many signal wires. In addition, one needs special extra field-shaping electrodes to keep the field uniform over the drift region. The usual drift chamber cannot

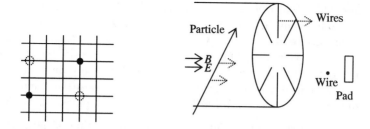

(a) "Ghost tracks" in two crossed chambers. (b) Time projection chamber.

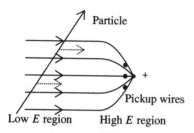

(c) Time expansion chamber.

Figure 3.9 Other ionization detectors.

take extremely high rates of events. A system of bundles of very small drift chambers (straws), each a few millimeters in diameter, can take rates up to 20 MHz.

With care this device can measure positions to accuracies of 50–300 μm or even better. The resolution time can be 2 ns and a typical dead time is 100 ns. Although gas-filled chambers are most common, chambers using liquids (argon, TMP) and even solids (neon) have been used or are showing promise as PWC and drift chambers.

The drift chamber measures distances in only one direction. If one uses two drift chambers at right angles, there is the problem of "ghost tracks" if two or more particles pass through. If two particles pass through the chambers at positions x_1, y_1 and x_2, y_2, respectively, is the position x_1, y_1 and x_2, y_2 or x_1, y_2 and x_2, y_1 (see Figure 3.9a)? Chambers in the diagonal directions (u and v chambers) are often used to resolve the ambiguity. There is also a twofold ambiguity in that it is not clear whether the drift is from the right or left in a single chamber.

The time projection chamber (TPC) is another variant. This is typically a large chamber with parallel E and B fields in, say, the z-direction (see

Figure 3.9b). The electrons drift in the z direction in the electric field. This is a long drift length, typically 1 m or more. The drift time measures z. The electrons arrive at a series of wires with pads behind them. The pads pick up signals capacitively. One obtains x, y from the position of the electrons at the detecting wires and pads and, in addition, obtains information on the ionization of the track from the pads. The position accuracy can be ≈ 200 μm for r, ϕ, and 750 μm for z. The time resolution is ≈ 15 ns.

The time expansion chamber (TEC) is still another variant. Here there are fewer signal wires. There is a large region of low constant field and then a region of high field (see Figure 3.9c). A single wire looks at the electrons from many places on the track. Discrimination is obtained by looking not just at the time of arrival but at the pulse shape using a sequence of very fast analogue to digital converters ("flash ADC's"). This method no longer seems as promising as some of the others.

3.3 b. Other Types of Particle Detectors

All of the preceding detectors used controlled amplification of the signals, i.e., proportional to the input signal. Another type of detector uses still higher amplification and makes a spark or glow discharge. An early particle detector, the Geiger–Müller counter, mentioned previously, was of this kind. Another simple detector of this type is the flash chamber. Here there are a number of tubes filled with a neon–helium mixture and a large electric field is placed across them. A glow discharge is formed when a charged particle goes through. Typically the walls of the tube are semi-insulating and the discharge is quenched by the charges set up on the walls. These charges then leak off slowly allowing the device to become sensitive again. One can either pick up the very large electronic signal or look at the light produced. One can effectively take a picture of one projection of an interaction.

A more sophisticated form of this device is the streamer chamber. This is a gas-filled device in which a short, strong electric field pulse is introduced. The discharge is formed but is only able to grow to about 2–3 mm within the duration of the pulse. A typical device has a measurement accuracy of about 300 μm, a resolution time of 2 μs, and a dead time of about 200 ns.

Another promising variant of this device is the resistive plate chamber. Here a spark is allowed to occur in a parallel plate device, but the plates are made of a very high resistance material. The discharge damps itself after discharging an area of the plate of the order of a few square millimeters. The signals have a very fast (ns) rise time, and are very large eliminating the need for amplifiers. The device is capable of reasonably large rates and is fairly radiation resistant. The time resolution can be ≤ 0.5 ns. The position resolution is limited by the size of the plates which are often several centimeters on a side.

The scintillation counter uses a different effect to detect particles, luminescence from ionization. A particle passes through a special plastic that converts some fraction of the energy lost to visible light. In plastic scintillators this fraction is of the order of 1–2%. In NaI, which is very efficient, it is about 8%. The light is detected by a photomultiplier tube (see Figure 3.10a). The light collection efficiency into the phototube is typically $\leq 10\%$. In this tube a bi-alkali cathode uses the photoelectric effect to convert the light to electrons (with perhaps 20% efficiency). The electrons are accelerated and crash into a metal plate knocking out several more electrons (secondary emission). This process is repeated at each of 10–14 stages. Signals with widths (after clipping) of 5–10 ns can be obtained and time resolutions of better than 150 ps can be obtained for small counters. Typical dead times are 10 ns.

Silicon photodiodes can give close to 70% efficiency in light conversion for photons in the visible region. For these latter devices care must be taken that the signal produced by the scintillator light is larger than the signal produced by particles traversing the photodiodes.

There are a number of kinds of scintillators. Plastic scintillators such as polystyrene are doped with various dyes (e.g., p-terphenyl plus tetraphenyl butadiene). In addition there are liquid scintillators, crystal scintillators, and scintillating glasses. Some examples of crystal scintillators are sodium iodide (doped with thallium), cesium iodide, barium flouride, and bismuth germanate (BGO). Scintillating plastic fibers with diameters of 1 mm or less are in use. The potential resolution is of the order of the fiber diameter/$\sqrt{12}$ if only one fiber is hit, and is better if several are hit, allowing a center of gravity to be found.

The bubble chamber was a mainstay of particle physics detectors for many years after its invention by Don Glaser at the University of Michigan in the 1950s. It has now largely been phased out of modern experiments, but for 20 years it dominated the field. In this detector a liquid is under pressure and at a temperature near its boiling point. The accelerator is set to give a short beam spill. Just before the beam comes, a piston is moved which expands the volume of the chamber and drops the pressure (see Figure 3.10b). At this new low pressure the temperature of the liquid is above its boiling point and it wants to boil. Just at this point the beam of particles comes through. Each charged particle provides a heat spike to the liquid along its trajectory, and the boiling initiates at these trajectories first. After a short delay to allow the bubbles to grow, a light is flashed and stereo pictures of interactions or decays occurring in the apparatus are taken. The piston is then recompressed. The expansion–recompression cycle lasts from about 0.1 to 100 ms depending on the size of the chamber and the sensitive time varies from 10 μs to about 2 ms. The chamber sizes have varied from about 10 cm to about 4 m in diameter. Liquids used have included hydrogen, deuterium, and helium as well as heavier

50   3. Particle Accelerators and Detectors

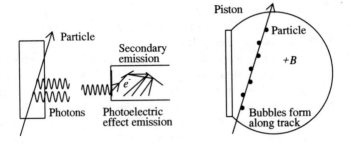

(a) Scintillator and photomultiplier.   (b) Bubble chamber.

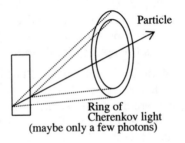

(c) Ring-imaging Cherenkov counter.

Figure 3.10 Other particle detectors.

liquids. The resolutions for many years were of the order of 150 μm-1 mm, but with holography in large chambers and with small precision chambers resolutions of 10 μm or better have been obtained.

The disadvantages of the device have been that it is sensitive for only a small fraction of the total time and that the storage medium is photographic film, which meant that it has been hard to collect many more than 10^6 events. Most experiments today are high rate experiments or colliding beam experiments.

There are several varieties of Cherenkov counter using the Cherenkov effect and measuring the resultant light. The threshold Cherenkov counter is simply a yes/no device. If a particle has velocity above threshold, it gives a signal.

The differential Cherenkov counter is also a yes/no device. Here the incoming particle direction is fixed and only the light traveling in a given

Color Plates

Figure 3.1 Aerial view of Fermilab. The large circle is the Fermilab Tevatron. The 16-story main laboratory building (Wilson Hall) is seen at the center left. Below it, booster rings and a cooling ring for anti-protons can be seen. Above it are various beam lines for fixed target experiments. (*Photograph courtesy of Fermilab Visual Media Services.*)

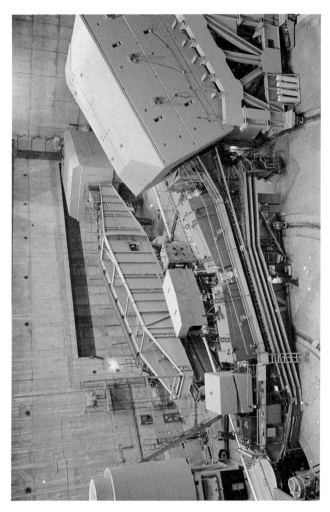

Figure 3.3 View of one of the end stations at the SLAC linear accelerator. *(Photograph courtesy of Stanford Linear Accelerator Center.)*

Figure 3.6 Aerial view of the CERN site showing the position of the LEP ring. The Geneva airport is seen in the lower left. On the middle left part of the picture, the main CERN campus is seen. The dotted line is the French-Swiss border. *(Photograph courtesy of the CERN Photographic Service.)*

Figure 3.13 The CDF $\bar{p}p$ collider detector. The black arches removed to the side contain modules of the central calorimeter, which, in operation, surround the region of the detector where the collisions of protons and anti-protons take place. (*Photograph courtesy of Fermilab Visual Media Services.*)

Figure 3.14 A view of the muon detectors for the CDF $\bar{p}p$ collider detector at Fermilab. (*Photograph courtesy of Fermilab Visual Media Services.*)

Figure 3.15 The UA1 detector at CERN. The Z and W intermediate bosons were discovered in this detector. *(Photograph courtesy of the CERN Photographic Service.)*

Figure 6.13 The SLAC apparatus, used for the discovery of charm and the τ lepton. *(Photograph courtesy of Stanford Linear Accelerator Center.)*

Figure 6.19 An example of the production of a $t\bar{t}$ pair in $p\bar{p}$ collisions in the CDF detector at Fermilab. In this example, $t \to b+W^+$; $W^+ \to e^+\nu$; $\bar{t} \to \bar{b}+W^-$; $W^- \to q\bar{q} \to 2$ jets, where the q and $\bar{q}$ are jets 2 and 3 and the b and $\bar{b}$ are jets 4 and 1 respectively. The B hadrons containing the b quarks travel a short distance before decaying. The decay vertices are clearly seen. The two jets from the W^- have an effective mass of about 79.5 GeV fully consistent with the 80 GeV mass of the W. *Drawing courtesy of the CDF group.*

Fig. 6.13 ▲

e + 4 jet event
40758_44414
24-September, 1992

TWO jets tagged by SVX

fit top mass is 170 +- 10 GeV

e^+, Missing E_t, jet #4 from top
jets 1,2,3 from top (2&3 from W)

LEGO view

Tracking View

Two Vertex Views
(note scales)

Fig. 6.19 ▶

Figure 13.7 The SLD detector at the SLAC SLC accelerator. *(Photograph courtesy of Stanford Linear Accelerator Center.)*

direction is looked at. An example of the use of this device would be to select out kaons for a fixed momentum incoming beam, vetoing pions and protons.

The ring-imaging Cherenkov counter (RICH) attempts to have an array of detectors after the Cherenkov region to measure the whole ring of Cherenkov light, obtaining information on the velocity and the direction of the particle (see Figure 3.10c). This device can handle several particles simultaneously and does not require that all the particles travel in parallel paths.

Another kind of detector is the transition radiation detector (TRD).[8] When a charged particle traverses the interface between substances of different dielectric properties, the particle produces a time-varying electric field and a transient polarization of the media that yields optical and x-ray photons. If a charged particle passes through a large set of foils, low energy photons are produced which can be detected by a drift tube or other detector. The signal is sensitive to the γ of the incident particle, and, if the particle momentum is known, this device can provide particle identification for particles in the 100 GeV range and beyond.

There are sets of new solid state detectors such as silicon strip detectors, charge coupled devices (CCD's), etc. These can take high rates and can give precisions in position accuracy of about 4–20 μm. Time resolution of some designs is expected to be ≤ 15 ns. They are generally rather small detectors so far. However, they are already very valuable and show great future promise. Silicon pixel detectors have resolutions of about 2 μm.

The best position accuracy is still obtained with old fashioned nuclear photographic emulsion, a heavily silvered emulsion in which accuracies of under 0.5 μm are possible. These detectors are still used in some specialized experiments. The density and radiation lengths of a typical nuclear emulsion are $\rho = 2.83$ g/cm^2, $X_\circ = 2.9$ cm.

Thus there are a panoply of detectors corresponding to the many varied needs of experimenters in designing experiments.

3.3 c.   Calorimeters

Calorimeters are combinations of the preceding detectors used to measure the total energy of particles. They are often sandwiches, consisting of a number of layers of absorbing material interspersed with layers of counters of one or another of the types described previously. First consider incident photons or electrons. Suppose an electron is incident on a large volume of matter. It radiates, emitting a photon. That photon is converted, producing an e^+e^- pair. Now there are three particles that can radiate. The process continues until there are many thousands of electrons, positrons,

52   3. Particle Accelerators and Detectors

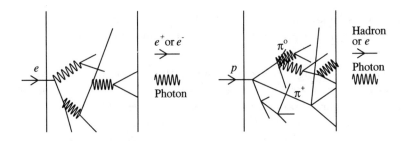

(a) Shower multiplication for an electromagnetic shower.

(b) Shower multiplication for a hadronic shower.

Figure 3.11 Particle showers in calorimeters.

and photons. Eventually these are all low energy and cannot produce more. This is called a shower (see Figure 3.11a). Typically the distance to the shower maximum varies with energy E as $\ln E$ (GeV) and is about 6–7 radiation lengths for 6 GeV particles. The size of the shower in the transverse direction is determined principally by multiple scattering considerations and is known as the Molière radius. The Molière radius is given by $R_M = X_\circ E_s/E_c$, where $E_s = 21$ MeV, and E_c is the critical energy, defined in Section 2.3.

An electromagnetic shower detector can be made from a large volume of material all of which gives a signal when radiation passes through it, or it can be a sampling device. The variance (σ) of the signal from a sampling calorimeter is typically dominated by sampling fluctuations leading to a fractional resolution (σ/E) varying as $1/\sqrt{E}$. There is often, in addition, a constant term. A common arrangement is to have a sandwich built up of layers of a high-Z material (Pb, U, W) and scintillators or PWC's (gas or liquid). The radiation length of Pb is 0.56 cm. A typical resolution of sandwich-type devices is $\sigma/E \simeq 0.1$–$0.3/\sqrt{E}$ GeV. A good totally sensitive, but expensive, device is bismuth germanate (BGO) crystals, which can have resolutions of 1% above 1 GeV and 5% at 100 MeV. Other totally sensitive devices (with somewhat less resolution than BGO) are sodium iodide (NaI) crystals, lead-glass Cherenkov counters, and scintillating glass. CsI (cesium iodide) and BFl (barium fluoride) have also been used. Each of these has advantages and disadvantages in terms of resolution, cost, radiation hardness, speed, etc. A spaghetti calorimeter made of scintillating fibers immersed in a heavy metal base gives resolutions of $\Delta E/E \simeq 0.1 - 0.15/\sqrt{E}$ GeV intermediate between sandwich and totally sensitive devices.

Next turn to measuring the energy of strongly interacting particles such as pions or protons. Again a shower can be measured, this time the nuclear

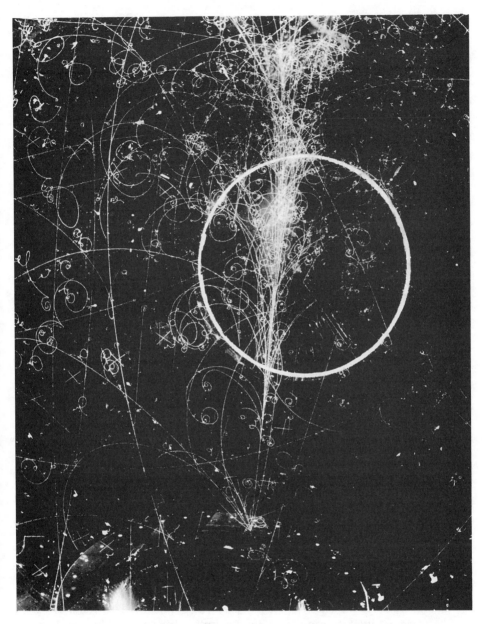

Figure 3.12 A neutrino interaction in the Fermilab 15-foot bubble chamber filled with a neon-hydrogen mixture. Note the nuclear and electromagnetic showers in this relatively short radiation and interaction length liquid. *(Photograph courtesy of the Fermilab Visual Media Services.)*

shower. Here the characteristic distance is a nuclear interaction length not a radiation length. There are typically many fewer particles, which means that fluctuations are more important and the accuracy less than with electromagnetic showers (see Figure 3.11b). Typical sandwich-type detectors are made similar to the electromagnetic calorimeters (but often with less high-Z absorber materials) and have resolutions in the range $\Delta E/E \simeq 0.35 - 1.0/\sqrt{E}$ GeV.

In Figure 3.12, examples of both electromagnetic and hadronic showers can be seen. Views of a large particle detector, the CDF detector at Fermilab, are given in Figures 3.13 and 3.14 in the color insert. The CERN UA1 detector is seen in Figure 3.15, also in the color insert.

In this chapter a very brief survey of some of the modern particle accelerators and detectors has been given. This is a very active area of research as indicated by the fact that the 1992 Nobel Prize in physics was awarded to Georges Charpak for his work in detector design. New ideas for detectors and combinations of detectors arise continually corresponding to the ever increasing needs of ever more complex experiments.

3.4 Exercises

3.1 The magnetic field in the bending magnets at the Fermilab collider is 4.4 T for 1 TeV protons, and the circumference of the collider is 6.28 km. The magnetic field of the bending magnets at LEP for 45.5 GeV electrons is 0.05 T, and the circumference of LEP is 26.7 km. For each machine calculate the fraction of the circumference that is filled with bending magnets.

3.2 A charged particle travels a curved path in a magnetic field. It is, therefore, being accelerated and, classically, emits electromagnetic radiation. This radiation is observed in particle accelerators and is known as synchrotron radiation. The peak of this radiation occurs at a frequency corresponding to a photon energy of $E(\text{GeV}) = \hbar c \gamma^3 / 2R(\text{in m}) =$ (for electrons) $0.74 \times 10^{-6} E(\text{GeV})^3 / R(\text{in m})$, and the radiation per turn is given by $4\pi e^2 \beta^3 \gamma^4 / 3R(\text{in m})$. For high-energy electrons this equals $8.85 \times 10^{-5} E$ (in GeV)$^4 / R(\text{in m})$. Find the energy of photons at the peak of the synchrotron radiation, and the average energy/turn lost for the Fermilab collider and for LEP. (See the previous problem.) This radiation is the principal limitation on the size of circular electron machines and the reason the LEP magnetic field is kept so low.

3.3 What is the frequency of revolution for particles at Fermilab and at LEP? If LEP runs eight bunches of particles in each beam, what is the time between collisions of bunches? LEP runs its rf accelerating cavities at about 352 MHz. To what harmonic of the fundamental revolution frequency does this correspond?

3.4 For colliding beam machines, the crucial intensity parameter is the luminosity L defined such that the rate in events/second for a process with cross section σ is given by $L\sigma$. Suppose for an idealized colliding beam machine, the current in each beam is I amperes, each beam is in a single bunch so collisions occur only in the interaction region, and each beam is uniform in cross section with radius r. Let the period of revolution for the beam in the machine be τ. Find the luminosity in terms of I, r assuming the beams collide head on. If each beam has n bunches, but the total current of all the bunches in each beam remains I, and the bunches are made to collide only at the one intersection region, what effect does this have on the luminosity?

3.5 The primary standard focusing elements in large accelerators are quadrupoles, and for these quadrupoles, to first order, simple lens optics equations can be used. Unfortunately, if a quadrupole magnetic field is focusing in the x-direction, it is defocusing in the y-direction, i.e., its focal length is f in one direction and $-f$ in the other direction. Show that if two quadrupoles, each of focal length f, are set so one is focusing and the other defocusing in the x-direction, then if they are a short distance D apart, the combination can be focusing in both directions. If the object and image distances are large with respect to f, and f is large with respect to D, find the effective focal length in each direction of the resulting quadrupole pair.

3.6 Two scintillators are used in coincidence in an accelerator experiment with a beam that is almost continuous over a large time interval ("slow beam spill"). The scintillators have pulses that, after shaping by electronics, are 5 ns (5×10^{-9} s) in width. Besides the desired coincidences, each of the two scintillators is individually hit by random particles at a rate of 2 MHz. Assume that if there is any overlap in time between the pulses from the two scintillators, it will be counted as a coincidence. What is the rate for accidental coincidences in this experiment?

3.7 A 10 GeV/c momentum muon travels 1 m in vacuo in a 2 T magnetic field. The position at the beginning, end, and center of this trajectory is measured with a 1 mm standard deviation in the transverse direction for each measurement. To what accuracy can the momentum of the particle be measured? [Hint: this is most easily approached by asking to what accuracy the sagitta (see problem 1.12) can be measured.]

3.8 Consider the particle and magnetic field of Problem 3.7. However, this time imagine that the particle is traveling within a solid iron magnet. The momentum measurement is limited by both measurement accuracy and multiple scattering. Find the fractional momentum uncertainty due to multiple scattering and show that it is independent of the incoming particle momentum. Ignore energy loss in the iron. However, you will need to include the correlations in the multiple scattering at the three points. For iron, $X_0 = 1.76$ cm.

3.9 Show that if the beams in a high-energy colliding beam machine collide at an angle θ instead of head-on (where $\theta = 0$ is head-on), then the center-of-mass energy squared is reduced to approximately $(1 + \cos\theta)/2$ of the head-on value.

3.10 High-energy protons are incident on a target. Following the target, a set of slits and magnets is used to define a small diameter beam of 10 GeV/c momentum positive particles. The beam contains appreciable amounts of e, μ, π, K, p. It is desired that threshold Cherenkov counters be added to tag which particles are K^+. How many Cherenkov counters are needed? What β threshold should each Cherenkov have?

3.11 In Problem 3.10, it is desired to emit about 100 photons between 400 and 700 nm. What length should each of the Cherenkov counters be? If the transverse dimensions of the beam are small, over what radius will these photons be spread at the end of the Cherenkov counter? What gas might be suitable for each, i.e., usable at not too unreasonable a pressure? (Look at the table on atomic and nuclear properties of materials in the *Particle Data Tables Handbook*.)

3.12 What path length is required to distinguish charged π's from charged K's, both with momentum 2 GeV/c, at the 90% confidence level, if the time resolution is 300 ps (300×10^{-12} s)?

4
Invariance, Symmetries, and Conserved Quantities

4.1 PRELIMINARY DISCUSSION Symmetries, invariances, and conserved quantities point the way to a complete theory, strongly limiting the choice of possible forms. They help one to see fundamental features of interactions hidden within vast amounts of phenomena and formalism.

4.1 a. Invariance Formalisms

In this section a general relation between symmetries and conserved quantities is discussed.

First *Hermitian* operators will be defined. A Hermitian operator, R, is an operator that satisfies the relation $R^\dagger = R$, where $R^\dagger$ is the complex conjugate of R with all operators taken in reverse order. Observables in quantum mechanics are associated with Hermitian operators. The Hermitian property is needed for the observables, which are the eigenvalues of the operator, to be real. $R^\dagger$ is called the Hermitian conjugate of R.

Next define *unitary* operators. Suppose a transform R exists such that for operators $A \to A' = RAR^{-1}$, and for wave functions $\Psi \to \Psi' = R\Psi$. Suppose the world is invariant under this transform, i.e., all matrix elements retain their same values: $\langle\Psi|A|\Phi\rangle = \langle\Psi'|A'|\Phi'\rangle$ for all Ψ, Φ, A. Then R is said to be unitary. It is seen that $\langle\Psi'|A'|\Phi'\rangle = \langle\Psi|R^\dagger(RAR^{-1})R|\Phi\rangle = \langle\Psi|A|\Phi\rangle$. Since $R^{-1}R = 1$ by definition, $R^\dagger R = 1$ for unitary operators.

In the discussion of the Heisenberg picture in Appendix A, it is noted that, if $RHR^{-1} = H$ and $\partial R/\partial t = 0$, then $dR/dt = [H, R] = 0$. Thus, an operator not depending on time explicitly is a constant if it commutes with the Hamiltonian. Hence, conserved quantum numbers are associated with Hermitian operators commuting with H.

If R is Hermitian, and in addition unitary, then $R^2 = 1$ and the only eigenvalues are ± 1. In this case it is a discrete operator, i.e., not continuous valued. If R is unitary but not Hermitian, and commutes with H, then let $R = e^{iW}$. W will be Hermitian and corresponds to a constant observable.

If a symmetry transformation commutes with both the free particle-Hamiltonian and the Hamiltonian with interactions, then the transformation properties of the physical state can be taken to be the same as for the corresponding free-particle state, since the transformation will commute with the U matrix introduced in Appendix A.

There is a quite generally valid theorem concerning symmetries known as Noether's theorem. It states that every continuous symmetry of a Lagrangian, i.e., every invariance under a continuous transformation, implies that there is a conserved "current." A conserved current is a quantity, j_μ, such that $\partial j_\mu/\partial x_\mu \equiv \partial^\mu j_\mu = 0$. Here a summation over the indices from $\mu = 0$ to 3 is implied. The proof of this theorem is given in Appendix B. In the next section it will be shown that a conserved current implies a conserved "charge," and thus a conserved quantum number.

Thus, symmetries imply conservation laws, a very important and fundamental result. The theorem does not necessarily imply the converse, but whenever a conserved quantity in physics is found, physicists look very hard for a corresponding symmetry.

4.1 b.   Conserved Electromagnetic Current

Conserved currents can be illustrated by using electromagnetism in classical physics. Consider the electromagnetic charge $Q = \int \rho dV$. From examining Figure 4.1 it can be seen that the current leaving a small section of the surface is $\vec{J} \cdot \hat{n} dS$, where $\vec{J}$ is the current per unit area and $\hat{n}$ is the unit outward normal to the surface element dS. The total current leaving the closed surface S is then $\oint \vec{J} \cdot \hat{n} dS$. By conservation of charge this must equal $-dQ/dt$. Thus

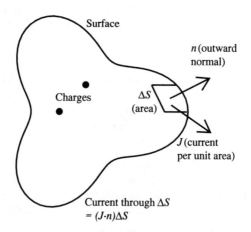

Figure 4.1 Surface surrounding a charge.

$$\frac{d}{dt}\int \rho dV + \oint \vec{J} \cdot \hat{n} dS = 0. \tag{4.1}$$

Next apply the divergence theorem $\oint \vec{J} \cdot \hat{n} dS = \int \text{div}(J) dV$. Using this, $\int [d\rho/dt + \text{div}(J)] dV = 0$ over any volume. Hence conservation of charge implies

$$\frac{d\rho}{dt} + \text{div}(J) = 0. \tag{4.2}$$

Now introduce the relativistic four-vectors $J = (\rho, J_x, J_y, J_z)$ and $x = (t, x, y, z)$. In terms of these vectors the conservation of charge relation can be written in the elegant form (summation over repeated indices is implied):

$$\partial_\mu J^\mu \equiv \frac{\partial J^\mu}{\partial x^\mu} = 0. \tag{4.3}$$

When there is a quantity J^μ obeying this equation, the current J^μ is said to be conserved. The derivation can be worked backward. If there is a conserved current, a conserved charge can be defined, the space integral of the fourth component of the current.

It will be found in Section 4.9 that there is a symmetry operation on the electromagnetic Hamiltonian whose presence requires that the electromagnetic current J^μ be conserved. As previously seen, this then implies that the electric charge is conserved. This is one of the important examples of a close relationship between symmetry principles and conserved quantities.

4.2 TRANSLATION INVARIANCE

As a first example of the principles discussed previously, assume that the Hamiltonian is invariant under translations. This invariance states that the laws of physics look the same regardless of where the origin of coordinates is placed. Consider an infinitesimal translation along x,

$$\Psi' = \Psi(x + \delta x) = \Psi(x) + \delta x \frac{\partial \Psi(x)}{\partial x} = \left(1 + \delta x \frac{\partial}{\partial x}\right) \Psi = \delta D \Psi, \tag{4.4}$$

$$\delta D = \left(1 + \delta x \frac{\partial}{\partial x}\right) = (1 - \delta x p_x / i), \tag{4.5}$$

where the fact has been used that in quantum mechanics $p_x = -i\partial/\partial x$.

60   4. Invariance, Symmetries, and Conserved Quantities

For a finite translation D, many (n) steps in succession can be made. Let $\Delta x = n\delta x$. Then $D = \delta D^n$, or in the limit,

$$D = \lim_{n\to\infty} [1 - \delta x p_x/i]^n \equiv e^{ip_x \Delta x}. \tag{4.6}$$

D is unitary, $D^\dagger D = 1$. p_x is called the generator of the operator D of space x-translations. Suppose H is independent of such space translations. Then $[D, H] = 0$, which implies that $[p_x, H] = 0$.

This result can also be obtained classically. Consider the potential energy V between two particles that form an isolated system. If this energy is dependent only on the relative positions of the two particles, not their absolute positions, then $V = V(\vec{r_1} - \vec{r_2})$. But then the total force on the system $-(\nabla_1 + \nabla_2)V = 0$, and the total momentum is conserved.

Hence, the very deep and fundamental theorem has been proven that conservation of momentum can be considered a consequence of the invariance of the laws of nature under space translation.

4.3   Rotation Invariance

Next assume that the Hamiltonian is invariant under rotations. Let ϕ be the azimuthal angle and consider an infinitesimal rotation about the azimuthal direction:

$$\Psi' = \Psi(\phi - \delta\phi) = \left(1 - \delta\phi \frac{\partial}{\partial\phi}\right)\Psi = \delta D \Psi, \tag{4.7}$$

$$\delta D = 1 - \delta\phi \frac{\partial}{\partial\phi} = 1 + iL_z \delta\phi, \tag{4.8}$$

where the fact that the z component of angular momentum is given by $L_z = i\partial/\partial\phi$ has been used. By an exactly similar procedure to that used for translation, it can be seen that for a finite rotation of $\Delta\phi$ about ϕ the operator is

$$D = e^{iL_z \Delta\phi}. \tag{4.9}$$

If H is invariant under such space rotations, then $[L_z, H] = 0$ and angular momentum is conserved. Conservation of angular momentum follows from the invariance of the laws of physics under space rotation.

L_z can also be written as

$$L_z = i\frac{\partial}{\partial \phi} \equiv -i\left(x\frac{\partial}{\partial y} - y\frac{\partial}{\partial x}\right). \qquad (4.10)$$

For future reference note that just by cyclically changing $x \to y \to z \to x$ one obtains

$$L_x = -i\left(y\frac{\partial}{\partial z} - z\frac{\partial}{\partial y}\right), \qquad (4.11)$$

$$L_y = -i\left(z\frac{\partial}{\partial x} - x\frac{\partial}{\partial z}\right). \qquad (4.12)$$

L_x, L_y, L_z are the generators of the operator D for space rotations. Note that L_x, L_y, L_z do not commute among themselves. The rotations correspond to a nontrivial group (SU_2). They are discussed further in Appendix C.

4.4 Time Invariance

Apply the Heisenberg equation of motion to the Hamiltonian itself, i.e., set $Q = H$. It is then seen that $idH/dt = i\partial H/\partial t$. Thus, if H is not explicitly dependent on time, it has constant eigenvalues. These are energy eigenvalues. If the laws of physics are not time dependent, then energy is conserved.

Thus, it has now been shown that the great conservation laws for momentum, energy, and angular momentum are related to the uniformity of space–time at a very fundamental level.

4.5 Parity

Next apply these invariance considerations to a discrete rather than a continuous transform. Are the equations of motion the same in a mirror world? The parity transform corresponds to space inversion,

$$x, y, z \to -x, -y, -z \text{ or } \vec{r} \to -\vec{r} \text{ or } [\theta \to \pi - \theta; \phi \to \pi + \phi],$$

$$P\Psi(\vec{r}) = \Psi(-\vec{r}). \qquad (4.13)$$

For mesons, $P^2 = 1$, since with P^2 one returns to the initial state. For fermions, $P^2 = \pm 1$. Fermions appear in pairs in matrix elements, since fermion number is conserved. Operating twice on a system of a pair of fermions involves P^4, i.e., P^2 for each fermion. Thus, $P^2 = -1$ is a possibility even for matrix elements conserving parity. Another way to see this is by analogy with spin angular momentum. Spin angular momentum can be shown for spin one-half particles to induce a two to one representation in which a rotation of the wavefunction by 2π gives minus the wavefunction, and a rotation of 4π is needed to get back to plus the original function. However, for all known particles $P^2 = +1$ seems to work. (There is a proposed possibility for neutrinos known as majorana neutrinos for which $P^2 = -1$.)

Assuming $P^2 = +1$, then P is a unitary operator with eigenvalues ± 1. Space inversion can be looked at as a mirror transformation followed by a 180^0 rotation. The mirror transformation takes z to $-z$, and a 180^0 rotation around the z axis then takes x, y to $-x, -y$.

A wavefunction does not have to have a definite parity. However, if $[P, H] = 0$, then, if the system starts in a well-defined parity state, it will stay in it. Evidence shows that the strong and electromagnetic interactions conserve parity. (See Section 4.12.)

Consider states that are spherical harmonics, $\propto Y_l^m(\theta, \phi)$. Using the operators discussed in Appendix C, these states can easily be shown to be states with orbital angular momentum l and z-component of orbital angular momentum m:

$$Y_l^m(\theta, \phi) \propto P_l^m(\cos\theta)e^{im\phi}. \tag{4.14}$$

Here P_l^m is the Legendre polynomial. The parity operation on this state gives

$$PY_l^m(\theta, \phi) = P_l^m(\cos(\pi - \theta))e^{im(\phi+\pi)} = (-1)^{l+m}P_l^m(\cos\theta)(-1)^m e^{im\phi}$$

$$= (-1)^l Y_l^m(\theta, \phi). \tag{4.15}$$

It is seen that the parity of the spherical harmonic, the behavior of the function under space inversion, is $(-1)^l$.

Next consider the "intrinsic parity" of a particle. Suppose a π-meson is produced. One can ask whether $P(\Phi_\pi) = \pm \Phi_\pi$ after all the explicit space dependence (the Y_m^l) is taken out, or whether it might even be mixed. Since probabilities involve absolute values of matrix elements squared, only the relative parity is measurable, the parity compared to something else as a standard. The proton (p) or neutron (n) is taken as that standard.

In Section 4.7 it will be found that the intrinsic parity of the pion is negative, $P(\Phi_\pi) = -\Phi_\pi$. It also will be shown in Section 7.1 that, for fermions, the parity of an anti-fermion is equal to minus the parity of the fermion in question.

4.6 Time Reversal Invariance

Define time reversal as changing t to $-t$ but leaving the space dimensions unchanged (i.e., $\vec{r} \longrightarrow \vec{r}$). Under T,

$\vec{p} \longrightarrow -\vec{p}$.
$\vec{\sigma} \longrightarrow -\vec{\sigma}$.
$\vec{E} \longrightarrow \vec{E}$.
$A^k \longrightarrow -A^k$, for the vector potential, $k = 1, 2, 3$, (since $\vec{B} \longrightarrow -\vec{B}$).
$A^0 \longrightarrow A^0$, where $A^0 = c\phi$ and $\phi =$ the scalar electric potential.
$\Psi(\vec{r}, t) \longrightarrow \Psi^\star(\vec{r}, -t)$ (Schrödinger equation).

Why is this relation for Ψ appropriate? Apply time reversal to the Schrödinger equation:

$$T\left[i\frac{\partial}{\partial t}\Psi(\vec{r},t) = H\Psi(\vec{r},t)\right] \longrightarrow -i\frac{\partial}{\partial t}\Psi(\vec{r},-t) = H\Psi(\vec{r},-t). \quad (4.16)$$

This is not in the form of a Schrödinger equation because of the extra minus sign on the left-hand side of the equation. (The energy spectrum of the time reversed state should remain positive.) Take the complex conjugate of Equation 4.16:

$$i\frac{\partial}{\partial t}\Psi^\star(\vec{r},-t) = H^\star \Psi^\star(\vec{r},-t). \quad (4.17)$$

Now this is in the form of a Schrödinger equation with Ψ transformed as shown and with H replaced by $H^\star$. If H is not explicitly time dependent, then H is invariant under time reversal if and only if H is real.

It is seen from the preceding that time reversal is not a unitary operator. T is a unitary operator $\times$ complex conjugation, $T = U \times K$, where K is the complex conjugation operator.

4.6 a. Detailed Balance

A very important theorem known as "detailed balance" will be proven here. Suppose the Heisenberg representation is assumed. Then the matrix element for transitions between states α and β will be essentially $\langle \Psi_\alpha | \Psi_\beta \rangle$. First a lemma will be established:

$$\langle T\Psi_\alpha | T\Psi_\beta \rangle = \langle \Psi_\alpha | \Psi_\beta \rangle^\star = \langle \Psi_\beta | \Psi_\alpha \rangle. \quad (4.18)$$

Expand in orthonormal base vectors $|n\rangle$:

$$|\Psi_\alpha\rangle = \Sigma |n\rangle\langle n|\Psi_\alpha\rangle, \quad (4.19)$$

$$T|\Psi_\alpha\rangle = \Sigma U|n\rangle^\star \langle n|\Psi_\alpha\rangle^\star, \quad (4.20)$$

$$\langle T\Psi_\alpha | T\Psi_\beta \rangle = \Sigma_n \Sigma_m \langle \Psi_\alpha | n \rangle^\star \langle n | U^\dagger U | m \rangle^\star \langle m | \Psi_\beta \rangle^\star$$

$$= \Sigma_n \langle \Psi_\alpha \mid n \rangle^* \langle n \mid \Psi_\beta \rangle^*, \qquad (4.21)$$

$$\langle T\Psi_\alpha \mid T\Psi_\beta \rangle = \langle \Psi_\alpha \mid \Psi_\beta \rangle^*, \qquad (4.22)$$

where the facts have been used that since U is unitary $U^\dagger U = 1$ and therefore $\langle n \mid U^\dagger U \mid m \rangle = \delta_{nm}$. Thus, under T, matrix elements transform into the complex conjugate of their previous value.

Suppose now that the Hamiltonian is invariant under T and P. Call the transition matrix element $M_{\alpha\beta}$. Then $|TP(M_{\alpha\beta})| = |M_{\alpha\beta}|$. T reverses initial and final states, reverses p, and reverses σ. P reverses p but not σ. Thus,

$$|M_{\alpha\beta}|^2_\sigma = |M_{\beta\alpha}|^2_{-\sigma} \text{ (if } T \text{ and } P \text{ invariant)}. \qquad (4.23)$$

This result is known as "Detailed Balance." Its use in particle physics will be shown shortly. It also plays a fundamental role in statistical mechanics considerations. Experiments show that the strong and electromagnetic interactions are T and P invariant. (See Section 4.12.)

4.7 Illustration of Uses of Invariance Principles: Pions

Pions can be used to illustrate some of the concepts introduced in the previous sections and to show how the properties of the fundamental particles can be established. There exist both charged and neutral pions. $m_{\pi^\pm} = 139.6$ MeV/c^2, $m_{\pi^0} = 135$ MeV/c^2. The pions are the lightest known strongly interacting particles. They were first identified in 1947.[9] In older theories they were the glue binding the n and p in nuclei. Exchanges of virtual photons can be looked at as the cause of the Coulomb force and exchanges of pions in a similar manner were imagined to be the cause of the nuclear force (see Figure 4.2). It will be seen somewhat later that exchanges of another set of particles, the gluons, are now considered to be the fundamental cause of nuclear binding.

Since the charged and neutral pions have slightly different masses (at least partially due to the Coulomb energy), why is it believed that they are different charge states of the same particle? The answer is that all their quantum numbers and their interactions are consistent with this.

In the following sections, determinations of the spin and parity of charged and neutral pions are discussed using the symmetry principles formulated above.

4.7 Illustration of Uses of Invariance Principles: Pions

Photons as electromagnetic glue

Pions as nuclear glue

Figure 4.2

4.7 a. Pion Spin

Charged Pions

Consider the process $A + B \to C + D$. The cross section for this process is

$$\sigma_{A+B \to C+D} = \text{constants} \times \text{phase space} \times |\text{matrix element}|^2. \quad (4.24)$$

The phase space just corresponds to the integrals over outgoing momenta, etc. Concentrate for now on the matrix elements. If A and B are unpolarized, it is necessary to average over the initial spins. If the final spins are not specifically observed, then one must sum over the possible final spins. (The final spins that are possible may differ depending on the initial state.) The matrix element can thus be written

$$|M|^2 = \frac{1}{(2S_A+1)(2S_B+1)} \sum_{m_A=-S_A}^{S_A} \sum_{m_B=-S_B}^{S_B} \sum_{\text{final spins}} |M|^2_{A+B \to C+D}, \quad (4.25)$$

$$|M|^2 = (2S_C+1)(2S_D+1)|M|^2_{A+B \to C+D} \text{ (averaged)}. \quad (4.26)$$

Here by "averaged" is meant $|M|^2$ averaged over both initial and final states.

Next consider the process in the reverse direction, $C + D \to A + B$:

$$\sigma_{C+D \to A+B} = \text{constants} \times \text{phase space} \times |\text{matrix element}|^2, \quad (4.27)$$

where for this process:

$$|M|^2 = \frac{1}{(2S_C+1)(2S_D+1)} \sum_{m_C=-S_C}^{S_C} \sum_{m_D=-S_D}^{S_D} \sum_{\text{final spins}} |M|^2_{C+D \to A+B}, \quad (4.28)$$

$$|M|^2 = (2S_A + 1)(2S_B + 1)|M|^2_{C+D \to A+B}(\text{averaged}). \tag{4.29}$$

For strong interactions it is believed that H is invariant under T and P. The combined effect of time reversal and space inversion invariance is, then, that the detailed balance relation holds:

$$|M|^2_{A+B \to C+D}(\text{averaged}) = |M|^2_{C+D \to A+B}(\text{averaged}). \tag{4.30}$$

Therefore, compare the two reactions

$$p + p \longleftrightarrow \pi^+ + d. \tag{4.31}$$

The matrix element for the reaction in which the final state is $2p$ has an additional factor of $1/2$ because of the anti-symmetry of the final state, i.e., effectively, one integrates only over $1/2$ of the total solid angle. The "d" is the deuteron, the nucleus of the heavy hydrogen atom. It is a nucleus made up of one proton and one neutron. It is known to have spin 1. Since $(2S_\pi + 1) = 1$ for pion spin 0 and 3 for pion spin 1, it is easily established that the $\pi^\pm$ has spin 0 from the relative cross sections for $p + p \to \pi^+ + d$ and $\pi^+ + d \to p + p$.

Neutral Pions

The major decay mode of the π^0 is $\pi^0 \to \gamma + \gamma$. The photon (γ) is a spin 1 particle. It is seen immediately that the spin of the pion is integral (not half-integral), since it decays into two integral spin particles.

Suppose the z-direction is the direction of propagation of the photon. It then turns out that only $m_z = \pm 1$ is possible, i.e., because they are massless, for real photons only right-handed and left-handed photons are possible, not longitudinal ones. This is equivalent to the two allowed polarizations of light, since $\vec{E} \perp \vec{v}$. (Virtual photons are allowed to have longitudinal polarization and indeed even a polarization in the time dimension.)

Consider the decay in the center of mass of the pion. The spatial part of the angular momentum of the decay products must have $m_z = 0$ along the z-axis, the axis of emission of the photons. ($\vec{L} = \vec{r} \times \vec{p}$ and the cross product is 0 in a direction along the line of flight of the particle.) Suppose it is assumed (falsely) that the spin of the pion is 1. Then only the states with the spins of the photons having a total $m = 0$ along the axis of emission are possible, since the photon spin projections are only ± 1 each. Thus, the z-component of the total angular momentum is 0.

One needs appropriate linear combinations of the two photon states to obtain states with diagonal J and J_z. This process, which involves Clebsch–Gordan coefficients, is discussed in Appendix C. From looking at the Clebsch–Gordan coefficient tables one sees that, for $m = 0$, the spin wavefunction of the photons is symmetric under interchange for total photon spin, $S_\gamma = 0, 2$ and anti-symmetric for $S_\gamma = 1$. Since the photons are Bose particles, the total wavefunction, including the spatial part, must be symmetric under interchange.

Consider the center-of-mass system for the two photons. Let L and S now refer to the relative orbital angular momentum and total spin of the photons. For $J = 1$ one can have $(L, S) = (1,1)$, $(2,2)$, $(0,1)$, $(1,0)$, $(1,2)$, $(3,2)$, and $(2,1)$. All of the last five are odd under interchange of the photons and therefore forbidden because the photons obey Bose–Einstein statistics. However, the Clebsch–Gordan coefficients are zero for

$$|L = 1, L_z = 0\rangle + |S = 1, S_z = 0\rangle \longrightarrow |J = 1, J_z = 0\rangle,$$

$$|L = 2, L_z = 0\rangle + |S = 2, S_z = 0\rangle \longrightarrow |J = 1, J_z = 0\rangle.$$

Hence, a two-photon system cannot have $J = 1$.

Thus, the pion spin is either 0 or ≥ 2. Assuming it is less than 2, then the spin of the neutral pion is 0.

Furthermore, indirect evidence for this result comes from high-energy pp collisions. It is found there that one obtains approximately equal numbers of π^+, π^0, π^- or, more exactly, the number of π^0 is approximately equal to the average of the number of π^+ and π^-. This implies that the spin factors are the same and that the spin of the π^0 should be the same as the spin of the $\pi^\pm$.

4.7 b. Pion Intrinsic Parity

Charged Pions

Consider the process $\pi^- + d \to n + n$. One can look at this process for stopped pions; it is found that it occurs reasonably quickly, too quickly to occur by electromagnetic or weak interactions. Thus, experimentally, it is allowed by the strong interactions. It occurs from the S state ($l = 0$) of the pion–deuteron system. This is known because, as the pion stops, it is captured by the deuteron forming a "mesonic atom" with the pion taking the place of the electron. As the pion cascades down through the various allowed orbits, it emits photons. These photons are in the x-ray region and are known as mesonic x-rays. By looking at the strength of the various lines it is seen that most of the pions get down to the lowest S state before the preceding interaction occurs.

The spins of the pion and deuteron are known to be 0 and 1, respectively. Since the interaction occurs in the S state, the total angular momentum, $J = 1$.

The two neutrons in the final state are identical Fermi–Dirac particles. The wavefunctions, therefore, must be overall anti-symmetric under exchange of the particles by the Pauli principle. The space (Y_L^m) wavefunction gives a factor $(-1)^L$; interchanging the two neutrons changes x, y, z to $-x, -y, -z$ and it was noted previously that $PY_L^m = (-1)^L Y_L^m$. What does the spin function do?

For two spin 1/2 particles, it can be seen from the Clebsch–Gordan tables (Appendix C, Figure C.1) that the possible total spin wavefunctions are

$$S = 1, \ m = 1 \quad \uparrow\uparrow,$$

$$S = 1, \ m = 0 \quad \frac{1}{\sqrt{2}}(\uparrow\downarrow + \downarrow\uparrow),$$

$$S = 1, \ m = -1 \quad \downarrow\downarrow,$$

$$S = 0, \ m = 0 \quad \frac{1}{\sqrt{2}}(\uparrow\downarrow - \downarrow\uparrow).$$

Thus, the symmetry of the spin function under particle exchange is $(-1)^{S+1}$. The overall symmetry of the two-neutron system under exchange then is $(-1)^{L+S+1}$. Fermi–Dirac statistics require that this must equal -1 forcing $L + S$ to be an even number.

A $J = 1$ two-neutron state can be built from $(L = 1, S = 1)$, $(L = 1, S = 0)$, $(L = 0, S = 1)$, or $(L = 2, S = 1)$. From the preceding discussion, all but the first possibility are forbidden because of Fermi–Dirac statistics. The two neutrons have $L = 1$.

The parity of this final state of two neutrons is $(-1)^L = -1$. Since parity is conserved in the strong interaction, the initial state is also a negative parity state. The deuteron is a bound state of a neutron and a proton. It is known to be a $J = 1$ state and it is known that the orbital angular momentum is mostly $L = 0$ (S-wave) with a few percent $L = 2$ (D-wave). Thus, the deuteron has positive parity. The initial state of $\pi^- d$ is S wave. Hence, the parity of the $\pi^- d$ system is equal to the intrinsic parity of the π^-. The intrinsic parity of the charged pion is negative.

Neutral Pions

Consider the pion decay into two photons in the center of mass of the pion. Looking at Figure 4.3 there are four possible states for the two photons.

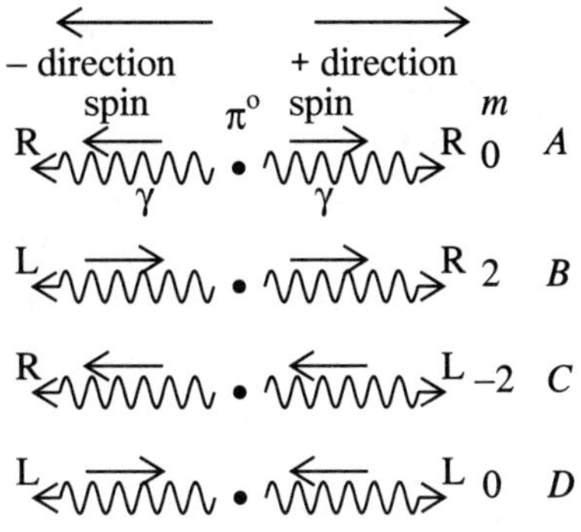

4.7 Ilustration of Uses of Invariance Principles: Pions

```
                ←                →
           − direction       + direction
              spin      π°   spin          m
         R ⇐∿∿∿ • ∿∿∿⇒ R         0      A
              γ              γ

         L ⇒∿∿∿ • ∿∿∿⇒ R         2      B

         R ⇐∿∿∿ • ∿∿∿⇐ L        −2      C

         L ⇒∿∿∿ • ∿∿∿⇐ L         0      D
```

R = right-handed, i.e., spin parallel to momentum.
L = left-handed, i.e., spin anti-parallel to momentum.

Figure 4.3 $\pi^0 \to \gamma + \gamma$ decay.

Only states A and D are possible, since the pion has spin 0. Under reflection the states A and D are reversed, since the spins reverse under reflection. (A right-handed spin around an axis parallel to the plane of the mirror goes to a left-handed spin.) Thus, $A' = D$; $D' = A$. This means that the states invariant under the parity operation are $A \pm D$. These states have $\pm$ parity, respectively.

Consider an electromagnetic plane wave. There are two states of polarization for a wave in the z-direction, $\vec{E}_x$ and $\vec{E}_y$. Right and left circularly polarized waves correspond to $\frac{1}{\sqrt{2}}(\vec{E}_x \pm i\vec{E}_y)$, respectively, and these can be shown also to correspond to the $m = \pm 1$ states of the photon. Then

$$A \pm D = R^1 R^2 \pm L^1 L^2$$

$$= \frac{1}{2}[(\vec{E}_x^1 + i\vec{E}_y^1)(\vec{E}_x^2 + i\vec{E}_y^2) \pm (\vec{E}_x^1 - i\vec{E}_y^1)(\vec{E}_x^2 - i\vec{E}_y^2)], \quad (4.32)$$

$$A + D = [\vec{E}_x^1 \vec{E}_x^2 - \vec{E}_y^1 \vec{E}_y^2] \text{ (planes of polarization parallel)}, \quad (4.33)$$

$$A - D = i[\vec{E}_x^1 \vec{E}_y^2 + \vec{E}_y^1 \vec{E}_x^2] \text{ (planes of polarization perpendicular)}. \quad (4.34)$$

Occasionally in the decay of the π^0 to two photons, the Feynman diagram is more complicated with the photons both converting to $e^+ e^-$ pairs within the amplitude, so that the photons are virtual. One can look at these pairs and ask whether their planes tend to be parallel ($A + D$, + parity) or perpendicular ($A - D$, − parity). Experimentally they tend to be perpendicular. The parity of the neutral pion is negative.

There is an alternative argument leading to the same polarization prediction. It is a very general argument and will be used further in this text. Consider the set of four-vectors available in the problem. There are $(\vec{E})^1, (\vec{E})^2$, and $\vec{k}$ (= momentum vector of photon 1 = − momentum vector of photon 2). Note that $(\vec{E})^1$ and $(\vec{E})^2$ are perpendicular to $\vec{k}$.

Next consider the scalars, involving both $(\vec{E})^1$ and $(\vec{E})^2$, that can be made from these vectors. There are $(\vec{E})^1 \cdot (\vec{E})^2$, which is a scalar, and $((\vec{E})^1 \times (\vec{E})^2) \cdot \vec{k}$, which is a pseudo-scalar i.e., has negative parity. If the pion is a pseudo-scalar particle and parity is conserved in the strong and electromagnetic interactions, only the second product can occur, and again it is expected that the two planes of polarization are perpendicular to each other.

For future reference note that the same argument works for positronium (the bound states of e^+e^-). Consider the state $L = 0$, $S = 0$. The parity is negative, since, as will be shown in Section 7.1, the parity of a Fermion anti-particle is opposite to that of the particle. Thus, the planes of polarization of the photons emitted in the decay (positronium → two photons) are perpendicular to each other.

4.8 CHARGE CONSERVATION

Charge conservation can be shown to follow directly from an invariance principle. The absolute value of electric potential can be set arbitrarily; only gradients of potential are expected to influence physical events. The invariance of a system to the absolute value of the electric potential, ϕ, and conservation of energy will now be shown to imply that there is a conserved charge. The proof is due to E.P. Wigner.

Consider a transformation $\phi \longrightarrow \phi + C$, where C is a fixed constant, and suppose the theory is invariant under this transformation. Suppose conservation of charge is not valid and that a charge is created at a point $\vec{r}$. The work to create the charge is W_c. W_c must be independent of ϕ, since the absolute value of ϕ at any point is arbitrary. Suppose the charge is moved to another point, where the potential is ϕ', and then destroyed. An energy W_d is returned, where W_d is independent of ϕ'. However, a net amount of energy $W_d - W_c + q(\phi - \phi')$ has now been gained. Since the difference $\phi - \phi'$ can be arranged to be any value depending on what batteries are used, conservation of energy would be violated. The proof is complete.

4.9 Gauge Invariance

Next a very deep relation of invariance to the form of the electromagnetic equations will be examined. Classically, in Gaussian units, the electromagnetic fields are related to the potentials by

$$\vec{B} = \nabla \times \vec{A}; \quad \vec{E} = -\nabla\phi - \frac{\partial \vec{A}}{\partial t}. \tag{4.35}$$

$\vec{A}$ and ϕ are not unique. Let χ be any scalar function of $(t, \vec{x})$. $\vec{B}$ and $\vec{E}$ remain the same under the transformation

$$\vec{A}' = \vec{A} + \nabla\chi; \quad \phi' = \phi - \frac{\partial \chi}{\partial t}. \tag{4.36}$$

Define $A^\mu \equiv (\phi, \vec{A})$. Then the preceding transformation can be written as

$$(A')^\mu = A^\mu + \frac{\partial \chi}{\partial x_\mu}. \tag{4.37}$$

This is called "local gauge invariance." It is local because χ is a function of $(t, \vec{x})$. If χ were a constant with respect to the three-vector $\vec{x}$, then this would be called "global gauge invariance." In practice a particular choice of gauge is often made. For example, a popular choice is to pick χ such that

$$\frac{\partial A^\mu}{\partial x^\mu} = 0.$$

This is known as the Lorentz gauge or the Maxwell gauge.

A four-vector notation was introduced above. In terms of four-vectors the Maxwell equations can be written in the succinct form

$$\frac{\partial F^{\mu\nu}}{\partial x^\mu} = 4\pi j^\nu, \tag{4.38}$$

where

$$F^{\mu\nu} \equiv \frac{\partial A^\nu}{\partial x_\mu} - \frac{\partial A^\mu}{\partial x_\nu}; \quad j^\nu \equiv (\rho, \vec{j}). \tag{4.39}$$

Under the local gauge transformation $(F')^{\mu\nu} = F^{\mu\nu}$.

Look next at the Schrödinger equation. The classical Hamiltonian for electromagnetism is

$$H = \frac{1}{2m}\left(\vec{p} - \frac{q\vec{A}}{c}\right)^2 + q\phi. \tag{4.40}$$

The Schrödinger equation then is

$$\left(\frac{1}{2m}\left(-i\nabla - \frac{q\vec{A}}{c}\right)^2 + q\phi\right)\Psi = i\frac{\partial \Psi}{\partial t}. \tag{4.41}$$

The quantity

$$\mathcal{D}^\mu \equiv \frac{\partial}{\partial x_\mu} - i\frac{qA^\mu}{c}$$

is known as the "covariant derivative." For the first three components $-i\mathcal{D}^\mu = (-i\nabla - \frac{qA}{c})^\mu$. In Section 11.1 it will be shown that, for the relativistic wave equations, $\mathcal{D}^\mu$ will enter in a natural way.

Suppose the preceding local gauge transformation is now made. It is seen that the Schrödinger equation remains invariant if one changes Ψ to $e^{iq\chi}\Psi$. Similar transformations also work for the relativistic quantum equations (the Dirac and Klein–Gordon equations) as well as for the Schrödinger equation. Gauge transformations for the Dirac equation will be discussed in Section 11.1.

Consider now the very important reverse argument. Ask that the theory be invariant under the transformation

$$\Psi \longrightarrow \Psi' = e^{iq\chi}\Psi.$$

First ask that this be true for a global gauge invariance function $\chi = at$, where a is a constant. The exponential becomes e^{iqat}. This corresponds to an extra constant energy factor in the exponential, $E_c = qa$. If the theory is invariant under this, it means that adding a constant "a" to the electric potential, ϕ, does not affect the theory. This is just the condition for the Wigner theorem of the previous section to apply, and it is seen that global gauge invariance guarantees conservation of charge. There is a problem with this argument. State vectors of a system do not have to have the full symmetry of the Lagrangian and, hence, may not be gauge invariant. A fuller discussion of this kind of symmetry breaking is given in Chapter 11.

Next consider a local gauge invariance function. It turns out that for linear fields the preceding forms for $\vec{A}$ and ϕ are essentially determined and they must obey the Maxwell equations. (See Section 11.1.) Local gauge invariance demands that A_μ exists! The entire electromagnetic interaction can be viewed as a consequence of local gauge invariance!

Local gauge invariance demands a zero-mass photon, since a photon mass term spoils the invariance. There have been many experiments checking that the photon mass is zero. They are usually based on looking at macroscopic phenomena and attempting to detect deviations from the Coulomb force law, i.e., $F \propto 1/r^{2+\epsilon}$. If ϵ were nonzero, this would induce a violation of Gauss's law. Very good evidence comes from satellite observations of Jupiter's magnetic field. The results imply[10] $m_\gamma < 6 \times 10^{-16}$ eV.

Gauge invariance has been a seminal idea for modern particle physics theory. Are all interactions between particles the result of symmetry principles?

4.10 CHARGE CONJUGATION

Charge conjugation is a symmetry operation (C) that changes particles to their anti-particles. It changes q to $-q$ and magnetic moments to minus themselves, since any currents change sign.

It changes baryon numbers (the number of p, n, and other strongly interacting unit charge fermions minus the number of their anti-particles) to minus themselves. It changes the lepton number for each generation to minus itself. (For the first generation, for example, the lepton number is the number of electrons and electron neutrinos minus the number of positrons and electron anti-neutrinos.)

Experimentally, the strong and electromagnetic interactions are invariant under C. Evidence for this invariance is cited in Section 4.12.

A fermion is not an eigenstate of C, since the particle and anti-particle are distinct. It, therefore, is a matter of convention whether C is defined to take a particle wavefunction into plus or minus the anti-particle wavefunction. For later convenience define

$$Cp = -\bar{p}, \; Cn = -\bar{n}, \; Cu = -\bar{u}, \; Cd = -\bar{d}, \; Ce^- = -e^+. \qquad (4.42)$$

The other generations of particles are treated similarly. Note that $C^2 = 1$ defines signs for C on anti-particles.

What happens for bosons?

$$C\left|\pi^{\pm}\right\rangle = \eta_{\pi^{\pm}}\left|\pi^{\mp}\right\rangle, \tag{4.43}$$

$$C\left|\gamma\right\rangle = \eta_{\gamma}\left|\gamma\right\rangle. \tag{4.44}$$

Since pions can be produced singly, by applying C twice it is seen that $\eta_{\pi^{\pm}}^2 = 1$ is necessary. However, the phase of $\eta_{\pi^{\pm}}$ is arbitrary since π^+ and π^- are not eigenstates of C. Choose $\eta_{\pi^{\pm}} = -1$. This is a useful choice for further quantum number discussions (G-parity).

For the photon $\eta_{\gamma}^2 = 1$. Electromagnetic fields are produced by charges and currents that change sign under C. Thus, $CA_{\mu} = -A_{\mu}$, where A_{μ} is the vector potential. The electric and magnetic fields also change sign. $\eta_{\gamma} = -1$. No C-conserving matrix element should connect a state that consists exclusively of an even number of photons to a state that consists exclusively of an odd number of photons. This is true to all orders of perturbation theory and is known as Furry's Theorem.

What about the π^0?

$$C\left|\pi^0\right\rangle = \eta_{\pi^0}\left|\pi^0\right\rangle. \tag{4.45}$$

Here again $\eta_{\pi^0}^2 = 1$. However, $\pi^0 \to \gamma + \gamma$ by strong and electromagnetic interactions, which conserve C. Therefore, $\eta_{\pi^0} = +1$.

Note that π^0 decays into an odd number of photons are forbidden. For example, $C(3\gamma) = (-1)^3(3\gamma) = -1 \times (3\gamma)$. Data show that (rate for $\pi^0 \to 3\gamma$)/(rate for $\pi^0 \to 2\gamma$) $< 3.1 \times 10^{-8}$ at the 90% confidence level.[11]

To summarize,

$$C\left|\pi^{\pm}\right\rangle = -\left|\pi^{\mp}\right\rangle;\ C\left|\gamma\right\rangle = -\left|\gamma\right\rangle;\ C\left|\pi^0\right\rangle = +\left|\pi^0\right\rangle. \tag{4.46}$$

In the production of mesons via the process $e^+e^- \to$ virtual gamma $\to$ mesons, the final mesons must have $C = -1$, $P = -1$, and spin 1. For spin 1, the parity is minus as the photon is a 1^- particle. In fact, for some electromagnetic interactions, there is a 0^+ part because, although $A_i (i = 1, 3)$ is a vector, A_0 is a scalar. However, one can show that, because the electromagnetic current is conserved, the meson cannot be spin 0. (See Exercise 4.18.)

4.11 Illustration of Charge Conjugation: Particle–Anti-Particle Pairs

Consider a particle–anti-particle pair in a state of definite orbital angular momentum L, spin S, and charge conjugation C. (If it is in an eigenstate of L, S, then it will be in an eigenstate of C.)

The extended Bose–Einstein symmetry principle states that for bosons the wavefunction of the particle–anti-particle pair must go to plus itself under total exchange of the particles.

The extended Pauli principle states that for Fermi–Dirac particles the wavefunction of the particle–anti-particle pair must go to minus itself under total exchange of the particles. The consequences of these principles will be discussed subsequently. They, in fact, follow from the relativistic equations of state in which the particle and anti-particle are treated in one equation. In Chapter 7, a brief discussion of the Dirac equation, the relativistic equation for spin one-half fermions, will be given.

Bosons
Consider the pair $\pi_1^+ \pi_2^-$. Exchange particles 1 and 2. A factor of $(-1)^L$ is obtained from the orbital angular momentum. There is no spin term. A factor C is obtained from charge conjugation.

$$(-1)^L C \Psi(\pi^+\pi^-) = +\Psi(\pi^+\pi^-) \Longrightarrow C\Psi(\pi^+\pi^-) = (-1)^L \Psi(\pi^+\pi^-). \tag{4.47}$$

Fermions
Consider the pair $e_1^+ e_2^-$. Exchange particles 1 and 2. A factor of $(-1)^L$ is obtained from the orbital angular momentum. There is a factor of $(-1)^{S+1}$ from the spin angular momentum (as discussed in section 4.7b) and a factor C from charge conjugation:

$$(-1)^L (-1)^{S+1} C\Psi(e^+e^-) = -\Psi(e^+e^-); \quad C\Psi(e^+e^-) = (-1)^{L+S} \Psi(e^+e^-). \tag{4.48}$$

Apply this to $(e^+e^-) \to 2\gamma$ and 3γ. Most bound-state e^+e^- (positronium) annihilations come from the ground state where the orbital angular momentum L is 0. Consider the two cases $S = 0$ and 1.

$S = 0, L = 0 \Longrightarrow C = +1 \Longrightarrow 2\gamma$ is allowed. The experimental lifetime, τ, for this process is 1.25×10^{-10} s.

$S = 1, L = 0 \Longrightarrow C = -1 \Longrightarrow 3\gamma$ (or other odd number of photons) is allowed. The experimental lifetime, τ, for this process is slower, 1.5×10^{-7} s, because of the extra photon coupling needed.

This result can also be obtained without using C. In fact, it was already shown, when discussing the spin of the π^0, that a two-photon system cannot have $J = 1$. Thus, the $J = S = 1$ e^+e^- system cannot decay into two photons, even in the absence of C conservation.

76   4. Invariance, Symmetries, and Conserved Quantities

4.12   P, C, and T. WHO IS INVARIANT UNDER WHAT?

Consider the symmetry operations P, C, and T and the combined operations CP, and CPT.

One can show that under very general conditions physical theories will be invariant under the combined operation, CPT. The theorem is true for any local field theory possessing Lorentz invariance and having the usual spin-statistics relation. A local field theory is one in which the Lagrangian consists of products of operators taken at the same spacetime point. All known interactions are consistent with being invariant under CPT. CPT would be violated if the total lifetime of particles and anti-particles were different or if the masses of particles and anti-particles were different. The partial lifetime to a specific final channel is not constrained by CPT to be the same for particle and anti-particle. Indeed, this happens in the neutral kaon system discussed subsequently. If the final channel, however, consists of free particles with no mutual interactions, then CPT invariance constrains the particle and anti-particle partial lifetimes to be the same. CPT requires the charges and magnetic moments of protons and anti-protons to be equal in magnitude and opposite in sign.

P, C, and T are separately conserved by the strong and the electromagnetic interactions.

The weak interactions conserve only CPT exactly. They violate P and C separately in a maximal way. They almost conserve CP and, hence, almost conserve T. However, there is a small violation of CP invariance observed in the neutral kaon system. This will all be described.

Consider now some of the evidence for conservation of P, C, and T in the strong and electromagnetic interactions.

For P conservations one looks at transitions between various nuclear states. Parity violation is seen only at about the 10^{-7} level and is about the right size for the interference term in the matrix element between the weak interactions and the strong interactions.

For C it has been noted that the π^0 decay into an odd number of photons is found to be very highly forbidden. One can also look at the ratio of $+$ to $-$ particle spectra in $p + \bar{p} \to \pi^+ + \pi^- + \cdots$. The two spectra should be the same and are found to be the same experimentally.

The decay of the η, a neutral boson particle of mass $= 550$ MeV/c^2 provides another example of C conservation. The decay $\eta \to 2\gamma$ occurs with about a 39% branching fraction. This implies that $C = +1$ for the η. If this is so, the decay $\eta \to \pi^0 \gamma_{\text{virtual}} \to \pi^0 e^+ e^-$ is forbidden because the π^0 has $C = +1$, and the γ has $C = -1$. The experimental branching fraction is less than 5×10^{-4}. The η has been involved in other C tests. The η decays into $\pi^+ \pi^- \gamma$ and into $\pi^+ \pi^- \pi^0$. Within each reaction the spectra of the $+$ and $-$ pions have been carefully compared and found to be the same as required by C invariance.

T invariance has been checked in strong interactions by looking carefully at detailed balance relations. These tests find that T violating interactions are less than 0.3% of the T conserving interactions.

T (or P) invariance also has been carefully examined by looking for electric dipole moments of elementary particles. $\sigma \cdot \vec{E}$ is proportional to the energy of a dipole, where $\vec{E}$ is the electric field and σ is the spin operator. Note that $T(\text{ or }P)\sigma \cdot \vec{E} = -\sigma \cdot \vec{E}$. ($T\sigma = -\sigma$; $T\vec{E} = +\vec{E}$ while $P\sigma = +\sigma$; $P\vec{E} = -\vec{E}$, the latter following since $\vec{E}_i = -\partial \phi/\partial x_i$.)

Hence an elementary particle should not have an electric dipole moment unless (for P) there were an almost degenerate state of opposite P to mix with. In practice no dipole moments have been seen. The neutron electric dipole moment is less than 10^{-25}cm^2.

4.13 Isotopic Spin Invariance

The strong interactions of the neutron and proton appear to be the same, although the electromagnetic interactions are quite different. One can therefore find an invariance under the strong (but not under the electromagnetic or weak!) interactions by thinking of the p and n as two states of the same particle:

$$|\Psi_{\text{nucleon}}\rangle = \Psi(x)\xi_{\text{spin}}\eta_{\text{charge}}, \qquad (4.49)$$

where $\eta_+ = p$ and $\eta_- = n$ are the two base states of η. It will be required that the strong interaction matrix elements for a single nucleon be independent of the charge state of the nucleon, i.e., independent of whether η_+ or η_- is in the wavefunction.

Define τ_3 by $\tau_3 p = p$ and $\tau_3 n = -n$. Thus, in the space of n and p,

$$\tau_3 = \begin{pmatrix} 1 & 0 \\ 0 & -1 \end{pmatrix} \begin{pmatrix} p \\ n \end{pmatrix}; \qquad (4.50)$$

$$\eta_+ = \begin{pmatrix} 1 \\ 0 \end{pmatrix}; \ \eta_- = \begin{pmatrix} 0 \\ 1 \end{pmatrix}. \qquad (4.51)$$

This is a new space, Isotopic Spin Space or I-Spin Space. Next define

$$\tau_1 = \begin{pmatrix} 0 & 1 \\ 1 & 0 \end{pmatrix}; \ \tau_2 = \begin{pmatrix} 0 & -i \\ i & 0 \end{pmatrix}. \qquad (4.52)$$

If you have studied spin one-half particles, you know these are just the Pauli spin matrices.

It is easily seen that $\tau_k^2 = 1$. $\tau_k\tau_l = -\tau_l\tau_k$, for $l \neq k$. $\tau_k\tau_l - \tau_l\tau_k = 2i\tau_m$, for k, l, m a cyclic permutation of 1,2,3.

Define the Isotopic spin (or I-spin or Isospin) of the nucleon to be 1/2 in analogy with spin. Define the I-spin operator for 1 nucleon to be:

$$I_k = \frac{1}{2}\tau_k; \quad [I_k, I_l] = iI_m. \tag{4.53}$$

In the latter relation (which is true for any I-spin not just I-spin 1/2), k, l, m are a cyclical permutation of 1,2,3. This is just the same as the angular momentum commutation relations and the group structure is, therefore, $SU(2)$ as in angular momenta. This means exactly analogous relations will hold and the Clebsch–Gordan coefficient tables will work for I-spin as well as angular momentum.

I-spin invariance assumes that the strong interactions are invariant under I-spin rotations.

4.13 a.   The Two-Nucleon System and Multinucleons

Apply I-spin considerations to two and multinucleon states. There are four possible states for two nucleons: $p^1p^2; p^1n^2; n^1p^2; n^1n^2$. In analogy with angular momentum, define for two nucleons

$$I_k = \frac{1}{2}(\tau_k^1 + \tau_k^2); \quad I^2 = \Sigma_{k=1}^3 (I_k)^2. \tag{4.54}$$

From the angular momentum analogy it is known that the eigenvalues of the I^2 operator are $I(I+1)$. I ($= 0$ or 1 for the two-nucleon state) will be used to identify the state. Next build states that are eigenstates of I^2 and I_3. From the Clebsch–Gordan tables it is seen explicitly that the $|I, I_3\rangle$ states are

$$|1,1\rangle = p^1p^2; \quad |1,0\rangle = \frac{1}{\sqrt{2}}(p^1n^2 + n^1p^2);$$

$$|1,-1\rangle = n^1n^2; \quad |0,0\rangle = \frac{1}{\sqrt{2}}(p^1n^2 - n^1p^2). \tag{4.55}$$

As an example consider the deuteron, the bound n, p system. The ground state is known to be mainly an S state ($l = 0$) and it is, therefore, symmetric under space exchange. It has $J = 1$, which means that the spin state is symmetric under exchange.

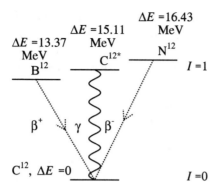

Figure 4.4 B–C–N isospin triplet of nuclei.

Note that it is not possible to have a two-proton or two-neutron state with the same spin and angular momentum as the preceding deuteron, because of the Pauli principle. Since only the np state is possible here, the deuteron ground state must be an $I = 0$ state. If the np were in a space-spin anti-symmetric state, the I-spin concept says that the similar pp and nn states must have the same interactions and thus they would be in an $I = 1$ state. The preceding argument is generally true for symmetric and anti-symmetric nucleon-nucleon states. Hence, a generalized Pauli principle obtains; the overall wavefunction, including I-spin must be anti-symmetric.

A neutron can change into a proton under the weak interaction. This further reinforces the concept of neutron and proton being different states of the same particle.

The same considerations generalize to multinucleons. Consider the triplet of nuclei B^{12}, C^{12} (excited), N^{12} where the superscript labels the number of nucleons. These three nuclei are believed to be an $I = 1$ triplet with 5,6,7 protons, respectively. They all have $J^P = 1^+$. They are observed to have almost the same binding energy. This can be seen experimentally as they all decay into the C^{12} ground state. (See Figure 4.4.)

4.14 Illustration of I-Spin Considerations: Pions and the Pion Nucleon System

I-spin considerations are now applied to pions. The pion has three charge states of almost the same mass. The strong interactions of these three states seem to be the same and it is desirable to add them to the I-spin picture. The pions are, therefore, placed in an $I = 1$ triplet:

$$|1,1\rangle = \pi^+; \quad |1,0\rangle = \pi^0; \quad |1,-1\rangle = \pi^-. \tag{4.56}$$

It is assumed that, with this definition, the strong interactions are again invariant under I-spin rotations. Note that for both pions and nucleons the charge can be written as $Q = (B + 2I_3)/2$, where B is the baryon number, 1 for p or n and -1 for their anti-particles. This formula will be modified when strangeness is considered.

Two pions can be in states of $I = 0, 1$, or 2 and the states can be found by using the Clebsch–Gordan tables. For example, for $I = 2$, the states of $|I, I_3\rangle$ are

$$|2, 2\rangle = |\pi^+\pi^+\rangle; \quad |2, 1\rangle = \frac{1}{\sqrt{2}}(|\pi^+\pi^0\rangle + |\pi^0\pi^+\rangle); \tag{4.57}$$

$$|2, 0\rangle = \frac{1}{\sqrt{6}}(2|\pi^0\pi^0\rangle + |\pi^+\pi-\rangle + |\pi^-\pi^+\rangle); \text{ etc.} \tag{4.58}$$

It is worth noting that I-spin independence does *not* imply charge independence for the two-pion system. Charge independence would say that the strong interactions are independent of the charge states of the particles. There are six states of two pions, $\pi^+\pi^+$, $\pi^+\pi^0$, $\pi^+\pi^-$, $\pi^0\pi^0$, $\pi^0\pi^-$, and $\pi^-\pi^-$. Suppose that the two pions have a symmetric space state. The strong interactions do not have to be the same in all six states, since the states are, in general, mixtures of $I = 0$ and $I = 2$, and I-spin invariance says nothing about the relative interactions in these two states and in fact they are not the same. I-spin invariance does not imply charge independence.

Consider the pion–nucleon system. Here there are two possible values, $I = 1/2$ or $3/2$. Thus,

$$|3/2, 3/2\rangle = |p\pi^+\rangle; \quad |3/2, 1/2\rangle = \frac{1}{\sqrt{3}}(|n\pi^+\rangle + \sqrt{2}|p\pi^0\rangle); \tag{4.59}$$

$$|3/2, -1/2\rangle = \frac{1}{\sqrt{3}}(|p\pi^-\rangle + \sqrt{2}|n\pi^0\rangle); \quad |3/2, -3/2\rangle = |n\pi^-\rangle; \tag{4.60}$$

$$|1/2, 1/2\rangle = \frac{1}{\sqrt{3}}(\sqrt{2}|n\pi^+\rangle - |p\pi^0\rangle); \quad |1/2, -1/2\rangle = \frac{1}{\sqrt{3}}(|n\pi^0\rangle - \sqrt{2}|p\pi^-\rangle). \tag{4.61}$$

Consider the two processes: $p + p \to d + \pi^+$ and $p + n \to d + \pi^0$. In the first process the two initial protons can only be in an $I = 1$ state. The final d and π^+ is also necessarily in an $I = 1$ state since the d has $I = 0$. In the second process the initial $p + n$ system can have $I = 0, 1$ while the final $d + \pi^0$ system necessarily has $I = 1$.

4.14 Pions and the Pion Nucleon System

I-spin is believed to be conserved in strong interactions. Thus, in the second interaction only the $I = 1$ component of the initial $p + n$ state can contribute to the reaction. $p^1 n^2 = \frac{1}{\sqrt{2}}(|1,0\rangle + |0,0\rangle)$. Two nucleons go to a deuteron and a pion in both reactions. Since cross sections are proportional to the square of the matrix elements and only 50% of the initial state of the second reaction can participate, one expects $\sigma_2/\sigma_1 = 1/2$ ignoring slight phase space effects, and this is indeed observed.

Next consider pion–nucleon scattering. There are six reactions that can be measured: $\pi^+ p \to \pi^+ p$; $\pi^- n \to \pi^- n$; $\pi^- p \to \pi^- p$; $\pi^- p \to \pi^0 n$; $\pi^+ n \to \pi^+ n$; $\pi^+ n \to \pi^0 p$. However, there are only two amplitudes, the $I = 1/2$ and $I = 3/2$ amplitudes, since I_3 shouldn't matter if the interactions are invariant under I-spin rotations. Previously the six I-spin states were written down in terms of pion–nucleon wavefunctions. Now the inverse is needed, the pion–nucleon functions in terms of the I-spin states. The inverse, also, can be read off the Clebsch–Gordan Coefficient Table, Figure C.1 in Appendix C:

$$|\pi^+ p\rangle = |3/2, 3/2\rangle; \quad |\pi^+ n\rangle = \frac{1}{\sqrt{3}}(|3/2, 1/2\rangle + \sqrt{2}|1/2, 1/2\rangle); \quad (4.62)$$

$$|\pi^0 p\rangle = \frac{1}{\sqrt{3}}(\sqrt{2}|3/2, 1/2\rangle - |1/2, 1/2\rangle); \quad (4.63)$$

$$|\pi^0 n\rangle = \frac{1}{\sqrt{3}}(\sqrt{2}|3/2, -1/2\rangle + |1/2, -1/2\rangle); \quad (4.64)$$

$$|\pi^- p\rangle = \frac{1}{\sqrt{3}}(-\sqrt{2}|1/2, -1/2\rangle + |3/2, -1/2\rangle); |\pi^- n\rangle = |3/2, -3/2\rangle. \quad (4.65)$$

Let M_3 be the $I = 3/2$ scattering amplitude, $\langle 3/2, x| A |3/2, x\rangle$ and M_1 be the $I = 1/2$ scattering amplitude, $\langle 1/2, x| A |1/2, x\rangle$. Using the preceding relations, the various pion–nucleon elastic scattering amplitudes can be described in terms of M_1 and M_3. The results are given in Tables 4.1 and 4.2.

There is a famous $I = 3/2$ pion nucleon resonance, the $\Delta(1232)$, which will be studied in the next chapter. (The 1232 is the approximate mass of the resonance in MeV.) The ratio of these cross sections at the resonance peak is in the 9:1:2 ratio predicted here indicating that the resonance is indeed $I = 3/2$.

Table 4.1 I-Spin Amplitudes for Pion–Nucleon Reactions

Reaction	Amplitude
$\pi^+ p \to \pi^+ p$	M_3
$\pi^+ n \to \pi^+ n$	$2/3 M_1 + 1/3 M_3$
$\pi^+ n \to \pi^0 p$	$\sqrt{2}/3(-M_1 + M_3)$
$\pi^- p \to \pi^0 n$	$\sqrt{2}/3(-M_1 + M_3)$
$\pi^- p \to \pi^- p$	$2/3 M_1 + 1/3 M_3$
$\pi^- n \to \pi^- n$	M_3

Table 4.2 I-Spin Coefficients for Pion–Nucleon Cross Sections

I-spin state	$\sigma(\pi^+ p \to \pi^+ p)$	$\sigma(\pi^- p \to \pi^- p)$	$\sigma(\pi^- p \to \pi^0 n)$
	M_3^2	$(2/3 M_1 + 1/3 M_3)^2$	$2/9 (M_3 - M_1)^2$
$I = 3/2$	1	1/9	2/9
$I = 1/2$	0	4/9	2/9

4.15 Further Use of I-Spin: Nucleon–Anti-Nucleon Pairs

The I-spin doublet for an anti-nucleon is similar to that for a nucleon. It is desirable to have the anti-particles of (p, n) also form an I-spin doublet, transforming in the same way as (p, n). Since the $\bar{p}$ has negative charge, it will be the $I_3 = -1/2$ member of the doublet, and $\bar{n}$ will be the $I_3 = +1/2$ member. This is almost, but not quite, correct, and the problem needs to be examined more carefully.

Start by applying the charge conjugation operator to the (p, n) doublet, obtaining $(\bar{p}, \bar{n})$. If CPT is conserved, C is equivalent to PT and, as noted in section 4.6, T involves complex conjugation. Thus, the pair $(\bar{p}, \bar{n})$ transforms as U^* for the same unitary I-spin transformation that U induces on (p, n). Let

$$U = e^{1/2 i \alpha \cdot \tau}; \quad U^* = e^{-1/2 i \alpha \cdot \tau^*}. \tag{4.66}$$

Here α is the I-spin vector defining the specific transformation desired.

Using the commutator relations for the Pauli matrices, it can be seen that

$$U^\star = (i\tau_2)U(-i\tau_2). \tag{4.67}$$

Let $\Psi = (\bar{p}, \bar{n})$ and $\Xi = (-i\tau_2)\Psi$.

$$U\Xi = U(-i\tau_2)\Psi = (-i\tau_2)(i\tau_2)U(-i\tau_2)\Psi = -i\tau_2 U^\star \Psi = -i\tau_2 \Psi' = \Xi',$$

where Ψ' and Ξ' are the transformed states under the operation. Thus, $-i\tau_2\Psi$ transforms according to U as desired. The appropriate I-spin doublet for the anti-particles is $-i\tau_2\Psi = (-\bar{n}, \bar{p})$, i.e., a minus sign is introduced into one of the states. Since either $\pm i\tau_2$ could have been used, the minus sign could have been introduced to either (but not both!) states. The standard choice is the one given previously. For a nucleon–anti-nucleon pair, the I-spin singlet state is then $1/\sqrt{2}(p\bar{p} + n\bar{n})$ and the $I=1$, $I_3 = 0$ state is $1/\sqrt{2}(p\bar{p} - n\bar{n})$.

4.16 ELECTROMAGNETISM AND ISOTOPIC SPIN

Electromagnetic interactions clearly are not invariant under I-spin. The proton is charged, the neutron is neutral. To express charge in terms of I-spin, note that the I-spin operator $(1 + \tau_3)/2$ has the property that it gives 1 for a proton and 0 for a neutron. However, for anti-particles, one needs $(-1 + \tau_3)/2$, since the anti-proton has $I_3 = -1/2$. In Section 4.18 the operator hypercharge, Y, will be introduced. This operator has the property of being 1 for protons, -1 for anti-protons, and 0 for pions. $(Y + 2I_3)/2$ is, thus, the appropriate charge operator, a generalization of the relation given in Section 4.16. Since mesons can be considered as nucleon–anti-nucleon (or later, quark–anti-quark) bound states, the same operator works for mesons.

Since $2I_3 = \Sigma\tau_3$ is a vector operator, this form has the important consequence that, although electromagnetic interactions are not invariant under I, an electromagnetic vertex can change I only by $\Delta I = 0$ or 1.

I-spin, as discussed previously, is a very important concept, considerably simplifying the discussion of strong interactions.

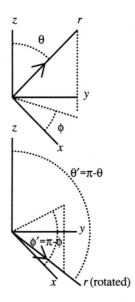

Figure 4.5 Rotation of 180° about the y-axis.

4.17   G-Parity

Another very useful quantum number for the strong interactions can be obtained by combining some of the symmetry operations discussed previously. Define G-parity:

$$G \equiv CR \equiv Ce^{i\pi I_2}. \tag{4.68}$$

The operator R corresponds to a 180° rotation about the y-axis (2-axis) in I-spin space. G-parity is conserved whenever C and I-spin are conserved; strong interactions conserve G-parity.

Recall that under C, $C\pi^0 = +\pi^0$, and $C\pi^\pm = -\pi^\mp$. What is $R\pi$? The analogy with angular momentum will be used, since the same $SU(2)$ group is involved. Consider the rotation properties of the spherical harmonics. From examining Figure 4.5, it can be seen that R corresponds to $\theta \to \pi - \theta$ and $\phi \to \pi - \phi$. Now $Y_1^0 = (3/4\pi)\cos\theta$ and $Y_1^{\pm 1} = \pm(3/8\pi)\sin\theta e^{\pm i\phi}$. Hence, $RY_1^0 = -Y_1^0$ [and generally $RY_l^0 = (-1)^l Y_l^0$] while $RY_1^1 = +Y_1^{-1}$.

$G\pi^0 = -\pi^0$ and $G\pi^\pm = -\pi^\pm$ and hence the G-parity of a pion is -1. G-parity is multiplicative. For n pions G gives $(-1)^n$.

As an example of its use note that the ρ meson is a meson of mass 770 MeV/c^2 with $I = 1$. It decays strongly into two pions, which implies that it has positive G-parity. This means that the three-pion decay mode of the ρ is forbidden by G-parity.

Next consider $\overline{N}N$ annihilations from rest, where the system has spin s and orbital angular momentum l. Consider, first, the neutral system $\bar{p}p$ or $\bar{n}n$. As for positronium, $C\left|\overline{N},N,l,s\right\rangle = (-1)^{l+s}\left|\overline{N},N,l,s\right\rangle$. For the neutral system, $I_3 = 0$ and since $RY_l^0 = (-1)^l Y_l^0$, one obtains $G\left|\overline{N},N,l,s\right\rangle = (-1)^{l+s+I}\left|\overline{N},N,l,s\right\rangle$.

Since the strong interactions are invariant under I-spin rotations, this relation must also hold if the system is not neutral, i.e., if $I_3 \neq 0$. Suppose now $\overline{N}N \to n\pi$. Then $(-1)^{l+s+I} = (-1)^n$.

Examine $\bar{p}n \to \pi^0\pi^-$. The initial state must have $I = 1$ since it is charged. This means that $l+s$ is odd. It will be shown in Section 7.1 that, for fermions, the intrinsic parity of an anti-particle is minus that of the particle. The parity of the $\bar{p}n$ system for $s = 0$ is then $(-1)^{l+1}$. However, the parity of the two-pion system is $(-1)^{l_\pi}$. For $s = 0$, $l = l_\pi$ and the reaction is forbidden. For $s = 1$, the S-state ($l = 0$) is allowed and the P-state ($l = 1$) is forbidden.

G-parity is, then, a simple multiplicative quantum number. The fact that the G-parity of a pion is negative leads to a number of restrictions on particle interactions and decays.

4.18 STRANGENESS

In the late 1940's[12] physicists saw some odd looking events in cloud chambers. A cosmic ray particle would interact in a plate in the cloud chamber and beneath it there would appear a two prong decay of a neutral particle (V^0) or one of the charged prongs coming from the interaction would break sharply indicating a decay ($V^\pm$). (See Figure 4.6.) These were known to be decays from the number of prongs. (For example, a π^-p interaction must give an even number of outgoing prongs by charge conservation.)

This created a major theoretical problem. These new particles were produced with millibarn cross sections characteristic of the strong interactions. (1 barn $= 10^{-24}$ cm^2; 1 millibarn $= 10^{-3}$ barn.) However, the particles lived about 10^{-10} s before decaying, whereas strong interaction lifetimes should have been about 10^{-22} s. After considerable initial confusion it was finally realized that these particles were being produced in pairs. M. Gell-Mann suggested that they had a new quantum number (S = strangeness) that was conserved in production by the strong or electromagnetic interactions but violated in decays. Strangeness could be conserved in production if two particles, one with $S = +1$ and one with $S = -1$, were created. The lifetime of the decays was characteristic of the weak interactions responsible for β-decays of radioactive nuclei. Table 4.3 lists the long-lived strange particles. It will be seen that there are many short-lived excited states of these particles and they will be related to various symmetry schemes.

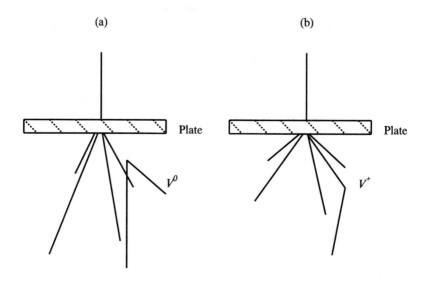

Figure 4.6 Strange particle decays.

Note that the masses of the I-spin multiplet partners are not exactly the same. This was also true for the pions. It is partly due to electromagnetic effects, but will be examined further in Chapter 6, when quarks are discussed.

The decays of these particles are quite rich. The strangeness -1 baryons decay into a pion and a nucleon with high probability and have rarer decay modes into lepton–neutrino–nucleon. The Σ^0 decay is dominated by the Λ–γ mode. The kaons decay into two and three pions, into lepton–neutrino and pi–lepton–neutrino modes, all with branching ratios greater than 1%. Note that the strange mesons, the kaons, occur as an isotopic-spin doublet, not a triplet as was the case for the pions.

The formula relating Q, B, I_3 must now be expanded to include S. The result is known as the Gell-Mann–Nishijima formula:

$$Q = I_3 + (B + S)/2 \qquad (4.69)$$

$(B + S)$ is known as "hypercharge," (Y) and the negative strangeness baryons are known as "hyperons." There are now two separate conserved quantities, I-spin and strangeness. Can they be put together to form a useful larger invariance group? The answer is they can and the group is called $SU(3)$. This is a very useful group, but, surprisingly, turns out not to have fundamental significance. This will be discussed in more detail in Chapter 6.

Table 4.3 Some Low Lying Strange Particle States

Particle	Mass MeV/c^2	Q	I	I_3	$Spin^P$	S
Baryons						
Λ	1115.6	0	0	0	$1/2^+$	-1
Σ^+	1189.4	+	1	+1	$1/2^+$	-1
Σ^0	1192	0	1	0	$1/2^+$	-1
Σ^-	1197	-	1	-1	$1/2^+$	-1
Ξ^0	1315	0	1/2	1/2	$1/2^+$	-2
Ξ^-	1321	-	1/2	-1/2	$1/2^+$	-2
Ω^-	1672	-	0	0	$3/2^+$	-3
Mesons						
K^+	493.7	+	1/2	1/2	0^-	+1
K^0	497.7	0	1/2	1/2	0^-	+1

The parity of the K^- has been measured to be that of the π^-, i.e., negative. Stopping K^- mesons[13] were absorbed by He^4, forming $\pi^- He^4_\Lambda$. He^4_Λ is a hypernucleus, a nucleus containing a negative strangeness baryon. He^4_Λ consists of $ppn\Lambda$. It is known to have $J^P = 0^+$ as does He^4. Thus the parity of the K^- must be the same as that of the π^-, since the orbital states are identical.

4.18 a. K^0 and $\overline{K}^0$ Decays

The K^0 and the $\overline{K}^0$ are anti-particles of each other, but what prevents them from mixing together to form a state that is partially K^0 and partially $\overline{K}^0$? They do differ by two units of strangeness. However, it has been noted that strangeness is violated in the weak decays. Since here $\Delta S = 2$, a violation has to be doubly weak, but this doesn't mean it is impossible. For example, a K^0 can virtually decay into two pions, which can then, still in the virtual state, recombine to form a $\overline{K}^0$.

This is important because of a peculiar fact. In production of these particles by the strong or electromagnetic interactions, S is conserved and the eigenstates are K^0 and $\overline{K}^0$. However, these are not the eigenstates for decay![14] Consider the effect of CP on the K^0 and $\overline{K}^0$. For now assume that CP is conserved in the weak decays. (It is conserved everywhere else and almost conserved in the weak decays.) The K^0 and $\overline{K}^0$ are not eigenstates of CP. In fact,

$$CP\left|K^0\right\rangle = \eta\left|\overline{K}^0\right\rangle; \text{ and } CP\left|\overline{K}^0\right\rangle = \eta'\left|K^0\right\rangle. \qquad (4.70)$$

Take the phase factors η and η' as $+1$. Call the CP eigenstates K_1 and K_2. Then,

$$|K_1\rangle = \frac{1}{\sqrt{2}}(K^0 + \overline{K}^0); \text{ and } |K_2\rangle = \frac{1}{\sqrt{2}}(K^0 - \overline{K}^0). \qquad (4.71)$$

The K_1 has $CP = +1$ and the K_2 has $CP = -1$.

Similarly,

$$\left|K^0\right\rangle = \frac{1}{\sqrt{2}}(K_1 + K_2) \text{ and } \left|\overline{K}^0\right\rangle = \frac{1}{\sqrt{2}}(K_1 - K_2). \qquad (4.72)$$

(This means that the initial kaon is produced in a mixed CP state even though CP is conserved. Suppose the initial reaction were $\pi^- p \to K^0 \Lambda$ in production. The kaon is in a mixed CP state because the initial $\pi^- p$ system is in a mixed state. One would have to have a mixture of p and $\bar{p}$ to have a pure CP initial state!)

Suppose CP is conserved in the decay and consider the decays $K^0 \to 2\pi$ and 3π. The kaon and the pions are spin 0 particles. When a kaon decays into two pions, the two pions are therefore in an S state ($l = 0$), which means that $P = +1$. By the extended Bose–Einstein statistics principle the wavefunction must be totally symmetric in the two pions. The I-spin function is then symmetric which implies $C = +1$, and hence $CP = +1$.

For the three-pion system from K^0 decay note that the kinetic energy is low. The Q value, which is $m_{\text{init}} - \Sigma m_{\text{final}} = m_K - \Sigma m_\pi$, is only 70 MeV. Suppose, then, that the pions are in an S state with no angular momentum between any pairs of pions. $P = -1$ as the pions have negative intrinsic parity (they are pseudoscalar). As with the two-pion example, the extended Bose–Einstein principle states that the wavefunction must be totally symmetric under interchange. Since the S wave space wavefunction is symmetric, the I-spin function is symmetric and $C = +1$. Hence for the three-pion system $CP = -1$.

An initial K^0 can be considered an equal mixture of K_1 and K_2. The K_1 component can decay into two pions, while the K_2 can decay into three pions. Because of the very small Q value for the three-pion decay and consequent small phase space, the rates for the two processes are very different. The lifetime for the two-pion component is 0.9×10^{-10} s, while, because of the reduced phase space, the lifetime for the three pion component is 0.5×10^{-7} s. Thus in decay the kaon behaves like a mixture of two particles with very different lifetimes!

4.18 b. K^0-$\overline{K}^0$ Oscillations

There is a strange and beautiful phenomenon associated with the neutral kaon mixture, namely, there is an oscillation between K^0 and $\overline{K}^0$.

There is a close analogy between this process and the motion of two identical pendulums that are loosely coupled. You will have studied this latter process in mechanics. There the eigenmodes are very slightly split by the coupling [corresponding here to $m(K_1) \neq m(K_2)$] and if one pendulum is started with the second at rest, some time later the first pendulum is at rest and all the energy is in the second.

Consider the kaon wavefunctions. Let $\Delta m = m(K_2) - m(K_1)$ and let Γ_1 and Γ_2 be the decay rates (1/lifetime) of the K_1 and K_2, respectively. It is the eigenstates that decay exponentially with definite lifetimes. If one starts with a K_1 at rest in vacuum, then

$$a_1(t) = a_1(0)e^{im_1 t}e^{-\Gamma_1 t/2} \text{ and Intensity } = I = a_1^\star a_1 = |a_1(0)|^2 e^{-\Gamma_1 t}. \quad (4.73)$$

Suppose now that one starts with a K^0 in vacuo in the rest system. One then has $a_1(0) = a_2(0) = 1/\sqrt{2}$. After a time t this becomes

$$\text{Intensity of } K^0 = I(K^0) = \frac{1}{\sqrt{2}}[a_1^\star(t) + a_2^\star(t)]\frac{1}{\sqrt{2}}[a_1(t) + a_2(t)], \quad (4.74)$$

$$\text{Intensity of } \overline{K}^0 = I(\overline{K}^0) = \frac{1}{\sqrt{2}}[a_1^\star(t) - a_2^\star(t)]\frac{1}{\sqrt{2}}[a_1(t) - a_2(t)], \quad (4.75)$$

$$I(K^0) = \frac{1}{4}(e^{-\Gamma_1 t} + e^{-\Gamma_2 t} + 2e^{-[(\Gamma_1 + \Gamma_2)/2]t}\cos \Delta mt), \quad (4.76)$$

$$I(\overline{K}^0) = \frac{1}{4}(e^{-\Gamma_1 t} + e^{-\Gamma_2 t} - 2e^{-[(\Gamma_1 + \Gamma_2)/2]t}\cos \Delta mt). \quad (4.77)$$

Thus the amplitudes oscillate with angular frequency Δm. This is actually seen. $\overline{K}^0$ can produce hyperons in interactions with matter, since the $\overline{K}^0$

and the hyperons have the same strangeness. K^0 can only produce hyperons in interactions if a new strange–anti-strange pair is made. Examining the production of hyperons as a function of distance then allows the oscillation to be examined. One finds $\Delta m = (+3.522\pm0.016)\times 10^{-6}$ eV $[m(K_2) > m(K_1)]$, $\Delta m\tau_1 = 0.477$. A mass difference with $\Delta m/m = 0.7\times 10^{-14}$ has been measured!

Regeneration is another phenomenon of the K^0 system. In this phenomenon the fact again is used that the K^0 and $\overline{K}^0$ interact differently with matter since they have different strangeness. The K^0 is allowed to decay until the K_1 component has vanished and only K_2 are left. Then the K_2 are allowed to interact in matter. Since they interact as K^0 and $\overline{K}^0$, it is found that after the interactions, K_1 decays again appear.

4.18 c. CP Violation

There is still another fascinating property of the neutral kaon system. In 1962 Christenson, Cronin, Fitch, and Turlay[15] found that a small part of the long-lived kaons decayed by the two-pion mode. This violates CP invariance! For this discovery J. Cronin and V. Fitch were awarded the 1980 Nobel Prize.

This means that the CP eigenstates K_1 and K_2 are not quite the eigenstates of the decay interaction. Let K_S and K_L be the eigenstates of the decay interaction. (S = short-lived and L = long-lived.) Then

$$|K_S\rangle = \frac{1}{\sqrt{2(1+|\epsilon|^2)}}[(1+\epsilon)|K^0\rangle + (1-\epsilon)|\overline{K}^0\rangle], \qquad (4.78)$$

$$|K_L\rangle = \frac{1}{\sqrt{2(1+|\epsilon|^2)}}[(1+\epsilon)|K^0\rangle - (1-\epsilon)|\overline{K}^0\rangle], \qquad (4.79)$$

where $|\epsilon| \ll 1$ and ϵ does not have to be real. K_S and K_L are the eigenstates of the Hamiltonian for the mass-decay matrix $H = M - i\Gamma$. M and Γ are individually hermitian since they correspond to observables (mass and lifetime). However, $M - i\Gamma$ does not have to be hermitian. In the K^0, $\overline{K}^0$ basis, $H_{11} = H_{22}$ from CPT invariance.

The magnitude and phase of ϵ are found from the experimentally determined mass-decay matrix. The phase ϕ of ϵ is given by $\phi \approx \tan^{-1}([2\Delta m\tau_s]/\hbar) = 43.59\pm 0.15°$. Unlike K_1 and K_2, K_S and K_L are not orthogonal states. In fact it is easily seen that $\langle K_S \mid K_L\rangle = 2\Re(\epsilon)/(1+|\epsilon|^2)$.

In the two-pion decay the pions can have $I = 0$ or 2, and there can be amplitudes, A_I, for each state. Furthermore, there are strong interaction "phase shifts," δ, as the pions interact with each other. That is, in the final state the two pions can interact and the final-state wavefunction is distorted. For a definite spin and I-spin state this is just a change in phase. In the next chapter the wave optical picture of particle scattering will be discussed. The phase changes for the present situation follow from the same line of argument to be discussed there. Let

$$\eta_{+-} = \frac{\text{Ampl.}\ (K_L \to \pi^+\pi^-)}{\text{Ampl.}\ (K_S \to \pi^+\pi^-)}; \quad \eta_{00} = \frac{\text{Ampl.}\ (K_L \to \pi^0\pi^0)}{\text{Ampl.}\ (K_S \to \pi^0\pi^0)}. \quad (4.80)$$

Then for the $\pi^+\pi^-$ decay:

$$I_{\pi^+\pi^-}(t) = I_{\pi^+\pi^-}(0)[e^{-\Gamma_S t} + |\eta_{+-}|^2 e^{-\Gamma_L t} + 2|\eta_{+-}|e^{-[(\Gamma_L+\Gamma_S)/2]t} \\ \times \cos(\Delta mt + \phi_{+-})]. \quad (4.81)$$

The formula for the $\pi^0\pi^0$ decay is exactly similar with $+-$ replaced by 00. One can write[16]

$$\eta_{+-} = \epsilon + \epsilon' = |\eta_{+-}|e^{i\phi_{+-}}, \quad \eta_{00} = \epsilon - 2\epsilon' = |\eta_{00}|e^{i\phi_{00}}, \quad (4.82)$$

where

$$\epsilon' \equiv \frac{i}{\sqrt{2}} \frac{\Im A_2}{A_0} e^{i(\delta_2 - \delta_0)} \quad (4.83)$$

using an obvious notation. (A_0 has been taken as real and positive here.) There is then, even in vacuo, an oscillation in the rate of 2π decay of the neutral kaons. CP violation is seen only in the kaon system so far, although it is hoped to see it in the B meson system in the future.

CP violation can only show up when there is an interference between two decay modes, here the $I = 0$ and $I = 2$ final-state modes leading to $\pi^+\pi^-$ and $\pi^0\pi^0$ decays. It is easy to see how this works. Let the decay matrix element for K^0 be $\sum_j g_j e^{i\delta_j}$, where δ_j represents the strong interaction phase shifts. Under charge conjugation, $g_j \to g_j^*$, since charge conjugation involves complex conjugation. However, $\delta_j \to \delta_j$, i.e., no complex conjugation, since the interaction between the anti-particles is the same as between the particles. When summing more than one g_j, the phase differences between the terms are then not the same for particles and anti-particles, leading to the possibility of CP violation.

Note that $K_L^0 \propto K_2^0 + \epsilon K_1^0$. CP violation can occur either from asymmetric mixing between the K^0 and its anti-particle or from direct CP violation in the decay process. The first process appears in the mass-decay matrix parameters and, as was seen previously, is parametrized by the parameter ϵ. The second process, the direct decay of $K_2^0 \to 2\pi$, is parametrized by ϵ' and can occur even if ϵ is zero.

Experimentally[17] it is found that $\eta_{+-} = (2.68 \pm 0.023) \times 10^{-3}$ and $\phi_{+-} = 46 \pm 1.2°$. There are two measurements of ϵ'/ϵ. They are not in very good agreement: $\epsilon'/\epsilon = [2.3 \pm 0.34 \pm$ (statistical) ± 0.65 (systematic) $] \times 10^{-3}$ and $[0.74 \pm 0.52$ (statistical) ± 0.29 (systematic) $] \times 10^{-3}$.[18] [19] [20] [21] Thus, the experiments have not yet settled the question of whether direct CP violation is occuring in the K decay process.

What causes CP violation? The answer is not yet clear. The mass-decay matrix violations, but not the direct mode violations, can be blamed on $\Delta S = 2$ processes, if one chooses to do so. It is worth noting that CPT conservation implies that if one partial decay mode is different for K^0 and $\overline{K}^0$, another must be different in the opposite direction to compensate, since CPT conservation implies the overall lifetimes must be the same.

A persistent question is "If it exists at all why is it so small?" One suggestion, due to L. Wolfenstein[22], is that there is a new superweak force connecting K^0 and $\overline{K}^0$ whose only observable physical effect is on the mass matrix. This theory predicts ϵ' to be zero, but as seen above, the experimental results are ambiguous on this possibility.

Within the Standard Model, which will be developed in this text, the size of CP violation can be interpreted in terms of the small size of the off-diagonal terms in the CKM matrix, which will be discussed in Section 7.12. Even there, a phase must be introduced. This is a parametrization, but does not serve as a deep explanation of the existence or size of this fascinating phenomenon.

In this chapter a number of standard symmetries and conserved quantities have been discussed: conservation of momentum, energy, and angular momentum have been related to basic invariance properties of space–time. Conservation of charge and gauge invariance have been shown to be deeply related. P, C, T symmetries have been introduced and their conservation by various forces discussed. The concepts of isotopic spin, G-parity, and strangeness have been discussed and their usefulness illustrated. In the course of the discussion, a number of the quantum numbers and symmetries of various subatomic particles were described.

4.19 Exercises

4.19 Exercises

4.1 Does a rotation R commute with the parity operation P?

4.2 It is observed experimentally that the reaction in which a π^- is captured by a d and produces two neutrons and a π^0 is inhibited, compared to the reaction with two neutrons only. This would be expected if the π^0 were forced to be in a nonzero angular momentum state because of the limited phase space (low energy for the pion). Show that this is expected if the parity of the π^0 is negative, but not if it is positive. Assume the parity of the charged pion is negative. Hint: follow the argument given for the experiment determining the parity of the charged pion.

4.3 The η meson is a neutral meson of 547.5 MeV that decays 39% of the time into two photons and most of the rest of the time into three pions (32% $3\pi^0$, 24% $\pi^+\pi^-\pi^0$). The decay angular distribution indicates that the pions are in an S state; all possible pairs of pions have $l = 0$. There are no charged partners of the η. Use some portion of the preceding information to find the spin and parity of the η.

4.4 If the three-pion decay of the η were a strong decay, what would be the G-parity of the η? However, note that the two-photon mode is competitive with the three-pion mode. The decay is not solely a strong decay, but must involve an electromagnetic vertex that can break G-parity conservation in the decay. In fact, from the information in problem 4.2, find the behavior of the η under C, the I-spin of the η, and, hence, the G-parity of the η.

4.5 The following is an alternate discussion of the argument that, given that the π^0 has a decay into two photons, it cannot have spin 1. Suppose the π^0 had spin 1.

a) Is the initial wavefunction a scalar, vector,...? (To save space, give these all the generic name "tensors.")

b) What physical tensors are available with which to construct a final-state amplitude? Look at this from the rest frame of the π^0 and use the relation that follows between the momentum of each gamma.

c) Construct three possible final-state tensors appropriate to match the tensor of part a given the physical tensors of part b. [Ignore parity. Assume that the final-state wavefunction must have some sort of products of the polarizations of each gamma (bilinear combinations).] That is, polarization one times polarization two is allowed, but polarization one plus polarization two is not. Note that cross and dot products must be considered.

d) What restriction, if any, must be imposed on the final-state tensors in part c by the spin-statistics theorem? (Consider the photons.) Apply it to each state of c if you find a restriction. At least one combination should survive.

e) What restriction, if any, must be imposed by the nature of the electromagnetic field, i.e., how does the polarization relate to the momentum of each photon? If there is a restriction, apply it to each state surviving part d. Does any state survive?

4.6 The decay $\pi^0 \to 3\gamma$ has not been seen, although the 2γ mode is the dominant mode. Why is the 3γ mode inhibited?

4.7 Write down the states of various I_3 for a two pion system with $I = 1$.

4.8 Suppose A, B, C, D are all spin 1/2 particles. Consider the reaction $A + B \to C + D$, where A and B are unpolarized. Show that parity conservation in this process guarantees that any polarization of C or D be perpendicular to the production plane.

4.9 Consider the reaction $\pi^- + p \to \Lambda + \pi^0$ followed by $\Lambda \to \pi^- + p$. Show that parity conservation forbids an up–down anisotropy of the decay π or p with respect to the production plane. Experimentally, such an anisotropy is observed, showing that this decay does not conserve parity.

4.10 Ignoring small phase space differences find the branching ratios expected for Σ^+ decay into $p\pi^0$ and $n\pi^+$. Assume first that the final state is pure $I = 3/2$, and then redo the calculation for a final state that is pure $I = 1/2$. Experimentally, the ratio between the decay rates for $p\pi^0$ and $n\pi^+$ is almost one.

4.11 What, if any, conclusions concerning the spin and parity of the K^0 can be drawn from the fact that the decay $K_S^0 \to 2\pi^0$ is a principal decay mode of the K_S^0?

4.12 When a K_L^0 beam is incident on a thin slab of matter, some K_S^0 decays appear afterward. Work out this phenomenon. Assume that if the wavefunction before the interaction for the $K^0 (\overline{K}^0)$ is $\psi_0 (\overline{\psi}_0)$, then after the interaction the wavefunctions are $\psi_0 e^{-a+i\delta} (\overline{\psi}_0 e^{-\bar{a}+i\bar{\delta}})$, where $a, \bar{a}, \delta, \bar{\delta}$ are real and $a, \bar{a}$ are positive. Hint: set $\epsilon = 0$ for this problem, since CP violation is not needed.

4.13 Show that the following "triangle relations" hold for the cross sections of $\pi p \to K\Sigma$:

$$\sqrt{\sigma(\pi^+ p \to K^+ \Sigma^+)} \leq \sqrt{\sigma(\pi^- p \to K^+ \Sigma^-)} + \sqrt{2\sigma(\pi^- p \to K^0 \Sigma^0)},$$

$$\sqrt{\sigma(\pi^- p \to K^+ \Sigma^-)} \leq \sqrt{\sigma(\pi^+ p \to K^+ \Sigma^+)} + \sqrt{2\sigma(\pi^- p \to K^0 \Sigma^0)},$$

$$\sqrt{2\sigma(\pi^- p \to K^0 \Sigma^0)} \leq \sqrt{\sigma(\pi^- p \to K^+ \Sigma^-)} + \sqrt{\sigma(\pi^+ p \to K^+ \Sigma^+)}.$$

Hint: if the amplitudes (complex numbers) obey a relation of the form $A_1 = A_2 + A_3$, then plotting them as vectors in the complex plane, the three amplitudes form a triangle, from which the inequalities follow. Find the amplitudes by resolving them, for each of the two particle states, into I-spin states and using conservation of I-spin for these strong interaction productions. Conservation implies that the matrix element will not change I-spin or I_3 and that the matrix element will not care about the projection, I_3, but will be only a function of $|I|$.)

4.14 Find the allowed I-spin states for $\pi^+\pi^-\pi^0$ and for $\pi^0\pi^0\pi^0$.

4.15 The ω meson has $J^{PG} = 1^{--}$, $I = 0$, $S = 0$, m=783 MeV.

a) What, if any, strong interaction decay modes are allowed for decays into pions (two or three body)? If strong interaction modes are not present list an electromagnetic mode you might expect to occur.

b) What sort of lifetime do you expect for this particle?

c) If it does go to two or three π mesons by strong interactions, what is the ratio of the various charge states of the pions?

4.16 The observation of any of the following would violate a regularity in particle physics. Name at least one violated "principle" for each of the following (one line or less, if possible), and indicate if the interaction or decay is expected to be absolutely forbidden or only inhibited. Hint: the values of $I^G(J^{PC})$ for the ω, π^0, and ϕ, respectively, are $0^-(1^{--})$, $1^-(0^{-+})$, and $0^-(1^{--})$, respectively. The masses of the three are 780, 135, and 1020 MeV respectively.

a) $\pi^+ + \pi^- \to 5\pi^0$.

b) $p \to ne^+\nu_e$.

c) An electric dipole moment for the Λ^0.

d) $\Lambda^0 \to K^0\overline{K}^0$.

e) $\Omega^- \to \Lambda^0\pi^-$.

f) $\phi \to \omega\pi^0$.

4.17 In this problem a simple experiment to observe K_S^0 regenerated from a K_L^0 beam impinging on matter will be designed. An actual experiment will be much more complicated, but this provides many of the essential considerations. The experiment is to be done at Fermilab. In the following when (type) appears insert decay, radiation, collision, or absorption as appropriate. (Consider absorption as interactions, such as nuclear interactions, which make the particle disappear or in some manner prevent it from penetrating further.) When (#) appears, insert an appropriate number. When ??? appears, insert the appropriate item, or position, etc.

a) Consider the K_L^0 production from the incident proton beam. The target should be (#) (type) lengths long. Justify (#). At a production angle of $\theta_p = 10^\circ$, particle production for K_L^0's with $\bar{p} = 2$ GeV/c is used.

96   4. Invariance, Symmetries, and Conserved Quantities

b) It is necessary to use a ??? to define the production angle and a ??? to remove charged particles from the beam.

c) To remove high-energy photons from the beam (#) (type) lengths of ??? (what material?) are placed in the beam. Justify (#) and material.

d) How far from the production target must the regenerator be placed if the K_S^0's initially produced are to be diminished at least by 10^{-6} (= e^{-14})?

e) After the regenerator, a ??? is placed to ensure that no charged particle exits the regenerator during a valid event.

f) What will be the typical half opening angle θ_h of symmetric $K_S^0 \to \pi^+\pi^-$ decays occurring behind the regenerator?

g) A spectrometer magnet then bends the $\pi^+\pi^-$ from symmetric decays so that they exit the magnet parallel to each other. The necessary field integral ($\int B dl$) is (#) T-m?

h) What mean $p_\perp$ due to multiple scattering is gained by a symmetric pion in traveling through a 1.3 m long air gap spectrometer magnet?

i) The two charged products of $K_L^0 \to \pi + e + \nu$ could mimic the desired decay $K_S^0 \to \pi^+ + \pi^-$. The $e\pi$ events can be discriminated against by a ??? placed ???. What aspect of this device insures low multiple scattering of the real $\pi^+\pi^-$ events?

j) Another background is caused by $K^0 \to \pi + \mu + \nu$ decay. To distinguish these events a particle is identified as a muon if it passes through a hadron filter of thickness (#) (type) lengths placed ???. Justify (#). Fe rather than Pb is used for this filter. Why?

k) In the data analysis how can the desired events be distinguished from $K_L^0 \to \pi^+\pi^-\pi^0$ events, where the π^0 is not observed in the apparatus. Draw a curve to illustrate your technique.

4.18 Show that in the reaction $e^+e^- \to$ virtual $\gamma \to$ meson, a spin-0 meson is not allowed. Hint: The matrix element for this reaction will involve the matrix element of the electromagnetic current operator between the meson and vacuum. Use the fact that the electromagnetic current is conserved and the relation that for any operator, $[P_\mu, O(x)] = -i\partial O/\partial x_\mu$, where P_μ is the momentum operator. Apply current conservation, $\partial J_\mu/\partial x_\mu = 0$. Note that $P_\mu |\text{vacuum}\rangle = 0$. If the final meson is spin 0 and the matrix element $\langle\text{meson}| J_\mu(0) |\text{vacuum}\rangle$ is a four-vector, find the most general four-vector dependence of $\langle\text{meson}| J_\mu(x) |\text{vacuum}\rangle$.

5
Hadron–Hadron Scattering

5.1 Wave Optics Discussion of Hadron Scattering

Scattering of hadrons can be discussed in a manner similar to scattering of light waves. For simplicity spinless particles will be assumed in the derivation, although the modifications for spin are not great. Imagine that there is an incoming plane wave normalized to one for a unit area in x,y:

$$\Psi_i = e^{ikz}e^{-i\omega t}, \tag{5.1}$$

where $k = 1/\lambda$ and $\lambda = \lambda/2\pi$; λ = the DeBroglie wavelength of the incident particle. In the following the factor $e^{i\omega t}$ will be left out.

Mathematically a plane wave can be written as a superposition of incoming and outgoing spherical waves if $kr \gg 1$. See Figure 5.1. Asymptotically,

$$\Psi_i = e^{ikz} = \frac{i}{2kr}\Sigma_l(2l+1)[(-1)^l e^{-ikr} - e^{ikr}]P_l(\cos\theta), \tag{5.2}$$

where the first term in the square brackets represents incoming waves and the second term represents outgoing waves. If kr is not in the asymptotic region, $kr \gg 1$, then the square bracket is replaced by $2kri^{l-1}j_l(kr)$, where j_l is the spherical Bessel function. P_l are the Legendre polynomials. The term l represents a spherical wave with orbital angular momentum l.

Suppose that this wavefunction represents the quantum mechanical wavefunction of a particle and that there is a scatterer at the origin. Scattering will affect only the outgoing waves. It can affect the magnitude and the phase of the wave. Therefore, multiply each outgoing wave by the factor $\eta_l e^{2i\delta_l}$, where the phase shift δ_l and the absorption factor η_l are both real and η_l is positive. Then including the effect of scattering:

$$\Psi_{\text{total}} = \frac{i}{2kr}\Sigma_l(2l+1)[(-1)^l e^{-ikr} - \eta_l e^{2i\delta_l}e^{ikr}]P_l(\cos\theta). \tag{5.3}$$

Now let $\Psi_{\text{scatt}} = \Psi_{\text{total}} - \Psi_i$. Then,

$$\Psi_{\text{scatt}} = \frac{e^{ikr}}{kr}\Sigma_l(2l+1)\frac{\eta_l e^{2i\delta_l}-1}{2i}P_l(\cos\theta) = \frac{e^{ikr}}{r}F(\theta), \tag{5.4}$$

where $F(\theta)$ is known as the scattering amplitude. It is simply related to

5. Hadron–Hadron Scattering

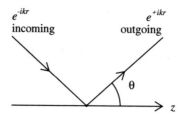

Figure 5.1 Incoming and outgoing waves.

the differential cross section as will be seen below. Note that $\Psi_{\text{total}} = e^{ikz} + \frac{e^{ikr}}{r}F(\theta)$ and that

$$F(\theta) = \frac{1}{k}\Sigma_l(2l+1)\frac{\eta_l e^{2i\delta_l} - 1}{2i}P_l(\cos\theta). \quad (5.5)$$

Define $f(l)$, the elastic scattering amplitude for the lth wave, to be

$$f(l) = \frac{\eta_l e^{2i\delta_l} - 1}{2i}. \quad (5.6)$$

$f(l)$ is a measure of how much, excluding the incident wave, of each wave l there is after the interaction with the initial "k" and hence it is a measure of the elastic scattering. For elastic scattering in the center-of-mass frame, $k_i = k_f$, which means $v_{\text{in}} = v_{\text{out}}$. Consider the outgoing flux through a large sphere. This flux is

$$\text{Flux}_{\text{out}} = v_{\text{out}}\Psi^{\star}_{\text{scatt}}\Psi_{\text{scatt}}r^2 d\Omega = v_{\text{out}}|F(\theta)|^2 d\Omega. \quad (5.7)$$

Next consider that the incident flux/unit area $\times d\sigma$ =outgoing flux through a sphere of radius R in a solid angle $d\Omega$. The incident flux/unit area $= v_{\text{in}}\Psi^{\star}_i\Psi_i$, and $\Psi^{\star}_i\Psi_i = 1$. Hence $\text{Flux}_{\text{out}} = v_{\text{in}}d\sigma$. Thus

$$\frac{d\sigma}{d\Omega} = |F(\theta)|^2. \quad (5.8)$$

Recall that the Legendre polynomials have the property that

$$\int P_l P_{l'} d\Omega = 4\pi\delta_{ll'}/(2l+1),$$

where $\delta_{ll'}$ is the Kronecker δ, which is 1 if $l = l'$ and is 0 otherwise. Then

$$\sigma_{\text{elastic}} = \frac{4\pi}{k^2}\Sigma_l(2l+1)\left|\frac{\eta_l e^{2i\delta_l} - 1}{2i}\right|^2 = 4\pi\lambda^2\Sigma_l(2l+1)\left|\frac{\eta_l e^{2i\delta_l} - 1}{2i}\right|^2. \quad (5.9)$$

Consider, first, the case of pure elastic scattering, i.e., no absorption,

$\eta_l = 1$. For this case:

$$\sigma_{\text{elastic}} = 4\pi\lambda^2 \Sigma_l (2l+1) \sin^2 \delta_l \qquad (5.10)$$

when it is noted that $|(e^{2i\delta_l} - 1)/2i|^2 = (1/4)|(\cos 2\delta_l - 1) + i \sin 2\delta_l|^2 = (1/4)[\cos^2 2\delta_l - 2\cos 2\delta_l + 1 + \sin^2 2\delta_l] = (1/2)[1 - \cos 2\delta_l] = \sin^2 \delta_l$.

Next consider the case where $\eta_l < 1$. From conservation of probability, the non-elastic part of the cross section, the reaction (or absorption) cross section, must be the difference of the incoming and outgoing spherical waves:

$$\sigma_{\text{reaction}} = \int (|\Psi_{\text{in}}|^2 - |\Psi_{\text{out}}|^2) r^2 d\Omega = \pi\lambda^2 \Sigma_l (2l+1)(1 - \eta_l^2), \qquad (5.11)$$

$$\sigma_{\text{total}} = \sigma_{\text{reaction}} + \sigma_{\text{elastic}}, \qquad (5.12)$$

$$\sigma_{\text{total}} = \pi\lambda^2 \Sigma_l (2l+1) 2(1 - \eta_l \cos 2\delta_l). \qquad (5.13)$$

From the preceding it is seen that, for given nonzero η_l, the maximum elastic scattering cross section possible for a given partial wave, l, occurs when $\delta_l = \pi/2$ and is

$$\sigma_{\text{elastic}}^{\max} = 4\pi\lambda^2 (2l+1)\eta_l^2. \qquad (5.14)$$

The maximum absorption cross section for a given partial wave occurs for $\eta_l = 0$. For this case the reaction and elastic scattering cross sections are

$$\sigma_{\text{reaction}}^{\max} = \pi\lambda^2 (2l+1) \; ; \; \sigma_{\text{elastic}} = \pi\lambda^2 (2l+1) \text{ if } \eta_l = 0.$$

Note that $P_l(\cos\theta) = +1$ for $\theta = 0$ (forward scattering). Thus,

$$\Im F(0) = 1/2k \Sigma_l (2l+1)(1 - \eta_l \cos 2\delta_l),$$

$$\Im F(0) = \frac{k}{4\pi} \sigma_{\text{total}}. \qquad (5.15)$$

This equation is known as the Optical Theorem. The imaginary part of the forward *elastic* scattering amplitude measures the total cross section. The imaginary part of the forward amplitude is in this sense the shadow of the total scattering.

5.2 BREIT–WIGNER RESONANCE FORMULA

Consider next what happens when one of the partial waves is near resonance, δ_l near $\pi/2$. Suppose there is no absorption, $\eta_l = 1$. Then,

$$f_l = \frac{\eta_l e^{2i\delta_l} - 1}{2i} = \frac{e^{i\delta_l}(e^{i\delta_l} - e^{-i\delta_l})}{2i} = e^{i\delta_l} \sin \delta_l.$$

Hence,

$$f_l = \frac{1}{\cot \delta_l - i}. \tag{5.16}$$

Near resonance δ_l is near $\pi/2$, which means $\cot \delta_l$ is near 0. Let E_R be the energy at resonance. Expand $\cot \delta_l$ in a power series around E_R:

$$\cot \delta_l(E) = \cot \delta_l(E_R) + (E - E_R)\left[\frac{d}{dE}\cot \delta_l(E)\right]_{E=E_R} + \cdots$$

$\cot \delta_l(E) \equiv -(E - E_R)(2/\Gamma)$ defines Γ. Then,

$$f_l(E) \cong \frac{\Gamma/2}{(E_R - E) - i\Gamma/2}. \tag{5.17}$$

The formula for elastic scattering (ignoring any other partial waves) then becomes

$$\sigma_{\text{elastic}} = 4\pi\lambda^2(2l+1)\frac{(\Gamma/2)^2}{(E - E_R)^2 + (\Gamma/2)^2}. \tag{5.18}$$

This is the Breit–Wigner resonance formula. The cross section at the peak measures l for the resonance. Γ is called the width of the resonance. Note that Γ is the full width of the resonance at half maximum (see Figure 5.2):

$$\sigma_{\text{elastic}}(E_R \pm \Gamma/2) = 0.5 \times \sigma_{\text{elastic}}(E_R). \tag{5.19}$$

The Breit–Wigner form is also the form for the width of an exponentially decaying state, an unstable particle decay. Let

$$\Psi(t) = \Psi(0)e^{-i\omega_R t}e^{-t/2\tau} \equiv \Psi(0)e^{-t[iE_R - \Gamma/2]}, \tag{5.20}$$

where $\Gamma = 1/\tau$ and $\tau =$ the decay lifetime. $I(t) = \Psi^\star(t)\Psi(t) = I(0)e^{-t/\tau}$.

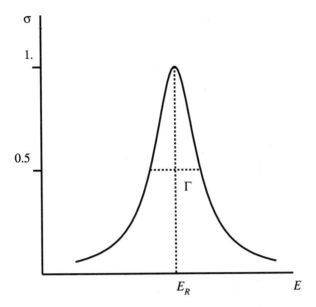

Figure 5.2 Breit–Wigner resonance curve.

Consider the Fourier transform of Ψ:

$$\chi(E) = \int_0^\infty \Psi(t) e^{iEt} dt = \Psi(0) \int e^{-t[\Gamma/2 + i(E_R - E)]} dt = \frac{-i\Psi(0)}{(E_R - E) - i\Gamma/2}. \tag{5.21}$$

Thus, the same characteristic form as a function of energy (frequency) is obtained. These resonances are thought of as new unstable particles.

If the initial particles have spin, the resonance formula is slightly modified. $l \longrightarrow J$ and for a resonance with a given J and parity one has, when averaged over spin states:

$$\sigma_{\text{elastic}} = \frac{4\pi \lambdabar^2 (2J+1)(\Gamma/2)^2}{(2s_a + 1)(2s_b + 1)[(E_R - E)^2 + (\Gamma/2)^2]}. \tag{5.22}$$

The reaction $\pi^+ + p \to \Delta(1232)$ is a fine example of the Breit–Wigner resonance form (Figure 5.3).

For particles with spin, the peak cross section determines J not l. The value of l can be obtained from the angular distribution. For example, the $\Delta(1232)$ is in a $J = 3/2$, $l = 1$ (P-wave) state. Thus, for the $\pi - p$ orbital state one has $\phi(l, m) = \phi(1, m)$ and the proton spin state is $\alpha(j, m) =$

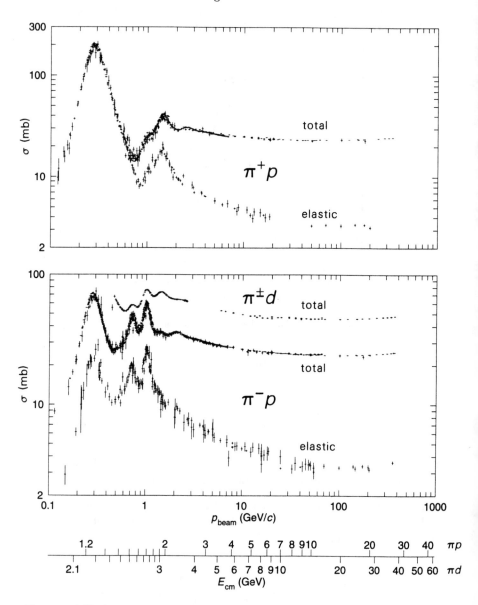

Figure 5.3 Hadronic total and elastic cross sections versus laboratory beam momentum and center-of-mass energy, showing clearly the $\Delta(1232)$ resonance and others.[17]

$\alpha(1/2, \pm 1/2)$. Initially, suppose the target protons are polarized. Then

$$\Psi(3/2, 1/2) = \sqrt{1/3}\phi(1,1)\alpha(1/2,-1/2) + \sqrt{2/3}\phi(1,0)\alpha(1/2,1/2), \quad (5.23)$$

where $\phi(1,1) = Y_1^1 = -\sqrt{3/(8\pi)}\sin\theta e^{i\phi}$ and $\phi(1,0) = Y_1^0 = \sqrt{3/(4\pi)}\cos\theta$. To find the intensity note that there are no cross terms because the proton has the opposite spin in each of these terms. Hence,

$$I(\theta) = \Psi^\star \Psi = \frac{1}{3}(Y_1^1)^2 + \frac{2}{3}(Y_1^0)^2 \propto \sin^2\theta + 4\cos^2\theta = 1 + 3\cos^2\theta. \quad (5.24)$$

One gets the same result for the other polarization and, hence, the same result for an unpolarized target, which is an incoherent mixture of spin forward and spin backward.

The wave optics picture of hadron scattering is very useful as is the Breit–Wigner form for resonances. The description of the many pion–nucleon resonances have conventionally used this formalism and the wave optics picture has continued to be useful for hadron scattering at the highest energies.

5.3 S-Matrix, Phase Space, and Dalitz Plots

In this section some rather formal definitions of matrix elements are given. There are several different terms used in the literature and (in at least one notation) some of the various scattering amplitudes you may read about are given here. Be cautioned that the normalizations vary from author to author. The normalizations here are those of Sakurai[23] and of Gasiorowicz.[24]

These definitions are needed to get to the standard definition of relativistic phase space. In later sections some of these formulas will be used to calculate cross sections. It is the intent, in this book, to provide abstract formalisms in (hopefully) bite-sized chunks rather than as a very long sequence.

Wavefunctions here are defined within a box, whose dimensions will eventually be taken to approach infinity. For wavefunctions defined within a box, the allowed momenta can be considered to be a large but discrete set of momenta determined by the boundary conditions.

The first concept to discuss is the S-matrix, which is named after the Swiss physicist, Stückelberg. In Appendix A, a definition of it is given using the non-relativistic Schrödinger equation. Here it will be defined in general terms. First a complete set of orthonormal "incoming" states $|p_1, \cdots, p_n; \text{in}\rangle \equiv |\alpha, \text{in}\rangle$ are defined. By these are meant a complete set of states that correspond to free particles as $t \to -\infty$. Next a complete orthonormal set of "outgoing" states are defined, $|\beta, \text{out}\rangle$, states that correspond to free particles as $t \to +\infty$.

5. Hadron–Hadron Scattering

Both of these sets are complete sets. Therefore, either can be expressed in terms of the other. The S-matrix is defined by

$$|\alpha, \text{in}\rangle \equiv S_{\beta\alpha} |\beta, \text{out}\rangle. \tag{5.25}$$

Note that both descriptions define states with the same total four-momentum. The S-matrix describes the time evolution of the states. It converts the state from an incoming to an outgoing state:

$$\langle\beta, \text{out} \mid \alpha, \text{in}\rangle \equiv S_{\gamma\alpha}\langle\beta, \text{out} \mid \gamma, \text{out}\rangle \equiv S_{\beta\alpha}, \tag{5.26}$$

where the fact has been used that the "out" states are orthonormal. Furthermore, since α and β are complete sets of states, S is a unitary transformation and $S^\dagger S = 1$.

The probability that $\alpha \to \beta$ is

$$\text{Prob}(\alpha \to \beta) = |\langle\beta, \text{out} \mid \alpha, \text{in}\rangle|^2 = |S_{\beta\alpha}|^2. \tag{5.27}$$

The transition rate, $d\omega$, is given by

$$d\omega = \frac{\text{Prob}(\alpha \to \beta)}{\Delta T \Delta V} \Pi_\beta \frac{d^3 p_\beta}{(2\pi)^3},$$

where the last term is the three-dimensional phase space factor, and $\Delta T \Delta V$ is the normalizing time and volume, which will be imagined here as approaching infinity. The normalization will be to one particle per unit volume.

It is desirable to remove the probability that nothing happens from S. Doing this gives the T matrix or M matrix, depending on the normalization chosen. Define

$$S_{\beta\alpha} \equiv \delta_{\beta\alpha} + (2\pi)^4 i \delta^4(p_\alpha - p_\beta) T_{\beta\alpha} \equiv \delta_{\beta\alpha} + \frac{i(2\pi)^4 \delta^4(p_\alpha - p_\beta)}{\sqrt{\Pi_\alpha(2E_\alpha)\Pi_\beta(2E_\beta)}} M_{\beta\alpha}. \tag{5.28}$$

The unitarity of the S matrix $[S^\dagger S = (S^\dagger)_{ik} S_{kf} = 1]$ can be expressed in terms of the M matrix. The intermediate states (index k) are to be summed over. This implies an integration $\int d^3 p_i/(2\pi)^3$ for $i = 1, \cdots, n$, where n is the number of particles in the intermediate state k. One obtains

$$\Im\langle f \mid M \mid i\rangle = \frac{1}{2} \sum_n (2\pi)^{4-3n} \int \Pi_{i=1}^n \frac{d^3 p_i}{2E_i} \delta^4(\sum_j p_j - P)$$
$$\times \langle p_1 \cdots p_n \mid M \mid f\rangle^* \langle p_1 \cdots p_n \mid M \mid i\rangle. \tag{5.29}$$

This equation is a generalization of the optical theorem, Equation 5.15.

When the M or T matrix is used for the transition probability, then after the matrix element is squared, there is a factor $[\delta^4(p_\alpha - p_\beta)]^2$. To handle this very singular function consider

$$(2\pi)^4[\delta^4(p_\alpha - p_\beta)]^2 = \int d^4x e^{i(p_\alpha - p_\beta)\cdot x}[\delta^4(p_\alpha - p_\beta)]$$

$$= \Delta T \Delta V [\delta^4(p_\alpha - p_\beta)].$$

Using this the arbitrary $\Delta T \Delta V$ in the transition rate definition can be removed:

$$d\omega_{\beta\alpha} = \frac{(2\pi)^4 \delta^4(p_\alpha - p_\beta)}{\Pi_\alpha(2E_\alpha)\Pi_\beta(2E_\beta)}|M_{\beta\alpha}|^2 \Pi_\beta \frac{d^3 p_\beta}{(2\pi)^3}. \tag{5.30}$$

Suppose this rate refers to the decay of a particle. When integrated over the final state momenta β,

$$\omega_{\beta\alpha} \equiv 1/\tau_\alpha \equiv \Gamma_\alpha, \tag{5.31}$$

where τ_α is the lifetime and Γ_α is the decay width of the particle.

Next examine the reaction $a + b \to c + d$. For this reaction $\alpha = a + b$ and $\beta = c + d$. In the coordinate system in which b is at rest, with the present normalization of one particle per unit volume, the transition rate should just equal the cross section times the velocity of a. The relativistic invariant that equals the relative velocity of the particles in the rest system of b is the Møller flux factor:

$$\frac{F_{ab}}{E_a E_b} = \frac{\sqrt{(p_a \cdot p_b)^2 - (m_a m_b)^2}}{E_a E_b}. \tag{5.32}$$

It can easily be verified that in the center-of-mass system the Møller flux factor is $|\vec{v}_a| + |\vec{v}_b|$. Using this flux factor, the transition rate is related to the cross section by

$$d\omega_{\beta\alpha} = d\sigma_{\beta\alpha} \frac{F_{ab}}{E_a E_b}. \tag{5.33}$$

In the literature the "invariant cross section," $E_{\beta_i} d^3\sigma/dp^3_{\beta_i}$ is often quoted.

The quantity "relativistic phase space" is defined to be

$$d\omega_{\beta\alpha} = (2\pi^4 |M_{\beta\alpha}|^2 / \Pi_\alpha (2E_\alpha)) \times \text{relativistic phase space}. \tag{5.34}$$

Let P.S. equal the relativistic phase space. Then, from the defining equation for $M_{\beta\alpha}$,

$$P.S. \equiv \Pi_{\beta_i} \frac{d^3 p_{\beta_i}}{(2\pi)^3 2 E_{\beta_i}} \delta^4(p_{\beta_1} + p_{\beta_2} + \cdots + p_{\beta_n} - p). \tag{5.35}$$

Although it will not be used in this text, note for reference it can be shown that (see Sakurai[23])

$$P.S. = \Pi_{\beta_i} \int_{(E>0)} \frac{2\pi \delta(p_{\beta_i}^2 + m_{\beta_i}^2) d^4 p}{(2\pi)^4} \delta^4(p_{\beta_1} + p_{\beta_2} + \cdots + p_{\beta_n} - p). \tag{5.36}$$

Consider now the dimensions of transition rates, phase space, and matrix elements. Cross sections have the dimensions of length squared. In the units in which $\hbar$ and c are one, cross sections can be considered to have units of $1/E^2$, since $\hbar c$ has units of energy times length. The Møller flux factor, $F_{ab}/E_a E_b$ is dimensionless as can be seen from Equation 5.32. Since the transition rate is related to the cross section by the Møller flux factor, this means that the transition rate also has the dimensions of $1/E^2$.

From examining the expressions given for relativistic phase space, (Equations 5.35 and 5.36), it can be seen that for two outgoing particles, phase space is dimensionless and, in general, phase space has dimensions E^{2n-4}, where n is the number of outgoing particles. The transition rate is proportional to $|M_{\beta\alpha}|^2 \times$ phase space$/\Pi E_\alpha$. From the preceding, the dimension of $M_{\beta\alpha}$ is E^{2-n} for an initial 2 body, final n body case.

Phase space can be expressed in some particularly useful forms for two- and three-body final states. For future reference some of these are quoted below. For two particles let $E_T = E_1 + E_2$ and $\vec{P}_T = \vec{p}_1 + \vec{p}_2$. Then

$$dP.S._2 = \frac{1}{4(2\pi)^6} \frac{p_1^3 d\Omega_1}{E_T p_1^2 - E_1(\vec{P}_T \cdot \vec{p}_1)} \text{(any frame)}, \tag{5.37}$$

$$dP.S._2 = \frac{p_1 d\Omega_1}{4(2\pi)^6 E_2} \text{(any frame; } m_2 \to \infty), \tag{5.38}$$

$$dP.S._2 = \frac{p_1 d\Omega_1}{4(2\pi)^6 E_T} \text{(center-of-mass frame; any } m_2). \tag{5.39}$$

5.3 S-Matrix, Phase Space, and Dalitz Plots

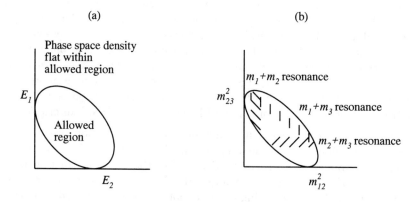

Figure 5.4 Dalitz plots.

For three particles let $d^3\Omega = d\alpha d\cos\beta d\gamma$, where α, β, γ are the Euler angles describing the decay plane. (If you don't know what Euler angles are, just realize that three angles are necessary to describe the orientation of a plane in space, two for the direction of the normal and one for the rotation of the plane around the normal.) Also let m_{12} = effective mass of body 1 + body 2, etc., and E_T, $\vec{p}_T$ be the sum of the energies and three-momenta, respectively, of the three particles. Then

$$dP.S._3 = \frac{1}{8(2\pi)^9 E_1} \left| \frac{p_2^3 p_1^2 dp_1 d\Omega_1 d\Omega_2}{P_2^2(E_T - E_1) - E_2 \vec{p}_2 \cdot (\vec{p}_T - \vec{p}_1)} \right| \text{ (any frame)}, \tag{5.40}$$

$$dP.S._3 = \frac{1}{8(2\pi)^9 E_1 E_3} [p_2 p_1^2 dp_1 d\Omega_1 d\Omega_2] \text{ (if } m_3 \to \infty\text{)}, \tag{5.41}$$

$$dP.S._3 = \frac{1}{8(2\pi)^9} d^3\Omega dE_1 dE_2$$

$$= \frac{1}{32 E_T^2 (2\pi)^9} d^3\Omega dm_{12}^2 dm_{23}^2 \text{ (center-of-mass frame)} . \tag{5.42}$$

For three-body decays the phase space relations above lead to a very convenient plot, the Dalitz plot. (See R.H. Dalitz,[25] Perkins[26] Ch. 4, Källen,[27] and Commins and Bucksbaum[28].) In the center-of-mass frame, the three-body phase space is proportional to $dE_1 dE_2$ or to $dm_{12}^2 dm_{23}^2$. If either of these two sets of variables is used as x and y axes, then, within the allowed region, there would be a flat distribution of probabilities if the matrix element were constant.

If the matrix element is not constant, then this shows up in the Dalitz plot. This plot was originally devised for $K \to 3\pi$ decays. The fact that the distribution of events within the plot was uniform was strong evidence that the decay had $L = 0$. The plot has been found to have much wider application. For example, if there is a resonance between particles 1 and 2 this shows as a band on the plot. See Figure 5.4. Furthermore, the distribution of events across this band gives information on the angular distribution of the decays of m_{12} and hence the spin. (See Exercise 5.4.)

In this chapter various formalisms for particle interactions were examined. In the first section the very useful phenomenological wave optical description was given. In the second section the more abstract, general description of scattering was expounded. The basic apparatus now has been established for calculating cross sections. Once it is described how to obtain matrix elements, the chain will be complete. This last element will be deferred to Section 7.5, after the quark model is described.

5.4 EXERCISES

5.1 A model for the cross section due to a black disk assumes that all partial waves are completely absorbed ($\eta_l = 0$) up through some integer L and after that point $\eta_l = 1$, $\delta_l = 0$. (Because angular momentum involves $r \times p$, the higher the angular momentum, the farther out in r the wavefunction is pushed.) The reaction cross section is defined as πr^2, which in turn, defines the value of r. Show that the total cross section in this model is $2\pi r^2$ and find r in terms of L and λ.

5.2 Suppose an elastic scattering cross section is parametrized as $e^{-B|t|+Ct^2}$, where t is the Mandelstam variable. Show that in the black disk model of problem 5.1, and for very small values of t, $B = r^2/4$, and $C = -r^4/192$. Assume that p is very large, that $L \gg 1$ for the black disk, and that $|t|$ is very small. Note that $\sum_{l=0}^{L} l = L(L+1)/2$; and $\sum_{l=0}^{L} l^n = L^{n+1}/(n+1)$ + terms of lower order in L.

5.3 Consider a grey disk model for the cross section in which η_l is a constant independent of l up through L and after that point $\eta_l = 1$, $\delta_l = 0$. Find the reaction cross section.

5.4 Suppose $A + B \to C + D + E$ at fixed bombarding energy. Consider the Dalitz plot of m_{CD}^2 versus m_{DE}^2. Let θ be the angle between D and E in the CD center-of-mass coordinate system. Suppose m_{CD} is held fixed. Show $m_{DE}^2 = \alpha - \beta \cos\theta$, where α and β are constants.

5.5 Find the expression for the energy of a γ-ray from the decay of a π^0 in terms of the mass of the π^0, the pion energy E, the pion velocity β in the laboratory, and the angle of γ emission in the center of mass system.

Show that, considering that the pion has zero spin, the energy distribution of the γ's will be uniform within the entire allowed range.

5.6 Show that the Møller flux factor (Equation 5.32) equals the velocity of a in the rest frame of b and $|\vec{v_a}| + |\vec{v_b}|$ in the center-of-mass system.

6
The Quark Model

6.1 INTRODUCTION

It is now known that quarks come in pairs and that there are three such pairs, three generations of quarks. Each quark is called a separate flavor of quark. However, the $u, d,$ and s quarks are distinguished from the others by having very low mass. Some important characteristics of these low lying quarks are given in the Table 6.1 (S = strangeness, B = baryon number).

The strong interactions of the quarks seem to be independent of quark flavor. Initially physicists took the wrong trail of assuming that these were a fundamental family of three. However, phenomenologically, because of the low masses, which make the particles degenerate for many purposes, the symmetry of three, $SU(3)$, is important even if it is not a fundamental symmetry. Color, considered later in this chapter, has $SU(3)$ as an exact symmetry. Furthermore, $SU(3)$ serves as a prototype of some of the higher symmetries proposed for particle physicists.

In a space in which the horizontal axis is I_3 and the vertical axis is $Y = S + B$, the quarks form a triangle consisting of the I-spin doublet (d, u) with $Y = 1/3$, and the I-spin singlet s with $Y = -2/3$. See Figure 6.1. The triplet is the fundamental basis of $SU(3)$.

A discussion of the evidence for the spin and fractional charge of quarks will be deferred until Chapter 9.

$SU(3)$ can be considered as the set of unitary 3×3 matrices with determinant $= +1$. The generators of the infinitesimal transformations may be taken to be any $3^2 - 1 = 8$ linearly independent traceless hermitian 3×3 matrices. The maximum number of mutually commuting generators is two. This is called the rank of the group [the rank of $SU(2)$ is one], and is also the number of Casimir operators, which are operators whose eigenvalues classify the various irreducible representations of the group. (L^2 is the Casimir operator for the rotation group.)

The generators of $SU(3)$ are conventionally given by the eight λ-matrices written down in Appendix D. These matrices are sometimes called the Gell-

6.1 Introduction 111

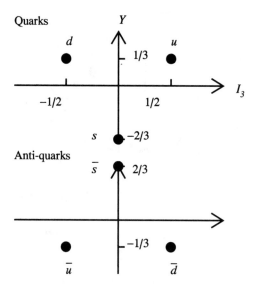

Figure 6.1 Quark and anti-quark triplets.

Table 6.1 Quark Quantum Numbers

Name	u (up)	d (down)	s (strange)
Q	$+2/3$	$-1/3$	$-1/3$
Spin	$1/2$	$1/2$	$1/2$
I	$1/2$	$1/2$	0
I_3	$1/2$	$-1/2$	0
S	0	0	-1
B	$1/3$	$1/3$	$1/3$
$Y = S + B$	$1/3$	$1/3$	$-2/3$
Mass	less than d	few MeV	≈ 150 MeV

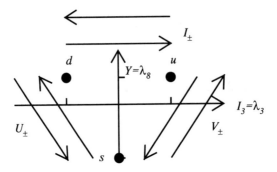

Figure 6.2 I-spin, U-spin, and V-spin on the quark triplet.

Mann matrices. They have the commutation relations:

$$\left[\frac{\lambda_i}{2}, \frac{\lambda_j}{2}\right] = i \sum_k f_{ijk} \frac{\lambda_k}{2}. \tag{6.1}$$

The f_{ijk} are known as structure constants and are also given in Appendix D. [The structure constants for $SU(2)$ consist of the totally anti-symmetric tensor ϵ_{ijk}.] The λ_1, λ_2, λ_3 matrices correspond to the three Pauli matrices of $SU(2)$ and form an $SU(2)$ subgroup of $SU(3)$. There are $SU(3)$ analogs of the L^+ and L^- operators. Consider the I-spin raising operator, I^+, for quarks. It changes a d quark into a u quark and gives zero on an s quark which has zero I-spin. If there were true symmetry between the three quarks, two new I-spin-like variables could be defined. Let U^+ take an s quark into a d quark, and V^+ take a u quark into an s quark. The "I-spins" associated with these two operators are called U-spin and V-spin, respectively.[29] They are exactly equivalent to I-spin, but in different directions in the quark space:

$$\frac{1}{2}(\lambda_1 \pm i\lambda_2) = I_\pm \quad \text{(I - spin)},$$

$$\frac{1}{2}(\lambda_4 \pm i\lambda_5) = V_\pm \quad \text{(V - spin)}, \tag{6.2}$$

$$\frac{1}{2}(\lambda_6 \pm i\lambda_7) = U_\pm \quad \text{(U - spin)}.$$

See Figure 6.2. λ_3 and λ_8 are diagonal matrices (the mutually commuting generators).

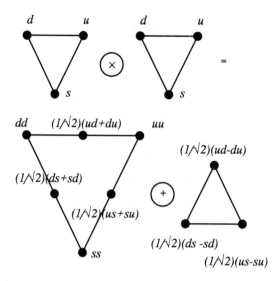

Figure 6.3 Combining two quarks, 3⊗3.

6.2 BARYONS

To make up a baryon, such as a proton, three quarks are needed to get $B = 1$. For low-lying baryons, each quark is one of u, d, s. In terms of group theory, there are three $SU(3)$'s multiplied together. The object resulting from the product of three $SU(3)$'s can be broken up as the sum of various, more fundamental representations. There are 27 ways to combine three quarks. First combine two quarks. See Figure 6.3. Represent the group representation by the number of particles in it. One sees that $3 \otimes 3 = 6 \oplus \bar{3}$.

Now add the third quark:

$$3 \otimes 3 \otimes 3 = (6 \otimes 3) \oplus (\bar{3} \otimes 3) = 10 \oplus 8 \oplus 8 \oplus 1. \tag{6.3}$$

This is the extension of the addition of angular momentum in $SU(2)$. The 10 is symmetric under interchanges of pairs of the three quarks, the 1 is completely anti-symmetric and the two 8's are of mixed symmetry. The 10 is called a "decuplet," the 8 is called an "octet," and the 1 is called a "singlet." The two octet states are called p_A and p_S. They are anti-symmetric or symmetric, respectively, under interchange of the first two quarks.

If it is assumed that, not only are the interactions the same between the three quarks, but that the interactions are spin independent, then there is a symmetry of 6, $SU(6)$, where one has spin up u, spin down u, spin up d, etc. The product of three of these quarks is

$$6 \otimes 6 \otimes 6 = 56 \oplus 70 \oplus 70 \oplus 20. \tag{6.4}$$

The 56 is symmetric under interchanges of pairs of the three quarks in the particle, the 20 is anti-symmetric, and the 70's are mixed. In terms of $SU(3)$, the $56 = (2,8) \oplus (4,10)$ where the first digit is the spin multiplicity and the second the $SU(3)$ representation. A spin multiplicity of 2 corresponds to spin 1/2 and a multiplicity of 4 to spin 3/2.

The 56 can be identified with physical particles. Presumably the other multiplets force spatial wavefunction symmetries making them higher mass and therefore not seen.

However, there is now a serious problem. The 56 is symmetric under particle exchange. Since the quarks are fermions, the overall wavefunction must be anti-symmetric. This seems to imply that the spatial wavefunction must be anti-symmetric under interchange. This, in turn, would imply that the wavefunction has nodes; it is positive in some places and negative in others. However, quantum mechanically $p^2 \propto \nabla^2$ and p^2 should, therefore, be larger for wavefunctions with nodes which have higher second derivatives. It is therefore surprising that a state with nodes should be the ground state. Although this is a simple argument, it was, in fact, very difficult to find a way around it.

The magic word is "color." It is now believed that there are three identical u quarks that are differentiated only by a new quantum number called color. There is a "red" u quark, a "yellow" u quark, and a "blue" u quark, and similarly for the others. This symmetry has to be completely unbroken and known particles are "colorless," i.e., have an equal mix of all colors.

How does this help the dilemma? Now there is a new function, the color function, to add to the space, spin, and $SU(3)$ functions. If the wavefunction is totally anti-symmetric in this new function, color (which is what is formally meant by "colorless"), then it can be symmetric in all the others as is required.

This is a very bold proposal to get out of a single difficulty. Further justifications of color will be delayed until later in this chapter, but it will be seen then that, in fact, there are a great many observational problems that the concept of color solves. For now assume it to be true and examine the 56, a totally symmetric state. As noted previously, the decuplet corresponds to spin 3/2 particles. The members of the lowest mass decuplet are given in Table 6.2.

One member of the decuplet is clearly the state with three spin up u quarks, $u \uparrow u \uparrow u \uparrow$. The other states in the 56, which belong to the $SU(3)$

Table 6.2 Baryon Decuplet States

State	Quark Structure	I	I_3	S
$\Delta^{++}(1232)$	uuu	3/2	3/2	0
Δ^+	$(uud + udu + duu)/\sqrt{3}$	3/2	1/2	0
Δ^0	$(udd + dud + ddu)/\sqrt{3}$	3/2	$-1/2$	0
Δ^-	ddd	3/2	$-3/2$	0
$\Sigma^+(1385)$	$(uus + sus + suu)/\sqrt{3}$	1	1	-1
Σ^0	$(uds + sud + dsu + sdu + dus + usd)/\sqrt{6}$	1	0	-1
Σ^-	$(dds + dsd + sdd)/\sqrt{3}$	1	-1	-1
$\Xi^0(1530)$	$(uss + sus + ssu)/\sqrt{3}$	1/2	1/2	-2
Ξ^-	$(dss + sds + ssd)/\sqrt{3}$	1/2	$-1/2$	-2
$\Omega^-(1672)$	sss	0	0	-3

10, can then be found by just taking ↑↑↑ and states symmetric in u, d, s as seen in Table 6.2. The particles in the baryon decuplet are all resonances decaying by the strong interaction with the exception of the Ω^- which decays only by the weak interaction.

Next consider the lowest mass baryon octet within the 56 representation. As seen previously this octet corresponds to spin 1/2 particles. These particles consist of "stable" baryons, the ones which either are really stable or decay by only the weak interaction (or for the Σ^0, the electromagnetic interaction). The particles in the octet are given in Table 6.3.

Figure 6.4 shows the baryon decuplet and octet where Y versus I_3 is plotted. These are the lowest mass baryons. The fact that they fit nicely into the 56 representation of $SU(6)$ is very important evidence that this classification is relevant. Historically, the existence of the Ω^- and its mass was predicted on the basis of this model. Experimenters then went out and found it.[30] (See Figure 6.5.)

The proton wavefunction and the other wavefunctions of the baryon octet are more complicated than those of the decuplet. The book of Frank

Table 6.3 Baryon Octet States

State	Quark Structure (Not symmetrized)	I	I_3	S
p	uud	1/2	+1/2	0
n	udd	1/2	−1/2	0
Σ^+	uus	1	+1	−1
Σ^0	uds	1	0	−1
Σ^-	dds	1	−1	−1
Λ	uds	0	0	−1
Ξ^0	uss	1/2	+1/2	−2
Ξ^-	dss	1/2	−1/2	−2

Close[31] on Partons has a good presentation of these wavefunctions, but the method of R.P. Feynman et al.[32] will be used here.

The proton has two u quarks and a d quark. Since it is known how to handle spin and I-spin two at a time using Clebsch–Gordan coefficients, arbitrarily divide the three quarks into two plus one. Call the two quarks the "core." The core can have an I-spin of 0 or 1. Denote I, I_3 of the core by $|I, I_3\rangle$.

Suppose the core has I-spin of 0. Note that for the core to have $I = 0$, it must have one u and one d. Thus for this case the wavefunction of uud is $\Psi_o = u\,|0,0\rangle$.

Suppose the core has I spin of 1. Using the Clebsch–Gordan tables it is seen $\Psi_1 = \sqrt{2/3}d\,|1,1\rangle - \sqrt{1/3}u\,|1,0\rangle$, where $|1,1\rangle = uu$ and $|1,0\rangle = \sqrt{1/2}(ud+du)$.

The spin can be constructed in a similar manner. The full wavefunction symmetric under interchange of any two quarks is then $\Phi = \Psi_o\chi_o + \Psi_1\chi_1$, where χ is the spin wavefunction. Written out explicitly the results for a spin up proton (normalized to 1) are

$$\Phi = \sqrt{1/8}[4/3uud \uparrow\uparrow\downarrow - 2/3uud \uparrow\downarrow\uparrow - 2/3uud \downarrow\uparrow\uparrow - 2/3udu \uparrow\uparrow\downarrow +$$

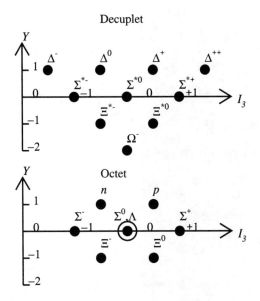

Figure 6.4 Baryon decuplet and octet states.

$$4/3udu \uparrow\downarrow\uparrow -2/3udu \downarrow\uparrow\uparrow -2/3duu \uparrow\uparrow\downarrow -2/3duu \uparrow\downarrow\uparrow +4/3duu \downarrow\uparrow\uparrow]. \quad (6.5)$$

By the same procedure the wavefunctions for the other particles in the octet can be constructed.

With this background an early experiment[33] giving important evidence for spin 1/2 quarks can be discussed. Examine the preceding wavefunction and calculate the probability of picking a quark at random and having it be spin up or spin down. Thus,

$$P(d \downarrow) = 1/24[(4/3)^2 + (4/3)^2 + (4/3)^2] = 2/9,$$

$$P(d \uparrow) = 1/24[(-2/3)^2 \times 6] = 1/9,$$

$$P(u \downarrow) = 1/24[(2/3)^2 \times 2 \times 3] = 1/9,$$

$$P(u \uparrow) = 1/24[(4/3)^2 \times 2 \times 3 + (2/3)^2 \times 2 \times 3] = 5/9.$$

The normalization is 1/24 not 1/8 as there are three quarks in the proton.

Consider forward or close to forward inelastic $e - p$ scattering. Imagine this corresponds to scattering off of a single quark in the proton (Figure 6.6). The exchanged photon is almost real and the longitudinal and timelike

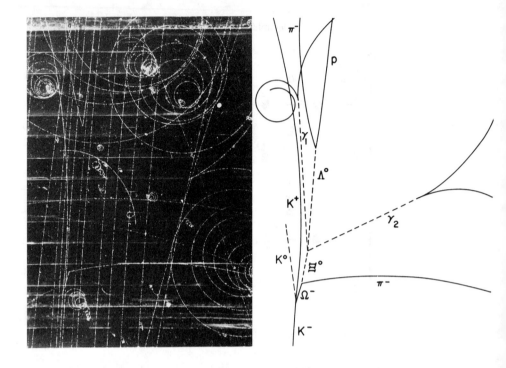

Figure 6.5 An example of the production of an Ω^- particle in the 80" Brookhaven bubble chamber filled with hydrogen. $K^- p \to \Omega^- K^+ K^0$; $\Omega^- \to \Xi^0 \pi^-$; $\Xi^0 \to \Lambda^0 \pi^0$; $\Lambda^0 \to p\pi^-$; $\pi^0 \to \gamma\gamma$. (Photograph courtesy Brookhaven National Laboratory.)

components can be ignored. To conserve angular momentum the spin of the electron and the spin of the quark must each flip in the process as shown in the figure. To produce a final state $e \uparrow$, only an initial $q \downarrow$ can be used. To produce a final state $e \downarrow$, only a $q \uparrow$ can be used. Thus:

$$\sigma_{e\uparrow p\uparrow} \propto \Sigma(\text{charge})^2 P(q \downarrow) = (1/3)^2 \times 2/9 + (2/3)^2 \times 1/9 = 6/81,$$

$$\sigma_{e\downarrow p\uparrow} \propto \Sigma(\text{charge})^2 P(q \uparrow) = (1/3)^2 \times 1/9 + (2/3)^2 \times 5/9 = 21/81,$$

$$A \equiv \frac{\sigma_{\uparrow\downarrow} - \sigma_{\uparrow\uparrow}}{\sigma_{\uparrow\downarrow} + \sigma_{\uparrow\uparrow}} = \frac{21 - 6}{21 + 6} = 5/9. \tag{6.6}$$

This is observed experimentally and corresponds to a test of quarks having spin 1/2.

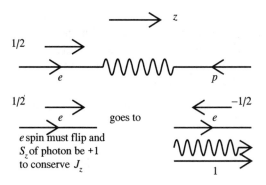

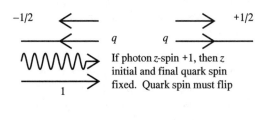

Figure 6.6 Forward inelastic $e - p$ scattering.

The strongest evidence for spin 1/2 quarks comes from the process $e^+e^- \to$ hadrons. The hadrons in this reaction often come out as two "jets" of particles. A jet is a set of particles which have low relative perpendicular momentum. At high energies jets appear as tightly grouped bundles of particles in space. The angular distribution of the jets with respect to the initial electron direction provides strong evidence for the picture $e^+e^- \to \gamma \to q\bar{q} \to$ 2 jets, with quark spin 1/2.[34]

The preceding picture is a bit oversimplified. The proton contains other components besides the three "valence quarks" discussed previously. An instantaneous snapshot would see a number of virtual quark–anti-quark pairs, which is called the "sea," and, also, a number of virtual gluons that are being exchanged between the quarks. If one looks at events with low q^2, the virtual four-momentum squared of the exchanged photon, then this corresponds to looking at protons with mostly small momentum quarks and if all momenta are small, the gluons and sea are small effects. The experiment described above corresponds to this regime.

If, however, one looks at events with high q^2, this means that at least one quark had a high momentum and there was a lot of phase space for the sea and the gluons. For these events not all of the spin need be carried by the

120   6. The Quark Model

valence quarks. A $\mu - p$ scattering experiment reported in 1987 the rather surprising result that, at high q^2, it seems that very little of the proton spin is carried by the valence quarks.[35] The meaning of this result is still under active discussion. Gluon polarization affects may have an influence on the interpretation of the data.

In general the baryons fit very well into the quark model picture given above.

6.3 MESONS

Turn now to quark meson states. Here there are mainly states of $q\bar{q}$ with $B = 0$. (There might also be states of $qq\bar{q}\bar{q}$. At present the question of whether these "exotic" states do actually occur is a matter of investigation both theoretically and experimentally.)

For $q\bar{q}$ the symmetry of $3 \otimes \bar{3}$ equals $8 \oplus 1$ in $SU(3)$. Again the lowest state is an $l = 0$ (S wave) spatial state. Experimentally, it turns out that it is an $s = 0$ (spin singlet) state also. The parity is negative as a fermion and anti-fermion have intrinsically opposite parity. This will be derived in Chapter 7. Hence, these are pseudoscalar mesons with spin parity = 0^-. For a more general representation, the parity is $P = -(-1)^L$, and the charge conjugation is $C = (-1)^{L+S}$, where L is the orbital angular momentum. All nine members of the octet and singlet (together called a "nonet") are found. Table 6.4 lists the pseudoscalar nonet mesons.

Note that the minus sign on the π^- wavefunction is needed so $C\pi^+ = -\pi^-$. This, in turn, as seen in Section 4.17, is desirable for the introduction of G-parity. The minus sign for $\overline{K}^0$ occurs for a similar reason to the minus sign in I-spin for $\bar{n}$ (See Section 4.15). η_8 and η_0 are the octet and singlet particles with $I = S = 0$. In reality the physical particles that are observed are mixtures of the two:

$$\eta(547) = \eta_0 \sin\theta - \eta_8 \cos\theta,$$

$$\eta'(958) = \eta_0 \cos\theta + \eta_8 \sin\theta.$$

Assume, rather arbitrarily, that for mesons $\langle \phi | H | \phi \rangle \propto$ mass^2. Using an obvious notation for the hamiltonian matrix,

$$M_\eta^2 = M_0^2 \sin^2\theta + M_8^2 \cos^2\theta - 2M_{08}^2 \sin\theta \cos\theta,$$

$$M_{\eta'}^2 = M_8^2 \sin^2\theta + M_0^2 \cos^2\theta + 2M_{08}^2 \sin\theta \cos\theta, \qquad (6.7)$$

6.3 Mesons

Table 6.4 Pseudoscalar Meson Nonet States

State	Quark Structure	I	I_3	S
K^+	$u\bar{s}$	1/2	+1/2	+1
K^0	$d\bar{s}$	1/2	−1/2	+1
π^+	$u\bar{d}$	1	+1	0
π^0	$1/\sqrt{2}(d\bar{d} - u\bar{u})$	1	0	0
π^-	$-d\bar{u}$	1	−1	0
η_8	$1/\sqrt{6}(d\bar{d} + u\bar{u} - 2s\bar{s})$	0	0	0
η_0	$1/\sqrt{3}(d\bar{d} + u\bar{u} + s\bar{s})$	0	0	0
$\overline{K}^0$	$-s\bar{d}$	1/2	+1/2	−1
$\overline{K}^-$	$s\bar{u}$	1/2	−1/2	−1

Table 6.5 Vector Meson Nonet States

State	Quark Structure	I	I_3	S
$(K^\star)^+(892)$	$u\bar{s}$	1/2	+1/2	+1
$(K^\star)^0$	$d\bar{s}$	1/2	−1/2	+1
$\rho^+(770)$	$u\bar{d}$	1	+1	0
ρ^0	$1/\sqrt{2}(d\bar{d} - u\bar{u})$	1	0	0
ρ^-	$-d\bar{u}$	1	−1	0
ϕ_8	$1/\sqrt{6}(d\bar{d} + u\bar{u} - 2s\bar{s})$	0	0	0
ϕ_0	$1/\sqrt{3}(d\bar{d} + u\bar{u} + s\bar{s})$	0	0	0
$(\overline{K}^\star)^0$	$-s\bar{d}$	1/2	+1/2	−1
$(\overline{K}^\star)^-$	$s\bar{u}$	1/2	−1/2	−1

$$M^2_{\eta\eta'} = 0 = (M_0^2 - M_8^2)\sin\theta\cos\theta + M_{08}^2(\sin^2\theta - \cos^2\theta).$$

The last equation results since η and η' are orthogonal states. These equations can be solved, yielding

$$\tan^2\theta = \frac{M_\eta^2 - M_8^2}{M_8^2 - M_{\eta'}^2}. \tag{6.8}$$

The mass of quarks will be discussed shortly. If it is assumed that the nucleon is the sum of the mass of three quarks of equal mass, then one might try $M_8 = (2/3)M_p = 625$ MeV, which yields $\theta = 10°$.

A second nonet of mesons occurs in an $s = 1$ (spin triplet) state, but with presumably the same $l = 0$ spatial state as the previous nonet. These mesons are then vector mesons with spin-parity $= 1^-$. The masses are higher than those in the pseudoscalar nonet and these mesons are all resonances. This mass difference gives information about the magnitude of the quark spin-spin interactions. The vector meson states are listed in Table 6.5.

Figure 6.7 shows the two meson nonets. In the vector nonet, ϕ_8 and ϕ_0 mix to form physical particles. The observed particles are:

$$\phi(1020) = \phi_0 \sin\theta - \phi_8 \cos\theta,$$

$$\omega(783) = \phi_0 \cos\theta + \phi_8 \sin\theta.$$

Here $\theta \approx 40°$. If θ were $35°$ ($\sin\theta = 1/\sqrt{3}$), then $\phi(1020) = s\bar{s}$, $\omega = (u\bar{u} + d\bar{d})/\sqrt{2}$. Since the actual angle is close to this, it might be expected that $\phi \to K\overline{K}$ would dominate the ϕ decay and indeed this has an 84% branching ratio, even though phase space favors the three pion modes. (The Q-value for the kaon decay is only 24 MeV.)

Why is it expected that $s\bar{s}$ goes predominantly into kaons? The diagrams shown in Figure 6.8 for the ϕ and ω decays are known as quark flow diagrams. The last diagram has the initial and final states disconnected. It is believed that such diagrams are forbidden to first order. This is known as "Zweig's rule." (This rule is probably due to the fact that the disconnected diagram requires multiple gluon exchange between the disconnected branches. For the ϕ it can be shown that at least three gluons are required.)

Thus both mesons and baryons are well described in the quark model.

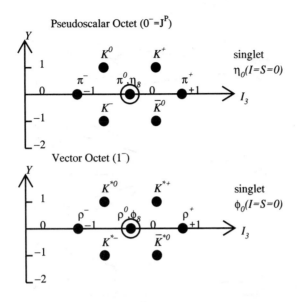

Figure 6.7 Pseudoscalar and vector meson nonets.

6.4 SIMPLE MODEL RELATIONS

Consider πN and NN scattering. Since there are two quarks in the pion and three in the nucleon, it might be expected that $\sigma_{\pi N}/\sigma_{NN} = 2/3$. This simple relation works surprisingly well.

In the Drell–Yan scattering process, shown in Figure 6.9, a pion scatters off a nucleus and a virtual gamma produces a muon pair. Choose an isoscalar nucleus, say C^{12}, so that there are an equal number of u and d quarks. Picture the process as an anti-quark in the pion annihilating with a quark in the nucleus, producing a virtual gamma, which then materializes as a muon pair. The cross section should be proportional to the square of the electric charges of the quarks. For π^- on carbon, a $\bar{u}$ of the π annihilates with a u in the carbon. For π^+ on carbon, a $\bar{d}$ of the π annihilates with a d in the carbon. The sea quarks and gluons are ignored here. Since the magnitude of the u-quark charge is twice that of the d-quark, it is then expected that $\sigma_{D-Y\pi^-}/\sigma_{D-Y\pi^+} = 4/1$. This is observed if one stays away from resonance regions in the muon mass distribution.

For vertices involving three octet particles such as $\pi^- + p \to$ virtual p, there are two ways of coupling the octets to give an $SU(3)$ invariant amplitude. These are called D and F couplings. Thus, the ratios of cross sections of octet partners are related with only the D/F ratio not fixed. The predictions are reasonable successful.

124   6. The Quark Model

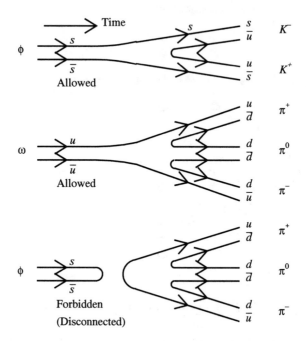

Figure 6.8 Quark flow diagrams for ϕ and ω decay.

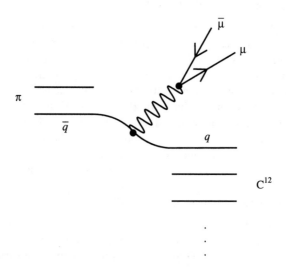

Figure 6.9 Drell-Yan scattering process.

6.5 MASS RELATIONS

Suppose that there is an approximate symmetry between all of the mesons or baryons in one multiplet. If this is so, what accounts for the mass differences of the particles within a multiplet? Some of this is attributable to the masses of the quarks themselves. One must be careful since the quark masses calculated this way are different than the quark masses that are obtained from looking at quark interactions. The masses here are constituent quark masses, or zero point energies for quarks bound by some potential. There are problems with the meaning of quark mass here since it is not possible to isolate free quarks. Here, from looking at ω, ρ masses, one estimates $m_u \approx m_d \approx 0.31$ GeV and from $\phi \approx s\bar{s}$, $m_s \approx 0.51$ GeV.

Look at the baryon decuplet. If the mass difference is attributed to the quark masses, then

$$\Sigma(1385) - \Delta(1232) = \Xi(1530) - \Sigma(1385) = \Omega^-(1672) - \Xi(1530).$$

Indeed these three differences in the measured masses are 152 MeV, 149 MeV, and 139 MeV.

In the baryon octet one would have

$$\Lambda(1116) - p(938) = \Xi(1315) - \Lambda(1116)$$

and the differences are 177 MeV, 203 MeV. One also expects $\Sigma(1193) = \Lambda(1116)$.

These are not bad first-order results, but it is clear that this is not the whole story. Suppose a term breaking the quark symmetry is now added. Work with U-spin and I-spin. In general, the elements of the U-spin operators will not commute with the I-spin operators. However, U^- and I^+ do commute. The proof of this statement will be left as an exercise in Chapter 7 (Exercise 7.1) after the concept of creation and destruction operators is introduced.

States that are eigenstates of the I-spin operators are not necessarily eigenstates of the U-spin operators. Note that U-spin does not change the charge of a quark and, hence, commutes with the charge operator Q. It does change strangeness. U-spin can only mix states of the same charge. In the baryon octet then, a U-spin triplet will be n, some mixture of Σ^0 and Λ, and Ξ^0. Assume the U-spin eigenstate mix of Σ^0 and Λ is given by $\alpha\Sigma^0 + \beta\Lambda$, where $\alpha^2 + \beta^2 = 1$ in order to normalize the state. Phases are ignored for this argument. Consider

$$U^- |n\rangle = \sqrt{2}(\alpha |\Sigma^0\rangle + \beta |\Lambda\rangle). \tag{6.9}$$

The $\sqrt{2}$ follows from the analogous relations for L^+ and L^- found in Appendix C (Equation C.8 and the preceding discussion) applied to a U-spin triplet. Operate on the result with I^+, i.e.,

$$I^+ U^- |n\rangle = 2\alpha |\Sigma^+\rangle.$$

The I-spin raising operator has given zero for the I-spin singlet Λ.

Next use the preceding fact that U^- and I^+ commute. Since the p and Σ^+ form a U-spin doublet,

$$U^- I^+ |n\rangle = U^- |p\rangle = |\Sigma^+\rangle.$$

This implies that $\alpha = 1/2$ and hence, from normalization, $\beta = \sqrt{3}/2$.

Now assume that the mass of the particles have one part that is a scalar in U-spin (S) and another part which is a vector in U-Spin (V), i.e., a part that is proportional to U_3. This is a physical assumption. In group theory terms it is equivalent to assuming that the mass difference has the symmetry of the $I = 0$ member of an octet representation. Other terms are possible, but ignored here. The $I = 0$ is necessary since the masses of the various members of I-spin multiplets are taken as the same in this approximation. Let M be the mass operator. For the U-spin triplet one then has

$$\langle \Xi^0 | M | \Xi^0 \rangle = S - V; \quad \left\langle \tfrac{1}{2}(\Sigma^0 + \sqrt{3}\Lambda) \middle| M \middle| \tfrac{1}{2}(\Sigma^0 + \sqrt{3}\Lambda) \right\rangle = S;$$
$$\langle n | M | n \rangle = S + V. \quad (6.10)$$

Since M conserves I-spin, there is no interference term in the second relation above and it becomes

$$\tfrac{1}{4} \langle \Sigma^0 | M | \Sigma^0 \rangle + \tfrac{3}{4} \langle \Lambda | M | \Lambda \rangle = S. \quad (6.11)$$

Combining these relations, one obtains

$$\tfrac{1}{2} \langle n | M | n \rangle + \tfrac{1}{2} \langle \Xi | M | \Xi \rangle - \tfrac{1}{4} \langle \Sigma | M | \Sigma \rangle - \tfrac{3}{4} \langle \Lambda | M | \Lambda \rangle = 0. \quad (6.12)$$

This relation is known as the Gell-Mann–Okubo[36][37] (GMO) mass relation. It is obeyed very well. The left-hand side is only -7.6 MeV.

6.5 Mass Relations

As an alternative to the GMO assumptions, suppose a spin dependence is added. Such a dependence is known to be present from the overall mass difference between the vector and pseudoscalar meson nonets. As a first-order estimate, add, therefore, a term proportional to $\sigma_i \cdot \sigma_j$ of the quarks. This gives a fit of better than 1% to the baryon octet and decuplet, fitting eight masses with four parameters. In Exercise 6.10, this model is applied to the meson states.

Until now the electromagnetic mass differences have been ignored. Considering baryons with similar quark charge structure it is expected that for electromagnetic mass shifts

$$\langle p| M |p\rangle = \langle \Sigma^+| M |\Sigma^+\rangle \, ; \, \langle \Sigma^-| M |\Sigma^-\rangle = \langle \Xi^-| M |\Xi^-\rangle \, ; \, \langle \Xi^0| M |\Xi^0\rangle = \langle n| M |n\rangle .$$

Combining these one obtains the Coleman–Glashow relation:

$$\langle p| M |p\rangle - \langle n| M |n\rangle = \langle \Sigma^+| M |\Sigma^+\rangle - \langle \Sigma^-| M |\Sigma^-\rangle + \langle \Xi^-| M |\Xi^-\rangle - \langle \Xi^0| M |\Xi^0\rangle .$$

This gives $-1.3 \text{MeV}/c^2$ on the right-hand side and $-8.0 \text{MeV}/c^2 + 6.4 \text{MeV}/c^2 = -1.6 \text{MeV}/c^2$ on the left-hand side.

This is reasonably good agreement. As previously seen, part of the mass differences in physical particles can be interpreted as due to a mass difference between u and d. An old puzzle is now understandable. For years people tried to calculate the $p - n$ mass difference. Not only could they not get the correct magnitude, they couldn't get the correct sign. This is easily seen. Since the proton is charged, one expects a positive potential energy from the net repulsion of different charge segments and one would therefore expect the proton to be heavier than the neutron, where nature has decreed the reverse. It is now believed that the neutron is heavier than the proton because the d is a little heavier than the u.

The meson multiplet masses will not be discussed here in detail. The pseudoscalar nonet is exceptional. The very low mass of the pions makes it difficult to calculate the masses. One would expect the isoscalar singlet (η, η') to be at about the same mass as the pion. However, the isoscalars are much heavier than the pion. This is known as the U_1 problem. It may be that part of the reason for the higher mass is that the η, η' can couple to two gluon intermediate states and the couplings have unusually large numerical coefficients due to internal symmetry. Since the gluon does not carry I-spin, a pion cannot have a two gluon intermediate state.

If the GMO assumptions are valid, the Gell-Mann–Okubo relation should work for the meson octet as well as for the baryon octet. One finds then that $3m_\eta - 4m_K + m_\pi = 0$. This does not work very well; the left-hand side is 0.21 GeV. However, if m is replaced by m^2, one obtains $3m_\eta^2 - 4m_K^2 + m_\pi^2 = 0$. This relation is obeyed fairly well; the left-hand side is -0.07 GeV^2.

But why should m be replaced by m^2 for mesons and not for baryons? Feynman argued intuitively that the relativistic equation for fermions is linear in m (see Section 7.1), while that for bosons involves m^2. T.D. Lee[1] did the perturbation theory using energy not mass and then used $\delta E = \delta \sqrt{p^2 + m^2} = \delta m^2/(2E)$, which leads naturally to a relation in m^2. Then for baryons, since $\delta m \ll m$, the equation can be linearized, $\delta m^2/E \approx 2m\delta m/E \approx 2\delta m$. However, neither justification is entirely persuasive.

It is easier to fit the mass differences for the vector nonet since that nonet is heavier. It agrees reasonably well with the quark picture.

Mass relations, except for calculational difficulties in the lowest mass meson nonet, are seen to be describable in the quark model.

6.6 BARYON MAGNETIC MOMENTS

Recall the $SU(6)$ proton wavefunction:

$$\Phi = \sqrt{1/8}[4/3 uud \uparrow\uparrow\downarrow -2/3 uud \uparrow\downarrow\uparrow -2/3 uud \downarrow\uparrow\uparrow -2/3 udu \uparrow\uparrow\downarrow + 4/3 udu \uparrow\downarrow\uparrow -2/3 udu \downarrow\uparrow\uparrow -2/3 duu \uparrow\uparrow\downarrow -2/3 duu \uparrow\downarrow\uparrow +4/3 duu \downarrow\uparrow\uparrow]. \quad (6.13)$$

Let μ_d and μ_u be the magnetic moments of the d and u quark, respectively. Then, assuming that the magnetic moment of the proton comes only from the quark magnetic moments:

$$\mu_p = \frac{1}{8}[(2\mu_u - \mu_d)(3 \times (4/3)^2) + \mu_d(6 \times (-2/3)^2] = [4/3\mu_u - 1/3\mu_d]. \quad (6.14)$$

The other octet baryons are done similarly.

Suppose one assumes the quarks to have point-like magnetic moments:

$$\vec{\mu}_i = Q_i \vec{\sigma}_i \frac{e}{2m_i}. \quad (6.15)$$

Here Q is the quark charge in units of the magnitude of the electron charge. Then from the proton and neutron magnetic moments, one estimates $m_u \approx 0.34$ GeV and from the Λ magnetic moment, one estimates $m_s \approx 0.51$ GeV, in good agreement with the estimates in the last section.

For a simple example, assume that $\mu_d = -(1/2)\mu_u$ as would naively be expected from the relative charges ignoring the mass differences. Then

$$\mu_p/\mu_n = [(4/3) - (1/3)(-1/2)]/[(4/3)(-1/2) - (1/3)] = -3/2.$$

The experimental value is -1.46. Some of the hyperon moments fit less well. This may be due to relativistic corrections. G. Karl and his coworkers have made many calculations of this effect and get reasonable fits to most of the data.[38]

Thus, regardless of the possible problems at high q^2 noted in Section 6.2, at zero q^2, the magnetic moment data are reasonably consistent with assuming that the spin of the nucleons is carried by the quarks.

6.7 HEAVY QUARKS

The heavy quarks c, b, t are considered here. It will be seen in Chapter 7 that the c quark was predicted before its discovery and the mass was guessed at. In fact at the 1974 International High Energy Physics Conference in London, the physicist J. Iliopoulos offered to bet a case of wine that it would be discovered within two years. He was right. The actual discovery was a remarkable confluence of events and initiated one of the most exciting times in recent particle physics history.

S. Ting and his group were doing an experiment on the East Coast of the United States, at Brookhaven Laboratory, looking at pN interactions and measuring, precisely, the effective mass of electron pairs coming out of the interactions. L. Lederman's group had previously observed a shoulder in the mass distribution which might indicate some non-uniformity. B. Richter and his group on the West Coast of the United States, at SLAC, had just built a storage ring and were looking at e^+e^- interactions.

Shortly before the announcements, H. Politzer and T. Appliquist had calculated that there might be bound $c\bar{c}$ states similar to positronium states in e^+e^-, and they might be quite narrow.

It is a matter of contention still, as to who was ready first, but it is clear that both experimental groups, even without the theoretical guidance, found enormous, narrow peaks, several hundred times larger than the nearby cross sections.[39][40] They announced these results on the same day in 1974. S. Ting and B. Richter shared the 1976 Nobel Prize for the discovery of this bound state now called the J/Ψ at 3.097 GeV. Some indication of their rivalry is indicated by the fact that the Ting group called it J and the Richter group called it Ψ. Neither would defer to the other and so the awkward J/Ψ designation remains. Figures 6.10 and 6.11 show the peaks they found.

130   6. The Quark Model

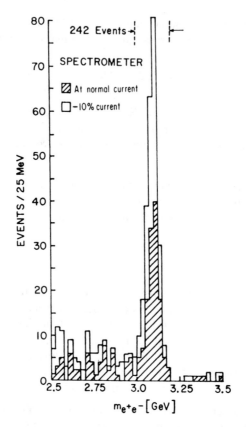

$pN \to e^+e^-X$. Mass spectrum of e^+e^- is shown.

Figure 6.10 Discovery of the J/Ψ particle at BNL.[39]

The SLAC experiment went on to become one of the most successful particle physics experiments in several decades. Additional bound "charmonium" states turned up. Besides the J/Ψ, another bound $c\bar{c}$ state, the Ψ', occurred at 3685 MeV. The charmed D mesons were found and evidence for a new generation appeared when a new lepton was found.

This latter discovery was completely unexpected. The SLAC group seemed to observe some interactions in which $e^+e^- \to \mu^+e^- +$ missing energy with no accompanying hadrons. A subgroup within the SLAC experiment lead by M. Perl argued that this might be due to pair production of new leptons, the τ leptons, which then decayed, one into $\mu\bar{\nu}_\mu\nu_\tau$ and the other into $e\bar{\nu}_e\nu_\tau$.[41] Eventually this proved true, and many other decay modes were seen of this τ particle of mass 1777 MeV.[42] For this discovery, M. Perl shared the 1995 Nobel Prize.

6.7 Heavy Quarks   131

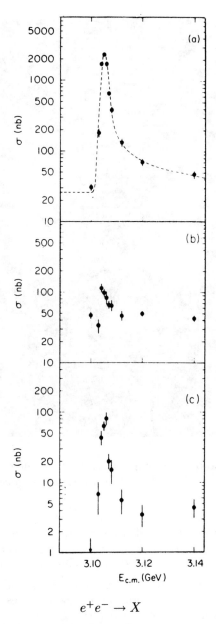

$$e^+e^- \to X$$

Cross section versus energy shown for
(a) multihadron final states,
(b) e^+e^- final states,
(c) $\mu^+\mu^-$, $\pi^+\pi^-$, K^+K^- final states.

Figure 6.11 Discovery of the J/Ψ particle at SLAC.[40]

Figure 6.12 The Brookhaven apparatus, used for the discovery of charm. *(Photograph courtesy of Brookhaven National Laboratory.)*

The τ spin was determined to be 1/2 from studies of the production cross section near threshold and from τ decay studies. (It will be noted in Chapter 13 that the Michel ρ parameter for the τ is equal to the value expected to high accuracy if spin 1/2 and the Standard Model is assumed.)

The J/Ψ peak was so sharp its natural width could not be directly measured. The experiments were limited by their resolutions and by a width due to a radiative mode in which the J/Ψ emits a photon and then decays. However, the width can be found from the total cross section integrated over the peak and the leptonic branching fraction. The width turns out to be 0.086 MeV.

The spin and parity of the J/Ψ was determined to be 1^- from the shape of the radiative part of the resonance curve and from the angular distributions of the e^+e^-. It was determined to be $I = 0$ from the branching ratios of various $J/\Psi \to \rho\pi$ decay modes.

The experimental equipment used by the Brookhaven and SLAC groups represent two contrasting styles, both of which have been quite important for particle physics work.

The SLAC group used an early form of an electronic detector covering much of the entire solid angle. This detector, although not as homogeneous as a bubble chamber, had enormous advantages in time resolution and ability to handle high rates. The SLAC detector covered 65% of the solid angle. More recent detectors, such as those described in Chapter 13, cover much larger fractions of the solid angle. The SLAC apparatus is shown in Figure 6.13 in the center color insert. There is a solenoidal magnet providing a 4 kG field in a 3 m long, 3 m diameter volume. A particle leaving the interaction region in the center first traverses a 0.15 mm stainless steel vacuum chamber needed for the circulating beams. It then encounters a pair of cylindrical scintillation counters that form an element of the trigger system. It next enters four sets of cylindrical wire spark chambers, with two gaps each, which allow the curved path of charged particles to be reconstructed. This is followed by a trigger hodoscope with time resolution of 0.5 ns. This, in conjunction with the initial counter, provides time of flight information allowing charged π/K separation up to 600 MeV/c momentum. The particle next passes through the one radiation length thick magnet coil and encounters an array of five radiation lengths of lead-scintillator shower counters that identify electrons. There is then a 20 cm thick iron magnet yoke that absorbs most of the hadrons, followed by a set of spark chambers to help identify muons. A hardware trigger required at least two particles with greater than 200 MeV/c momenta and hence did not record totally neutral final states.

The BNL detector took a completely different tack. The SLAC detector attempted to look at a large part of the solid angle. The BNL detector looked at a very small part of the solid angle, but looked at it with high precision. It looked at the $J/\psi \to e^+e^-$ mode. The apparatus consisted of two symmetric spectrometer arms, on opposite sides of the beam and each at 14.6° horizontally to the incident beam direction. Each spectrometer had two magnets, both with vertical bends to decouple the horizontal angle and the momentum determination. Each spectrometer had two threshold Cherenkov counters, one filled with one atmosphere of H_2 and the other with 0.8 atm of H_2. There were also scintillators providing time-of-flight information. The scintillators were followed by 11 planes of proportional wire chambers at 20° with respect to each other to minimize multiple track confusion. There were then eight vertical and eight horizontal scintillator hodoscopes followed, finally, by an array of 25 lead-glass Cherenkov counters, each three radiation lengths thick. As a check, spectra were taken with full magnetic field and with one tenth of full field. The gas Cherenkov counters and the lead-glass array provided independent electron identification tests. This apparatus is shown in Figure 6.12.

134   6. The Quark Model

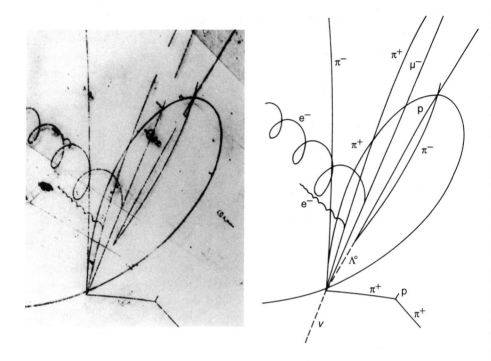

Figure 6.14 An example of charmed baryon production seen in the Brookhaven seven foot bubble chamber filled with hydrogen. $\nu p \to \mu^- \Sigma_c^{++}$; $\Sigma_c^{++} \to \Lambda_c^+ \pi^+$; $\Lambda_c^+ \to Y^{*+} \pi^+ \pi^-$; $Y^{*+} \to \Lambda^0 \pi^+$; $\Lambda^0 \to p \pi^-$. *(Photograph courtesy of Brookhaven National Laboratory.)*

Shortly after the discovery of charm a number of other experiments also were measuring properties of charmed particles. Figures 6.14–6.15 show two examples of charmed particle production seen in Brookhaven neutrino interaction experiments.

There are now known to be many bound states of charmonium and a minor industry within particle physics is concerned with measuring their properties and finding an appropriate heuristic potential to fit the spectrum of states. The $c\bar{c}$ system is interesting to study because the heavy c mass makes the system almost nonrelativistic. There are a number of $c\bar{c}$ states. For spin 0, the $^1S_0$ state η_c has been found. For spin 1, the $1^3S(\psi)$, $2^3S(\psi')$, $^3D_1(\psi'')$ states exist. $^3P_2$, $^3P_1$, $^3P_0$ states are seen as well as several other states.

There is considerable evidence that the spin of the c quark is 1/2. The pattern of charmonium states is in accord with spin 1/2. Furthermore, in $e^+e^- \to \overline{D}D$ production, where D is a meson with charm, the angular

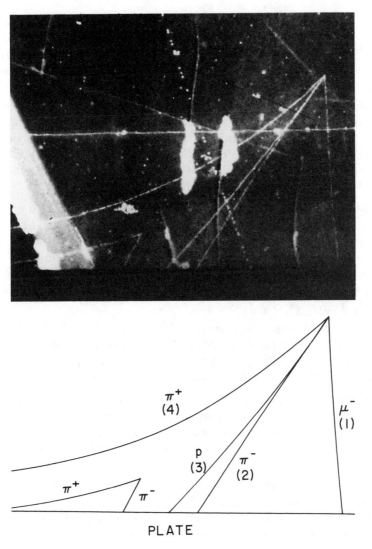

Figure 6.15 Another example of charmed baryon production seen in the Brookhaven seven foot bubble chamber, this time filled with deuterium. $\nu d \to \mu^- \Lambda_c^+ (p_{spectator})$; $\Lambda_c^+ \to pK^{*-}\pi^+$; $K^{*-} \to K^0 \pi^-$; $K^0 \to \pi^+\pi^-$. *(Photograph courtesy of Brookhaven National Laboratory.)*

distribution is observed to be $\sin^2 \theta$, where θ is the angle of the D with respect to the incident electron. This is expected if the D has spin 0. For an S-wave $\bar{c}d$, this would be expected if the spin of the c were 1/2. However, the spin of an S-wave $\bar{c}d$ state would necessarily be greater than zero if the spin of the c were greater than 1/2. In addition, deep inelastic scattering and Z decay branching ratios are all in accord with spin 1/2. In addition to D and D^* states, there are D_s states composed of $c\bar{s}$. Charmed baryon states such as Λ_c, Σ_c Ξ_c also exist.

In 1977 a still heavier quark was found, again by observing bound states. At Fermilab, a group lead by L. Lederman, stimulated by the discovery of the J/ψ, was again looking at lepton pair effective masses from pN collisions.[43] The apparatus and some of the physicists can be seen in Figure 6.16. This time their apparatus had sufficient resolution and they found two and perhaps three bound states which were labeled $\Upsilon, \Upsilon', \Upsilon''$ ("upsilons"). The masses were 9.4, 10.01, and 10.4 GeV/c^2. These turned out to be bound states of $\bar{b}b$ quark pairs. See Figure 6.17.

The first two Υ states were quickly confirmed by the DASP group working at the DESY e^+e^- storage ring DORIS.[44] Shortly afterward, Cornell completed an e^+e^- storage ring, CESR, in just the right energy region to explore this new physics. They confirmed the Υ'', found an Υ''', and found the various mesons with b quarks, the B mesons.[45] Some of their original results are shown in Figure 6.18. The Υ states are now known as $\Upsilon(1S)$, $\Upsilon(2S)$, $\Upsilon(3S)$, $\Upsilon(4S)$.

A plethora of b-states has now been found. In mesons, the B^+ (5.2787 ± 0.002 GeV), and B^0 (5.279 ± 0.002 GeV) consist of a $\bar{b}$-quark and a u or d quark, respectively. The B_s^0 (5.375 ± 0.006 GeV) consists of a $\bar{b}$-quark and an s quark. Higher mass B^* states are also known to exist.

The B mesons have a myriad of decay states. (Literally dozens of detected states are listed in the Particle Data Group tables.) These include decays into various D and D^* states, into J/ψ states, and kaon states. They are further subdivided into decays involving leptons (semi-leptonic) or only hadrons.

Mixing occurs; CP violation is expected to occur in B^0 decays, and, within the Standard Model, at a higher rate than in K^0 decays. This makes for a great deal of current interest in studying the B system. $B's$ will be discussed further in Section 7.12 and in Chapter 13.

So far, two complete quark doublets (ud), and (cs) have been found in addition to the b quark. Does the b have a doublet partner, the top, t, quark? There is good reason, from various theoretical calculations using properties of existing particles, to believe that the t exists and that its mass is less than 200 GeV.

Experimentally, firm evidence for the t has only recently been given. In 1994, the CDF group at the Fermilab $p\bar{p}$ collider announced that it was likely they were detecting events in which t quarks are produced.[46] The mass, from this direct observation, was about 174 GeV. In 1995, the CDF and D0 Fermilab groups simultaneously announced the discovery of the top quark. The CDF group[47] obtained a mass estimate, $m_t = 176 \pm 8($ stat.$) \pm$

Figure 6.16 A view of some of the experimenters and the apparatus used in the discovery of the Υ particle in a fixed target facility at Fermilab. *(Photograph courtesy of the Fermilab Visual Media Services.)*

10(*syst.*) GeV and the D0 group obtained $m_t = 199^{+19}_{-21}($ stat.$) \pm 22($ *syst.*$)$ GeV.[48] In Figure 6.19 in the center color insert, a fine example of the production of a $t\bar{t}$ pair in the CDF detector is shown. In this example, $t \to b + W^+$; $W^+ \to e^+\nu$; $\bar{t} \to \bar{b} + W^-$; $W^- \to q\bar{q} \to 2$ jets. The B hadrons containing the b quarks travel a short distance before decaying. The decay vertices are clearly seen. The two jets from the W^- have an effective mass of 79.5 GeV fully consistent with the 80 GeV mass of the W.

How many families are there? As noted in Chapter 1, a family (or generation) consists of a u-type quark, a d-type quark, a charged lepton, and a neutrino. It turns out that the number of families can be measured

138   6. The Quark Model

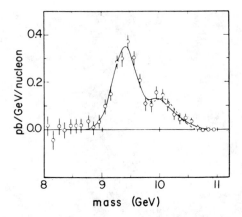

$pN \to \mu^+\mu^- X$. The dimuon mass spectrum is shown.

(a) Dimuon spectrum above 6 GeV.

(b) Excess of data over continuum fit. The solid curve is the three-peak fit; the dashed curve is the two peak fit.

Figure 6.17 Discovery of the Υ.[43]

as long as the neutrino in the family is low mass. As will be seen, the Z intermediate boson has a decay mode into $\nu\bar{\nu}$. The rate of this mode is proportional to the number of neutrinos, i.e., it has a certain rate into each species of neutrino. The width of the Z signal and the ratio of the width of the Z signal to the height of the visible Z peak have been measured. It has been found that there are 2.991 ± 0.016 families.[49] This mode has also been directly measured by measuring $e^+e^- \to Z^\star \to Z + \gamma \to \nu + \bar{\nu} + \gamma$. The two neutrinos cannot be detected but the γ signals the presence of the Z. A value of 3.14 ± 0.27 families was obtained.[50]

6.8   COLOR

It is now believed each quark u, d, c, s, t, b has a new quantum number, color, and that each quark comes in three colors, labeled red, blue, and yellow. Color symmetry is believed to be exact and all known particles are colorless, by which is meant a color singlet state, i.e., an equal mixture of all three colors.

As indicated in Section 6.2, one major justification for color comes from the need to have the baryon wavefunction anti-symmetric. Since there are three valence quarks in a baryon, each of which needs to be a different color, there must be (at least) three colors. The most natural way of constructing

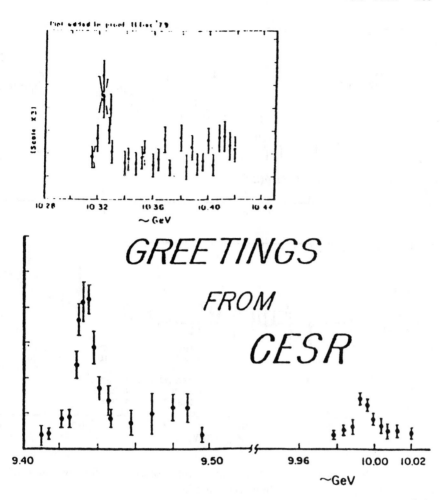

Figure 6.18 Cornell data on Υ states from 1979 CESR Christmas card.[45]

particles in the $SU(6)$ 56 leads to symmetric wavefunctions if color is not included. In this section other evidence for color will be examined.

Consider e^+e^- collisions and examine the ratio of cross sections with strongly interacting particles produced to the cross section for muon pairs. It will be shown that this latter cross section has the simple form $\sigma_{\mu\mu} = 4\pi\alpha^2/3s$, where s is the square of the energy of the e^+e^- system in the center of mass. Let

$$R = \sigma(e^+e^- \to \text{hadrons})/\sigma(e^+e^- \to \mu^+\mu^-). \qquad (6.16)$$

Figure 6.20 shows measured values of this quantity as a function of energy.

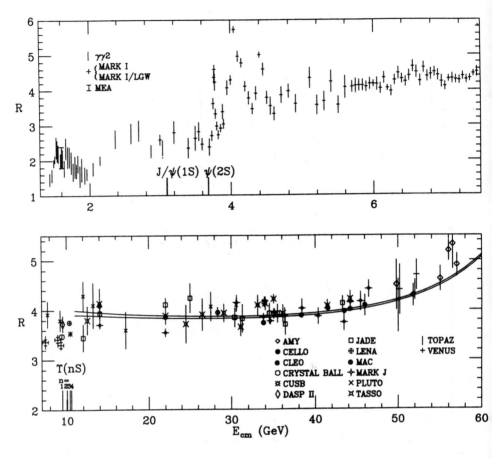

Radiative corrections and, where important, corrections for two-photon processes and τ production have been made.

Figure 6.20 $R = \sigma(ee \to \text{hadrons})/\sigma(ee \to \mu\mu)$ as a function of $\sqrt{s}$ in e^+e^- scattering.[17]

What is expected? Quarks are supposed to be point particles. Their electromagnetic interactions are just like those of leptons with the quark charge substituted for the leptonic charge. Consider the production of pairs of particles shown in Figure 6.21. "f" can be any of the elementary fermion particles. Far away from threshold effects and ignoring higher-order effects one has

$$R = \Sigma_{\text{quarks}} q_{\text{quark}}^2 / q_\mu^2. \tag{6.17}$$

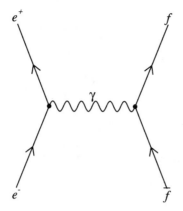

Figure 6.21 Feynman graph for $e^+e^- \to f\bar{f}$.

At around 2 GeV the u, d, and s are all able to be produced, but this energy is below the charm threshold. Without color

$$R(u,d,s; \text{no color}) = (q_u^2 + q_d^2 + q_s^2) = (2/3)^2 + (-1/3)^2 + (-1/3)^2 = 2/3. \quad (6.18)$$

The measured value of R is about 2.5 in this energy region, in disagreement with the preceding prediction.[51] If three colors are included, then there are three quarks of each kind:

$$R(u,d,s; \text{color}) = 3 \times 2/3 = 2. \quad (6.19)$$

This is in much closer agreement with the data. Above the charm threshold, but below the threshold for b and t quarks, one expects

$$R(u,d,s,c; \text{color}) = 3[(2/3)^2 + (-1/3)^2 + (-1/3)^2 + (2/3)^2] = 3.3. \quad (6.20)$$

The experimental value in this region must be corrected for the contribution of the τ particle. When this is done, a value of $R = 4.17\pm0.09$ (statistical)$\pm$ 0.42 (systematic) is obtained. There are known higher-order corrections that increase the expected value of R by about 0.5. Thus, with color, experiment and theory are in good agreement. Using the preceding first-order calculation it is expected that, above the b threshold, R should rise by 0.33 to 3.67. This rise is observed. Above the t threshold R is expected to rise further. (If purely electromagnetic interactions were involved, R would be expected to rise by 1.33 to 5. However, the mass of the t quark is large, even larger than the Z intermediate boson mass. Production of top pairs by the neutral weak force is then also important.)

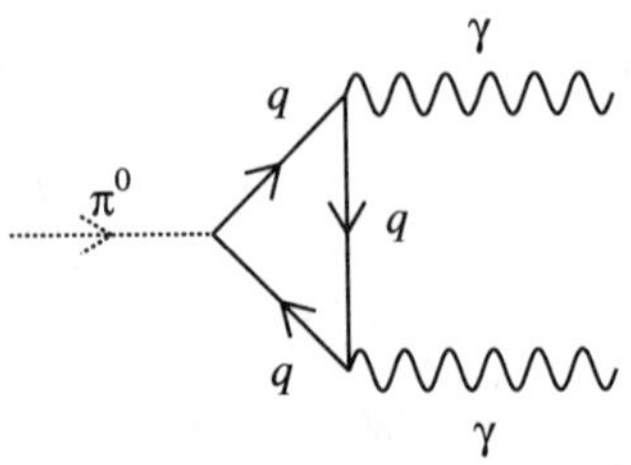

142   6. The Quark Model

Figure 6.22 Feynman graph for π^0 decay.

The preceding data seems also to be strong evidence for fractionally charged quarks. However, the question is subtle. In fact the preceding evidence only shows that an appropriate average over colors is fractional. This question will be discussed more fully in Chapter 9.

Another piece of evidence for color concerns π^0 decay. In the standard model to be developed in this text, the usual kind of diagrams cancel, giving no π^0 decay. π^0 decay is believed to occur largely through the anomalous diagram shown in Figure 6.22. There is a separate diagram for each quark species, q. These diagrams, known as the "triangle anomaly," have been evaluated and give a result too small to account for π^0 decay. The color hypothesis triples the number of diagrams. This would give a factor of nine increase. However, the color part of the pion wavefunction $((1/\sqrt{3})(r\bar{r}+y\bar{y}+b\bar{b}))$, when squared, introduces a factor of one-third. The combination of these effects gives an overall factor of three increase in the rate. The calculation with color is then in reasonable agreement with the measured lifetime of 0.87×10^{-16} s.[17]

Experimentally,[52] the lifetime of the π^0 is measured by looking at the vertex in a different process, shown in Figure 6.23. Forward π^0 photoproduction is dominated by the Feynman diagram shown. This is known as "the Primakoff effect." As is seen in this diagram, the $\pi^0 \to \gamma\gamma$ vertex can be measured by this means.

More evidence of color comes from the τ lepton decay branching ratios. Ignoring phase space corrections and assuming that the spin of the τ is $1/2$, the predicted branching ratios are given in Table 6.6.

Experimentally, the $e\nu\nu$ branching ratio is $17.7 \pm 0.4\%$ and the $\mu\nu\nu$ branching ratio is $17.8 \pm 0.4\%$,[17] essentially in agreement with the color hypothesis.

The evidence, presented in this section, for color seems very firm.

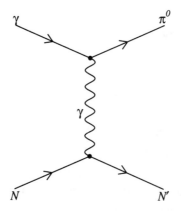

Figure 6.23 Primakoff effect Feynman diagram.

Table 6.6 Predicted τ Decay Branching Ratios

τ decay mode/	No color branching ratio	Color branching ratio
$e^+ \nu_e \bar{\nu}_\tau$	1/3	1/5
$\mu^+ \nu_\mu \bar{\nu}_\tau$	1/3	1/5
$(u\bar{d})\bar{\nu}_\tau$	1/3	3/5

6.9 FREE QUARKS

Free quarks have never been seen. It is strongly suspected that quarks cannot exist as free particles. There have been extensive searches in cosmic radiation and at accelerators looking for charge 2/3 or 1/3 particles. In addition, modern equivalents of the Millikan oil drop experiment have been performed at a number of places. Some of these gave initial positive results, but on careful re-examination, no evidence has been found at present for free quarks.

Theoretically, there are strong indications, a bit short of a proof, that quarks are not expected to exist in the free state.

In this chapter the very sizable body of evidence has been described which indicates that matter is composed of quarks, particles that can exist in bound, but not in free states. The new quantum number of color has been

144   6. The Quark Model

introduced. All interactions seem invariant under color transformations. Further discussion of quarks is given in Chapter 9.

6.10   EXERCISES

6.1 Assume that you are proposing an experiment to study the decays of the Λ^0.

a) How would you identify the Λ^0?

b) What apparatus would you need?

6.2 For each of the following reactions, draw three possible Feynman diagrams involving an intermediate state of one particle or an exchange of one particle. Determine the quantum numbers (Q, S, B) of the exchanged or intermediate particle. In each case, two diagrams involve the possible exchange of a known particle–identify it on the graph. The last diagram involves the exchange of an exotic particle and is strongly inhibited.

a) $\bar{p} + p \to K^- + K^+$.

b) $\pi^- + p \to K^+ + \Sigma^-$.

c) $\pi^- + p \to \bar{p} + d$.

6.3 Express the π^+ wavefunction in terms of the spin, flavor, and *color* of the component quarks.

6.4 It is a remarkable fact that the observed hadronic states are extremely limited as to the possible quantum numbers. For example, no doubly charged meson (e.g., 'π^{++}') has ever been seen. Such a state can *not* be made out of $q\bar{q}$ combinations of known quarks. Hence, its absence tends to confirm the simple quark model picture. Suggest

a) another mesonic state that can *not* be made out of known $q\bar{q}$ combinations, and

b) a possible baryonic state that can *not* be made out of known qqq combinations.

For each state in a and b, give a reaction in which the proposed state could, if it existed, be produced by the strong interaction, while respecting standard strong interaction conservation laws (e.g., $\pi^+ p \to$ 'π^{++}'$+ n$).

c) Examine the evidence for the existence of baryons with $S = +1$.

6.5 Find the valence quark content of

a) proton.

b) K^+, K^0, K^-.

c) π^+.

d) D^0.

e) ρ^+.

6.6 Using the simple quark model, calculate the magnetic moments of the particles in the lowest baryon octet in terms of the magnetic moments of the quarks, μ_u, μ_d, μ_s. Compare the measured values of the ratios of the magnetic moments to the proton's magnetic moment with the calculations. Assume the quarks have magnetic moments proportional to the simple point fermion magnetic moments and use estimates for the constituent quark masses given in Exercise 6.10 below.

6.7 At low energy, the cross section for K^--nucleon interactions is much greater than the cross section for K^+-nucleon interactions. Give a plausibility argument explaining why this should be expected.

6.8 The mean life of the Σ^0 is much less than the mean life of the $\Sigma^{\pm}$. Give a plausibility argument explaining why this should be expected.

6.9 The leptonic decay of neutral vector ($J^{PC} = 1^{--}$) mesons can be pictured as proceeding through a virtual photon:[53]

$$V(q\bar{q}) \to \gamma \to e^+ e^-.$$

The V-γ coupling is proportional to the charge of the quark q. Neglecting a possible dependence on the vector meson mass, show that the leptonic decay widths are in the ratios

$$\rho : \omega : \phi : \psi = 9 : 1 : 2 : 8.$$

Hint: Remember to add the amplitudes of the various pieces of the interaction before squaring.

6.10 As indicated in the text, a model for ground-state hadron masses can be made in which a QCD analogue of hyperfine splitting is used. In this scheme, meson masses are given by[53]

$$m(q_1 \bar{q}_2) = m_1 + m_2 + \frac{a(\sigma_1 \cdot \sigma_2)}{m_1 m_2},$$

where m_1 and m_2 are the masses of the quarks and a is a positive constant.

a) For the π (spin 0) and the K^* (spin 1), show that this expression gives

$$m(\pi) = m_u + m_d - \frac{3a}{m_u m_d},$$

$$m(K^*) = m_u + m_s + \frac{a}{m_u m_s}.$$

b) Calculate the masses of all the members of the 0^- and 1^- meson multiplets using

$$m_u = m_d = 0.31 \text{ GeV}; \quad m_s = 0.48 \text{ GeV}; \quad m_c = 1.65 \text{ GeV}.$$

Guess the value of a from looking at the π and K^* masses. Compare your results with the meson masses listed in the Particle Data Book.

c) Check how well the predictions work that

$$(\rho - \pi) = \frac{m_s}{m_u}(K^* - K) = \frac{m_c}{m_u}(D^* - D) = \frac{m_c m_s}{m_u^2}(D_s^* - D_s),$$

where the particle names are used to denote their masses.

7
Weak Interactions

In this chapter the weak interactions, responsible for β-decay and other decays, and for neutrino interactions will be discussed. New symmetries and conserved quantities will be found to occur. Two kinds of weak interactions, one mediated by charged and the other by neutral heavy bosons, will be described.

7.1 BRIEF OUTLINE OF DIRAC THEORY

To discuss the weak interactions, it is necessary to understand a little bit about relativistic equivalents of the Schrödinger equation for spin 1/2 particles. Only what is absolutely needed will be discussed here. $\hbar$ and c are taken as 1.

The non-relativistic Schrödinger equation is linear in energy and quadratic in momentum. One relativistic extension would be to use $E^2 = \vec{p}^{\,2} + m^2$ and make the usual quantum mechanical substitutions for E and $\vec{p}$ in terms of time and spatial derivatives. This leads to the Klein–Gordon equation and is quadratic in both $\vec{p}$ and E.

Dirac attempted to find an equation linear in $\vec{p}$ and E. Since the classical relation between $\vec{p}$ and E is not linear, this cannot be done with ordinary coefficients. However, Dirac had the insight to find that it was possible using coefficients that were operators and could be represented by matrices. In this manner he was able to obtain a relativistic theory for spin 1/2 particles.

In non-relativistic theory a spin 1/2 wavefunction has two components, corresponding to spin up and spin down. In Dirac theory the wavefunction has four components corresponding to the degrees of freedom of spin up, spin down and particle, anti–particle. Please do not confuse these four components with the four dimensions of space-time.

A useful set of four-dimensional matrices in this four-component spin space are the gamma matrices. There are four such matrices γ^μ. Each matrix is a matrix in component space. The index μ, on the other hand, *is* a spacetime index. The gamma matrices are defined through their (anti) commutation relations, which are

$$\gamma^\mu \gamma^\nu + \gamma^\nu \gamma^\mu = 2Ig^{\mu\nu}. \tag{7.1}$$

Here I is the 4×4 identity matrix and $g^{\mu\nu}$ is the metric tensor. μ and ν go from 0 to 3. The four-vector is $(\gamma^0, \gamma^1, \gamma^2, \gamma^3)$. Another useful matrix is

$$\gamma^5 \equiv i\gamma^0 \gamma^1 \gamma^2 \gamma^3. \tag{7.2}$$

$\gamma^5 \gamma^5 = I$ and γ^5 anti-commutes with γ^μ, $\mu = 0-3$, i.e., $\gamma^5 \gamma^\mu = -\gamma^\mu \gamma^5$.

To define the gamma matrices it is useful to again write down the 2×2 Pauli spin matrices:

$$\sigma_1 = \begin{pmatrix} 0 & 1 \\ 1 & 0 \end{pmatrix}; \; \sigma_2 = \begin{pmatrix} 0 & -i \\ i & 0 \end{pmatrix}; \; \sigma_3 = \begin{pmatrix} 1 & 0 \\ 0 & -1 \end{pmatrix}. \tag{7.3}$$

There are several representations of matrices satisfying these commutation relations, and different physicists use different ones. In this text, the gamma matrices are defined as

$$\gamma^k (k=1,2,3) = \begin{pmatrix} 0 & \sigma_k \\ -\sigma_k & 0 \end{pmatrix}; \; \gamma^0 = \begin{pmatrix} I & 0 \\ 0 & -I \end{pmatrix}; \; \gamma^5 = \begin{pmatrix} 0 & I \\ I & 0 \end{pmatrix}. \tag{7.4}$$

Here each entry is itself a 2×2 matrix. I is the 2×2 identity matrix. For example,

$$\gamma^5 = \begin{pmatrix} 0 & 0 & 1 & 0 \\ 0 & 0 & 0 & 1 \\ 1 & 0 & 0 & 0 \\ 0 & 1 & 0 & 0 \end{pmatrix}. \tag{7.5}$$

$$\gamma^{k\dagger}(k=1,2,3) \equiv (\gamma^{kT})^\star = -\gamma^k; \; \gamma^{0\dagger} = \gamma^0; \; \gamma^{5\dagger} = \gamma^5. \tag{7.6}$$

A free particle wavefunction in Dirac theory is

$$\Psi = u(p)e^{-ip\cdot x} + v(p)e^{ip\cdot x}. \tag{7.7}$$

Here $p \cdot x \equiv Et - p_1 x_1 - p_2 x_2 - p_3 x_3$, which is the four-dimensional scalar product. u and v are four-component objects known as "spinors." u and

7.1 Brief Outline of Dirac Theory

v correspond to particle and anti-particle, respectively. (Note that they correspond to the positive and negative frequency components of Ψ.) There is a spin up and a spin down u and similarly for v. Let r be the spin index and $r = 1$ be spin up, $r = 2$ be spin down. Then,

$$u_r = \sqrt{|E|+m}\begin{pmatrix}\chi_r^u \\ \frac{\sigma\cdot p}{|E|+m}\chi_r^u\end{pmatrix}, \text{ where } \chi_1^u = \begin{pmatrix}1\\0\end{pmatrix} \text{ and } \chi_2^u = \begin{pmatrix}0\\1\end{pmatrix}; \quad (7.8)$$

$$v_r = \sqrt{|E|+m}\begin{pmatrix}\frac{\sigma\cdot p}{|E|+m}\chi_r^v \\ \chi_r^v\end{pmatrix}, \text{ where } \chi_1^v = \begin{pmatrix}0\\-1\end{pmatrix} \text{ and } \chi_2^v = \begin{pmatrix}1\\0\end{pmatrix}. \quad (7.9)$$

Note that these spinors u and v have dimensions of $E^{1/2}$.

Define

$$\overline{\Psi} = \Psi^\dagger \gamma^0 \Rightarrow \overline{\Psi}_k = \Sigma_j \Psi_j^\star (\gamma^0)_{jk} \equiv \Psi_j^\star (\gamma^0)_{jk}, \quad (7.10)$$

where in the last expression the summation convention has been used: repeated indices in an expression are understood to be summed over. Also $\bar{u} \equiv u^\dagger \gamma^0$.

Let $\not{p} \equiv p_\nu \gamma^\nu = E\gamma^0 - p^1\gamma^1 - p^2\gamma^2 - p^3\gamma^3$ and $\not{\partial} \equiv \partial_\nu \gamma^\nu = \partial/\partial x^0 \gamma^0 + \partial/\partial x^1 \gamma^1 + \partial/\partial x^2 \gamma^2 + \partial/\partial x^3 \gamma^3$. The Dirac equation for a free particle is

$$(i\not{\partial} - m)\Psi = 0. \quad (7.11)$$

Then, for the spinors above, the Dirac equation is

$$(\not{p} - m)u = 0; \quad (\not{p} + m)v = 0; \quad \bar{u}(\not{p} - m) = 0; \quad \bar{v}(\not{p} + m) = 0. \quad (7.12)$$

The completeness relations easily follow (see Exercise 7.2):

$$\sum_{r=1}^{2} u_r \bar{u}_r = \not{p} + m; \quad \sum_{r=1}^{2} v_r \bar{v}_r = \not{p} - m. \quad (7.13)$$

Note that here $u\bar{u}$ is used. $\bar{u}u$ is a constant, the normalization; $u\bar{u}$ is a matrix.

Two useful operators are $\Lambda_\pm$ and $\Pi_n^\pm$:

$$\Lambda_\pm \equiv \frac{1}{2m}[\pm\not{p} + m]. \quad (7.14)$$

It is easily verified that $\Lambda_\pm$ projects out the $\pm$ frequencies of the solution. That is, if $\Psi = aue^{-ip\cdot x} + bve^{ip\cdot x}$, then $\Lambda_+ \Psi = aue^{-ip\cdot x}$. (See Exercise 7.2.)

7. Weak Interactions

To define the second operator first define n_μ as a generalization of the polarization direction. Let $n = (n_0, n_1, n_2, n_3)$. Furthermore, let $n^2 = 1$ and $n \cdot p = 0$ (4D). For a particle at rest, $n = (0, \hat{n})$. Then define

$$\Pi_n^\pm \equiv \frac{1}{2}[I \mp \gamma^5 \not{n}]. \tag{7.15}$$

$\Pi_n^\pm$ projects out the spin in the $\pm n$ direction. (See Exercise 7.3.) Helicity (the spin quantized along the direction of motion) commutes with H and p and hence can be simultaneously diagonalized. For helicity,

$$n = \left(\frac{|\vec{p}|}{m}, \frac{E}{m}\hat{p}\right).$$

Suppose that under the parity transformation $\Psi'(x') = A\Psi(x)$ where $x' = (x^0, -\vec{x})$. Then the Dirac equation for the parity-reversed state can be written

$$\left(i\vec{\partial}' \cdot \gamma + i\frac{\partial'}{\partial t}\gamma^0 - m\right)\Psi'(x') = \left(-i\vec{\partial} \cdot \gamma + i\frac{\partial}{\partial t}\gamma^0 - m\right)A\Psi(x) = 0.$$

Because they are derivatives with respect to x', the three-vector part of the partial derivatives has the wrong sign for the Dirac equation. To get back to the proper sign, it is necessary that $A = \eta\gamma^0$, where η is a constant. Then when A is brought to the left-hand side of the equation, using the appropriate γ commutation relations, the Dirac equation is recovered. From applying the parity operation twice it is seen that $|\eta| = 1$. As noted in Section 4.5, $\eta^2 = -1$ is a possibility; for the standard particles, $\eta^2 = 1$, but for a proposed neutrino model, the neutrinos are "majorana" particles. One aspect of this model is that $\eta^2 = -1$, i.e., a majorana neutrino has imaginary parity.

Note that the parity transformation involves γ^0. The lower components of γ^0 are -1, while the upper two are $+1$. Since, for a particle at rest, the upper two components refer to the particle components and the bottom two components refer to the anti-particle components of the wavefunction, the parity operator gives opposite sign results for particles and anti-particles. This is proof of the fact noted in Chapter 4 that the parity of a Fermi particle is opposite to that of its anti-particle.

7.1 a. Second Quantization

In field theory, quantization is taken one step further. The wavefunctions themselves are considered operators, creating and destroying particles on state vectors. This enables the handling of situations in which the number of particles is not constant.

Expand Ψ as a superposition of plane waves and let a be an operator destroying a particle in a state vector. Recall from Section 5.3 that the wavefunctions are normalized within a box and satisfy boundary conditions at the boundaries of the box; the allowed momenta form a large but discrete set:

$$\Psi = \sum_r \sum_{p(E>0)} \frac{1}{\sqrt{2|E|}} [a^{(r)}(p) u^{(r)}(p) e^{-ip\cdot x}]$$

$$+ \sum_{p(E<0)} \frac{1}{\sqrt{2|E|}} [a^{(r)\dagger}(p) u^{(r)}(p) e^{-ip\cdot x}]. \qquad (7.16)$$

In order to avoid negative energies, it is convenient to reinterpret the destruction of a negative energy particle of momentum p and spin σ as the creation of a positive energy anti-particle of momentum $-p$ and spin $-\sigma$. The creation operator for this anti-particle is defined as $b^\dagger$. $b^\dagger(\sigma, p) = a(-\sigma, -p)$.

The space–time dependent field operator for a Dirac field can then be written in the form

$$\Psi = \sum_r \sum_{p(E>0)} \frac{1}{\sqrt{2E}} [a^{(r)}(p) u^{(r)}(p) e^{-ip\cdot x} + b^{(r)\dagger}(p) v^{(r)}(p) e^{ip\cdot x}], \qquad (7.17)$$

where $a^\dagger$ is an operator creating a particle of momentum p,

$b^\dagger$ is an operator creating an anti – particle of momentum p,

a is an operator destroying a particle of momentum p,

b is an operator destroying an anti – particle of momentum p.

Thus Ψ destroys particles and creates anti-particles. $\overline{\Psi}$ destroys anti-particles and creates particles.

7. Weak Interactions

In order for spin 1/2 particles to have Fermi-Dirac statistics, i.e., in order that two identical spin 1/2 particles be forbidden to be in the same state, it is necessary following the classic analysis of Jordan and Wigner[54] to require that the field operators have anti-commutation relations. Use curly brackets ({}) to denote anti-commutation, i.e., $\{p, q\} = pq + qp$:

$$\{\Psi_\mu(\vec{r}, t), \Psi_\lambda^\dagger(\vec{r'}, t)\} = \delta^3(\vec{r} - \vec{r'})\delta_{\mu\lambda},$$

$$\{\Psi_\mu(\vec{r}, t), \Psi_\lambda(\vec{r'}, t)\} = 0, \qquad (7.18)$$

$$\{\Psi_\mu^\dagger(\vec{r}, t), \Psi_\lambda^\dagger(\vec{r'}, t)\} = 0.$$

Then this implies that all of the $a's$, $a^{\dagger's}$, $b's$, $b^{\dagger's}$ anti-commute, except

$$\{a^{(r)}(p), a^{(s)\dagger}(p')\} = \{b^{(r)}(p), b^{(s)\dagger}(p')\} = \delta_{rs}\delta^3(\vec{p} - \vec{p'}). \qquad (7.19)$$

For two dissimilar particles all of the creation and destruction operators anti-commute, i.e., a, $a^\dagger$, b, and $b^\dagger$ for an s quark anti-commutes with the similar operators for a u quark.

Consider $a^{(r)\dagger}(p)a^{(r)}(p)$ acting on a state with a particle of momentum p and spin state r, i.e., $a^{(r)\dagger}(p)|0\rangle$. From the anti-commutation relations,

$$a^{(r)\dagger}(p)a^{(r)}(p)\, a^{(r)\dagger}(p)|0\rangle = a^{(r)\dagger}(p)(-a^{(r)\dagger}(p)a^{(r)}(p) + 1)|0\rangle$$

$$= 1 \times a^{(r)\dagger}(p)|0\rangle,$$

since $a^{(r)}(p)|0\rangle = 0$. Thus the operator

$$A = \sum_r \sum_p a^{(r)\dagger}(p)a^{(r)}(p) \qquad (7.20)$$

counts all the particles in a state, i.e., $A|\text{state}\rangle = nA|\text{state}\rangle$, where n is the number of particles contained in the state.

If Γ is any 4×4 matrix whose elements are ordinary numbers (c-numbers), not operators, then it can be shown from the above anti-commutation relations that

$$\left[\int \Psi_\mu^\dagger(\vec{r'}, t)\Gamma^{\mu\nu}\Psi_\nu(\vec{r'}, t)d^3r',\ \Psi_\lambda(\vec{r}, t)\right] = -(\Gamma\Psi(\vec{r}, t))_\lambda \qquad (7.21)$$

Next consider a current of spin 1/2 fermions ($\overline{\Psi}O\Psi$), where O is some operator. These will become quite important for calculating matrix elements in future parts of this text. The creation and destruction operators in Ψ and $\overline{\Psi}$ must destroy the initial particles and create the final particles. In terms of the spinors then, from Equation 7.17, the results given in Table 7.1 apply. Thus, omitting the exponential, for $e^- \to e^-$, the form is $(\overline{u}Ou)$; for vacuum $\to e^+e^-$, the form is $(\overline{u}Ov)$.

7.2 Nuclear β-Decay

The prototypical nuclear β-decay is neutron decay, $n \to p + e^- + \overline{\nu}$, where the ν is known to be a massless or almost massless particle. Many nuclei β-decay, some emitting $e^-\overline{\nu}$ and some emitting $e^+\nu$. In the 1930s β-decay was thought to occur through a point interaction or almost a point interaction, i.e., an interaction depending on the overlap of all of the wavefunctions at each point in space–time.[55] Thus, the matrix element for $A + B \to C + D$ is

$$M \propto \Psi_C^\star(r)\Psi_D^\star(r)\Psi_A(r)\Psi_B(r). \tag{7.22}$$

(The point interaction picture will be modified later. The energies involved in β-decay are a few MeV, much smaller than the 80 GeV of the W intermediate boson. For β-decays, the interaction indeed appears pointlike.) Note that $n \to pe^-\overline{\nu}$, $\nu n \to pe^-$, $\overline{\nu}n \to \overline{p}e^+$, $p \to ne^+\nu$ (in nuclei), and $pe^- \to n\nu$ (K capture–a nucleus capturing a K-shell electron) are all different processes arising from the same basic interaction.

For A, B, C, D all spin 1/2, each Ψ has four components as seen in the last section and there are then $4^4 = 256$ possible terms in M. It is necessary to reduce and classify this multitude of terms by applying Lorentz invariance. The possible interaction forms should be covariant under a Lorentz transformation. Consider a bilinear term $\overline{\Psi}_1\Psi_2$, where, for example, one might have $1 = A$, $2 = C$. As noted previously, $\overline{\Psi}_1 = \Psi_1^\dagger \gamma^0$. There are five basic bilinear combinations of the form $\overline{\Psi}_1 O \Psi_2$, where O is an operator. Let $\sigma^{\mu\nu} = (i/2)[\gamma^\mu, \gamma^\nu]$. See Table 7.2.

These sum to 16 combinations $= 4^2$, which, therefore, exhaust the possibilities. For Lorentz covariance, when two bilinear combinations are put together, the number of indices must match, and the indices must be summed over, to get a Lorentz scalar result.

7.2 a. Parity Considerations

For VV or TT, etc., the matrix element will conserve parity because both terms do the same thing (go to plus or minus themselves) under the parity operator. For VA or SP, the matrix element goes to minus itself under the parity operator. However, transition probabilities depend on $|M|^2$ and hence the minus sign vanishes. These combinations also conserve parity.

Table 7.1 Use of u and v in Currents

	fermion	anti-fermion
initial state	u on right	$\bar{v}$ on left
final state	$\bar{u}$ on left	v on right

Table 7.2 Basic Bilinear $\bar{\Psi}_1 \Psi_2$ Terms

Combination	Name of Term	Number of Components
$\bar{\Psi}_1 \Psi_2$	Scalar (S)	1
$\bar{\Psi}_1 \gamma^\mu \Psi_2$	Vector (V)	4
$\bar{\Psi}_1 \sigma^{\mu\nu} \Psi_2$	Tensor (T)	6
$\bar{\Psi}_1 \gamma^5 \gamma^\mu \Psi_2$	Axial Vector (A)	4
$\bar{\Psi}_1 \gamma^5 \Psi_2$	Pseudoscalar (P)	1

Parity is violated if there is a mixture of parity types in the same bilinear combination. Terms like P(S+P) or (V−A)(V−A) will have parity violation in their interference terms and are forbidden if parity is conserved under the weak interactions.

But is parity conserved under the weak interactions? Until the 1950s it was thought that parity was absolutely conserved like momentum and angular momentum. But then a serious problem appeared. The K particle was still fairly new. Both $K^+ \to 2\pi$ and $K^+ \to 3\pi$ were seen in experiments. The decay rates were roughly of the right magnitude to be due to the weak interaction. The pion was already known to be a pseudoscalar particle. The three pion decay had very little phase space and Dalitz plots of the decay were flat. (The Dalitz plot was invented for this decay.) This seemed to indicate that the 3π were all in relative S-states, i.e., that the matrix element was constant. However, if this was the case, then the parity of the 2π state had to be opposite that of the 3π state because of the negative π intrinsic parity.

This was a very hard problem. At first it was thought that these were two different particles that just happened to be at almost the same mass. However, no matter how charged kaons were produced and no matter what absorbers they went through, the ratio of 2π to 3π decays remained the same. All sorts of models were proposed including opposite parity partners of all fundamental particles. Nothing seemed to fit well.

Finally, almost in desperation, T.D. Lee and C.N. Yang, following a suggestion of M. Block, examined the possibility that maybe parity was not conserved.[56] When they tried to examine in detail the experimental evidence for parity conservation in the weak interactions, they found there was none! The race among experimentalists was on. It was won by C.S. Wu[57] who found parity violation in decays of aligned Co^{60}. T.D. Lee and C.N. Yang received the 1957 Nobel Prize for their discovery.

Parity invariance is often discussed by asking whether one can distinguish the world from the world as seen in a mirror. Reflection in a mirror is simply parity inversion followed by a simple rotation. If parity is conserved there is no way to distinguish a world from a mirror world. If you contacted another civilization in outer space by radio, you would be unable to define for them "righthandedness."

Dr. Wu aligned the nuclei by placing her sample in a magnetic field at very low temperature. She found that the decay electrons mostly went opposite to the direction of nuclear spin. Under reflection in a mirror in the xy plane, a right circulating electric current in the xz plane becomes a left circulating current. This reverses the sign of B_y. However, the y component of momentum is unchanged. Thus, observation of a $\vec{B} \cdot \vec{p}$ term in this experiment distinguishes the world from the mirror world and hence violates parity invariance. This is more formally expressed by saying that p is a vector and B is a pseudovector and hence their dot product is a pseudoscalar. A pseudovector or pseudoscalar has the same spatial properties as the corresponding vector or scalar except that it has the opposite sign under the parity transformation. A pseudovector goes to plus itself under the parity transform and a pseudoscalar goes to minus itself.

It is now known that although P is violated, CP is almost conserved. Thus, the world is almost invariant if a parity transformation is made followed by interchanging particles and anti-particles. Thus, suppose you contact an outer space civilization, and agree to meet and shake right hands (or flippers). You have discussed P violation but not CP violation. You meet and the ambassador sticks out his left flipper. Don't shake!

7.2 b. Nuclear β-Decay Summary

Assume the matrix element is of the form

$$M = \frac{G}{\sqrt{2}} (\overline{\Psi}_p O_i \Psi_n)(\overline{\Psi}_e O'_i \Psi_\nu), \tag{7.23}$$

where O_i and O'_i are S,V,T,P,A operators or a combination of them times normalization coefficients C_S, C_V, C_T, C_A, C_P.

7. Weak Interactions

Table 7.3 Characteristics of Lowest Order β Decay Terms

Name of Term	Fermi (F)	Gamow–Teller (GT)
Spin Orientations	$n \uparrow \to p \uparrow + e^- \downarrow + \bar{\nu}_e \uparrow$	$n \uparrow \to p \downarrow + e^- \uparrow + \bar{\nu}_e \uparrow$
$\Delta \vec{J}$	0	1 ($\Delta J = 0, \pm 1$)
$J_i = 0 \longrightarrow J_f = 0$	Allowed	Forbidden
Lorentz terms	S,V	T,A
Non-Rel. term	$\langle | 1 | \rangle$	$\langle | \sigma | \rangle$

Here the hadron pair and the lepton pair have been taken as the two pairs. This is physically motivated, but an equally valid choice would have been the proton–electron and neutron–neutrino as the pairs. The physics is identical, and one formulation can be written in terms of the coefficients of the other. However, when the transformation is made, something which was pure vector in one system will have contributions from other symmetry operators in the other system. These transformations are known as Fierz transformations. As will be seen, with the present choice, an especially simple form results.

In the rest system of the decaying particle the hadrons are all very low energy and a non-relativistic approximation of the hadron wavefunctions can be used. The wavefunction is expanded in powers of v^2/c^2. The lowest-order term is called "allowed." The allowed terms require definite relative spin orientations and make definite statements about what is allowed for the spin of the initial nuclear state (J_i), the final nuclear state (J_f), and the change in angular momentum of the state ($\Delta \vec{J}$). (See Exercise 7.8.) There are two general classes of terms; they are shown in Table 7.3.

Specifically,[1] the decay rate for $Z \to Z' + e^- + \bar{\nu}_e$ is given by

$$\frac{1}{\tau} = \frac{m_e}{2\pi^3}(|C_V|^2 M_F^2 + |C_A|^2 M_{GT}^2) f\left(\frac{E_m}{m_e}\right), \tag{7.24}$$

where E_m is the maximum electron energy and

7.2 Nuclear β-Decay

Table 7.4 Characteristics of Weak Decay Terms
(for left-handed neutrino helicities; right-handed similar)

Term	Nuclear Recoil	Intensity $\propto$
S	Low energy	$1 - (v/c)\cos\phi$
A	Low energy	$1 - 1/3(v/c)\cos\phi$
T	High energy	$1 + 1/3(v/c)\cos\phi$
V	High energy	$1 + v/c\cos\phi$

$$f(x) = \frac{1}{60}(x^2 - 1)^{1/2}(2x^4 - 9x^2 - 8) + \frac{x}{4}\ln(x + \sqrt{x^2 - 1}),$$

$$M_F^2 = \left| \int \langle Z'| \Psi_p^\dagger \Psi_n |Z\rangle \, d^3r \right|^2, \tag{7.25}$$

$$M_{GT}^2 = \left| \int \langle Z'| \Psi_p^\dagger \vec{\sigma} \Psi_n |Z\rangle \, d^3r \right|^2.$$

If there is a $1 - \gamma^5$ in the lepton operator, the neutrino is left-handed (i.e., the spin is opposite to the direction of motion), and if there is a $1+\gamma^5$ in the operator, the neutrino is right-handed. For a left-handed neutrino, Tables 7.3 and 7.4 give the spin polarizations and the qualitative angular distributions for the S,T,V,A cases. ϕ is the angle between the electron and the neutrino and v is the velocity of the electron. P is hard to see here. It will be examined when pion decay is discussed. One can distinguish the other cases by measuring the energy distribution of the recoil nucleus or by measuring the $\pm$ sign on $(v/c)\cos\phi$. Both are difficult measurements, but have been done by clever and careful experimenters.

First the lepton terms $(\overline{\Psi}_e O'_i \Psi_\nu)$ will be considered, and afterward the hadron terms $(\overline{\Psi}_p O_i \Psi_n)$.

Lepton terms.

e^- helicity $-v/c$. [Co^{60} $\rightarrow$Ni^{60} + e^- + $\bar{\nu}$]. This is a $J = 5 \rightarrow J = 4$ transition. It was necessary to measure the longitudinal polarization of the electron. This was done by deflecting the electron by about $90°$ with an electric field that did not change the spin direction. The electron was then transversely polarized. It was next scattered from a high-Z material. Asymmetry in scattering measured the transverse polarization.[58][59]

e^+ helicity $+v/c$. [$Ga^{66} \to Zn^{66}+e^++\nu$]. This is a $0^+ \to 0^+$ transition and is pure F. The electron polarization was measured by e^+(flight)+ $e^- \to 2\gamma$ followed by scattering the γ off of polarized electrons.

$Na^{22} \to Ne^{22}+e^++\nu$. This is a $3^+ \to 2^+$ transition and is pure GT.[60] The polarization of the positrons was measured. This was done by having them captured by e^- to form positronium in the presence of a 1.0–1.5 T magnetic field. The ratio of triplet to singlet positronium was then measured.

ν helicity -1. [$Eu^{152} + e^-$(K capture) $\to Sm^{152}(1^-) + \nu$. The $Sm^{152}(1^-) \to Sm^{152}(0^+) + \gamma$]. This was a marvelous experiment.[61] Very briefly, by a trick one can measure the γ helicity and then measure the ν helicity. The trick is the following. Suppose one has a sample of $Sm^{152}(0^+)$. If E_γ is just right, one can get resonant absorption to excite $Sm^{152}(0^+)$ to $Sm^{152}(1^-)$. In the preceding reaction, the energy of the γ depends on the direction of the γ with respect to the direction of travel of the $Sm^{152}(1-)$ formed in the Eu^{152} decay. This, in turn, depends on the direction of the neutrino from the original Eu decay. Thus, the energy of the emitted γ depends on the direction the ν went and the γ direction. One then uses conservation of angular momentum to show that for the original Eu decay, the helicity of the recoil nucleus equals the helicity of the ν. It turns out that for the γ to have sufficient energy to be resonantly absorbed, the γ helicity is approximately equal to the recoil helicity of the $Sm^{152}(1-)$, and thus, if the γ helicity is measured, the ν helicity is known. One obtains the γ helicity by passing the γ's through magnetized iron before they impinge on the Sm^{152} absorber. The transmission is slightly different for + and − helicities.

In summary, it is known that O'_i has V,A only since the e^- and $\bar{\nu}$ have opposite helicities. Since the polarization is maximum, it is known that it has $(1 \pm \gamma^5)$ in it. Since the sign of the various helicities is known (the ν is left-handed), it is known that the minus sign is chosen. Thus $O'_i = \gamma^\mu(1 - \gamma^5)$.

Hadron terms. As just seen, examining these terms involves looking at correlations between the e and the ν.

$C_A \gg C_T$. [$He^6 \to Li^6 + e^- + \bar{\nu}$]. This is a $0^+ \to 1^+$ transition and is pure GT. There is no parity change of the nucleus. p_e and p_ν were found to have negative correlation.[62]

[$Co^{60} \to Ni^{60} + e^- + \bar{\nu}$]. This is a $J = 5 \to J = 4$ transition. It is pure GT. This was Ms. Wu's experiment.[57] Besides parity violation she observed from the spectra that e and ν tended to go in opposite directions as predicted by A. T predicted that they would tend to go in the same direction.

$C_V \gg C_S$. [Ar$^{35} \to$ Cl$^{35}+e^++\nu$]. This is almost a pure F transition. p_e and p_ν were found to have a negative correlation.[63]

C_A comparable with C_V. Sc^{46} decay was examined. This is a 4$^+ \to$ 4$^+$ decay followed by photon emission as the final state drops 4$^+ \to$ 2$^+ \to$ 0$^+$. Interference terms between A and V were seen implying that A and V are of similar size.

From the preceding experiments determining the hadron part, $O_i = \gamma^\mu(C_V + C_A\gamma^5)$. If the neutrino helicity is left free, the limits on the hadronic C_S/C_V and C_T/C_A are not exceedingly sharp having errors about ±0.2. This occurs because the experiments look at interference terms that can be zero if $C_S = -C'_S$; $C_T = -C'_T$, where the prime refers to the right-handed neutrino coupling. If the neutrino helicity is taken as fixed, these limits are quite good: $C_S/C_V = -0.001 \pm 0.006$; $C_T/C_A = -0.0004 \pm 0.0003$.

C_A and C_V are believed to be relatively real as time reversal invariance seems to be respected (no T odd correlations were found). It was shown in Section 4.6, that if H is not explicitly time dependent, then H is invariant under time reversal, if, and only if, H is real. Finally from measuring O^{14} (pure F) and He^6 (pure GT) it is found that $-C_A/C_V = 1.26 \pm .02$:

$$M = \frac{4G}{\sqrt{2}} \left[\overline{\Psi}_p \gamma^\mu \left(\frac{1-1.26\gamma^5}{2}\right)\Psi_n\right]\left[\overline{\Psi}_e\gamma^\mu\left(\frac{1-\gamma^5}{2}\right)\Psi_\nu\right], \quad (7.26)$$

where $GM_p^2 = 1.01 \times 10^{-5}$. The 4 in the first term of Equation 7.26 is a residue of an older notation in which no factor of two was included in the $(1-\gamma^5)$ term.

It is now believed that there is a universal V−A weak interaction. The 1.26 is due to strong interaction renormalizations (virtual cloud of pions, etc.) which can now be calculated in terms of πp cross sections. At quark level there is no 1.26 and one just has $[\overline{\Psi}_1\gamma^\mu(1/2)(1-\gamma^5)\Psi_2]$. The quark level interaction will be discussed later in this chapter.

The $(1-\gamma^5)/2$ term is responsible for the left-handedness of the weak interaction. It is not present in the electromagnetic interaction, which has only γ^μ, not the $\gamma^\mu(1-\gamma^5)/2$ of the weak interaction.

7.3 KURIE PLOTS; NEUTRINO MASS LIMITS

The matrix elements in β decay have been studied previously. Consider now the phase space. In Section 5.3 it was noted that the three-body final-state phase space could be written

$$dP.S._3 = \frac{1}{8(2\pi)^9 E_1 E_3}[p_2 p_1^2 dp_1 d\Omega_1 d\Omega_2] \quad (\text{if } m_3 \to \infty). \tag{7.27}$$

Let particle 1 be e and particle 2 be ν. Let Q be the total final-state kinetic energy for the decay, i.e., be the energy released into kinetic energy and T_e be the kinetic energy of the electron. Because of the large mass, the energy of the recoil nucleon is very low ($E = p^2/2M$). If this energy is ignored, then $|\vec{p_2}| = E_\nu \approx Q - T_e$ and $E_3 = M_N$:

$$dP.S._3 = \frac{1}{8(2\pi)^9 E_e M_N}[Q - T_e]p_e^2 dp_e d\Omega_1 d\Omega_2. \tag{7.28}$$

Suppose $|M|^2 \propto E_e E_\nu \approx E_e(Q - T_e)$. It can be shown (see next section) that this corresponds to the allowed Fermi case for the matrix elements of the last section. Then if $N(p_e)$ = the differential number of events at p_e:

$$N(p_e) dp_e \propto (Q - T_e)^2 p_e^2 dp_e. \tag{7.29}$$

If $\sqrt{[N(p_e)/p_e^2]}$ vs T_e is plotted, a straight line should be obtained cutting the axis at Q. This is known as a Kurie plot and the straight line is seen for many transitions.

If the electron is relativistic so that $p_e \approx E_e$, then the preceding form can be integrated for $N(p_e)$ to obtain

$$N_{tot} \propto Q^5. \tag{7.30}$$

This is known as Sargeant's rule. The rate is thus expected to be very strongly dependent on the "Q value."

There are some known corrections to the linear Kurie plot. The first is the Coulomb factor. When an e^+ is created near a nucleus, which like the e^+ is also positively charged, then the e^+ is accelerated as it leaves the nucleus. In a similar manner an e^- is decelerated as it leaves the nuclear environment. This distorts the energy spectrum.

The second correction is that, if m_ν is not zero, then, near the end point of the Kurie plot, there should be a deviation from straight line behavior. This is the basis for some of the most sensitive tests of neutrino mass. A few years ago an ITEP group[64] (Moscow, USSR) reported seeing an effect corresponding to a neutrino mass (see Figure 7.1). The central value was about 30 eV and they had limits from about 15 to 35 eV. This was a very

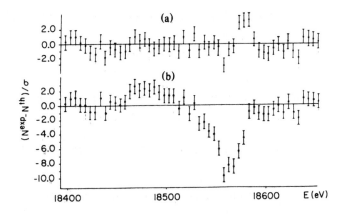

The end point region of the experimental/fitted spectra. a) $M_\nu^2 = 966\ eV^2$, $E_\circ = 18,580.8$ eV, b) $M_\nu^2 = 0\ eV^2$, $E_\circ = 18,576.3$ eV.

Figure 7.1 ITEP neutrino mass data.[64]

serious experiment done with great care. Nonetheless, the experiment is very difficult. Other researchers have not found an effect. The present upper limit[17] for the mass of the electron neutrino is 7.3 eV. This example illustrates some of the problems continually occurring in trying to understand phenomena. Physicists must deal with theoretical ideas that are not always clearly discerned and with data that are not always correct.

Limits on the mass of ν_μ and ν_τ are determined by looking at the decays of the μ and τ in a similar manner to the above. The 1994 Particle Data Group Tables[17] list these limits as $m_{\nu_\mu} < 0.27$ MeV/c^2, and $m_{\nu_\tau} < 31$ MeV/c^2.

7.4 ZERO MASS NEUTRINOS

If neutrinos have zero mass, the Dirac equation can be simplified:

$$(i\slashed{\partial} - m)\psi = 0 \Rightarrow i\slashed{\partial}\psi = 0.$$

Let w_U and w_L be the upper and lower two components, respectively, of either u or v. Then using the explicit forms for the γ-matrices given in Equation 7.5, one has

$$-(\vec{p} \cdot \vec{\sigma})w_L + Ew_U = 0; \quad (\vec{p} \cdot \vec{\sigma})w_U - Ew_L = 0. \tag{7.31}$$

Adding and subtracting these two equations, two decoupled equations

7. Weak Interactions

for the sum and difference of the upper and lower components are obtained:

$$\mp(\vec{p}\cdot\vec{\sigma})(w_U\pm w_L) + E(w_U\pm w_L) = 0. \tag{7.32}$$

Thus, there are two independent two component objects $(w_U\pm w_L)$. Call $\chi = u_U + u_L$ and $\phi = u_U - u_L$:

$$+(\vec{p}\cdot\vec{\sigma})\chi = E\chi; \quad -(\vec{p}\cdot\vec{\sigma})\phi = E\phi. \tag{7.33}$$

The Weyl representation corresponds to writing u as the four-component object

$$u = \begin{pmatrix} \chi \\ \phi \end{pmatrix}.$$

In this representation, also known as the chiral representation, one can take

$$\gamma^k (k=1,2,3) = \begin{pmatrix} 0 & \sigma_k \\ -\sigma_k & 0 \end{pmatrix}; \quad \gamma^0 = \begin{pmatrix} 0 & I \\ I & 0 \end{pmatrix}; \quad \gamma^5 = \begin{pmatrix} -I & 0 \\ 0 & I \end{pmatrix}. \tag{7.34}$$

The previous gamma matrix representation diagonalizes energy, whereas the Weyl representation diagonalizes the helicity (γ^5). γ^5 is called the chirality operator. $(1 \pm \gamma^5)/2$ is a projection operator, projecting out helicity ± 1 in the limit of energy much greater than mass, or for massless particles. (See Exercise 7.4.)

Each of the two equations for χ and ϕ is based on $E^2 = \vec{p}^2$; there is then one positive and one negative energy solution. For the second equation,

$$E\phi = |\vec{p}|\phi = -(\vec{p}\cdot\vec{\sigma})\phi \Rightarrow (\hat{p}\cdot\vec{\sigma})\phi = -\phi.$$

Thus, the positive energy particle has negative helicity, i.e., is left-handed, the spin pointing opposite to the direction of momentum. The negative energy solution, $E = -|\vec{p}|$, has positive helicity, i.e., is right-handed. This corresponds to an anti-neutrino of momentum $-\vec{p}$ with positive helicity. Call these solutions ν_L and $(\bar{\nu})_R$. Similarly the equation for χ describes ν_R and $(\bar{\nu})_L$. Nature might choose either or both of these two component pairs. It is necessary for experiment to decide among the alternatives, and, as has been seen, the choice is ν_L, $(\bar{\nu})_R$.

7.5 Evaluating Matrix Elements; Traces

This section provides the necessary final link so that, in principle at least, you will know how to go from the assumed interaction forms to transition rates and cross sections for first-order electromagnetic and weak processes.

Suppose one has matrix elements with initial and final state fermions. In this section it will be seen how to evaluate $|M|^2$.

Often, the spins of particles initially in an experimental situation are randomly oriented and one averages over possible initial states. It is also often the case that the spins are not detected in the final state. Then, one wishes to sum over the allowed final states and average over the initial states. Assume this to be the case.

Consider a bilinear operator, $\langle \overline{\Psi}_f | O | \Psi_i \rangle$. Leaving out the $e^{i\omega t}$ factors, this becomes $M = \langle \overline{u}(p_f) | O | u(p_i) \rangle$. Since there are two initial possible spin states, the average over initial states is $1/2$ the sum over initial states. Considering, in addition, the sum over final states, the quantity to evaluate is $(1/2)\Sigma_{\text{init}}\Sigma_{\text{final}}|M|^2$:

$$|M|^2 = \overline{u}(p_f)Ou(p_i)[\overline{u}(p_f)Ou(p_i)]^\dagger = \overline{u}(p_f)Ou(p_i)u^\dagger(p_i)O^\dagger \overline{u}^\dagger(p_f). \quad (7.35)$$

Include now the spin indices r and s that were suppressed above; also note that $(\gamma^0)^2 = 1$ and let $O' = \gamma^0 O^\dagger \gamma^0$:

$$|M|^2 = \overline{u}_s(p_f)Ou_r(p_i)\overline{u}_r(p_i)O'u_s(p_f). \quad (7.36)$$

Now average over initial states. It can be shown that (see Exercise 7.2)

$$\Sigma_{r=1}^2 u_r \overline{u}_r \equiv \Lambda_+ = \not{p} + m$$

$$\Sigma_{r=1}^2 v_r \overline{v}_r \equiv \Lambda_- = \not{p} - m$$

One then obtains

$$|M|^2_{\text{ave}} = (1/2)\overline{u}_{s\gamma}(p_f)[O(\not{p}_i + m_i)O']^{\gamma\delta}u_{s\delta}(p_f). \quad (7.37)$$

The indices γ, δ refer to elements in the four-dimensional component space of the spinors. Now sum over final states. Note that from the preceding definitions

$$\Sigma_{s=1}^2 u_{s\delta}(p_f)\overline{u}_{s\gamma}(p_f) = (\Lambda_+)_{\delta\gamma} = (\not{p}_f + m_f)_{\delta\gamma}.$$

Thus

$$|M|^2_{\text{ave+sum}} = (1/2)\Sigma_{\gamma\delta}[O(\Lambda^i_+)O']^{\gamma\delta}(\Lambda^f_+)_{\delta\gamma}$$
$$= (1/2)\text{Tr}[O(\not{p}_i + m_i)\gamma^0 O^\dagger \gamma^0 (\not{p}_f + m_f)]. \qquad (7.38)$$

Here Tr means the trace, which is the sum of the diagonal elements of the matrix. If one does not wish to sum over initial or final states, i.e., if there is polarization, this can be handled by the same methodology. Introduce projection operators, Π, which, when operating on u, give 1 for the desired state and 0 for the others.

Thus, calculating matrix elements involves taking the trace of a large 4×4 matrix. A whole technology has been built up to assist in this task. Many books (e.g., Relativistic Quantum Fields by J.D. Bjorken and S. Drell[65]) have tables of useful relations for traces of various combinations of γ matrices. There are also computer programs that do this. An early and still powerful example of this is the program Schoonship written by M. Veltman.

The opportunity will now be taken to write down the results of the trace calculation for two interesting interactions.

1. Electromagnetic interaction. $M = \bar{u}_l(k')\gamma^\alpha u_l(k)$. The trace is

$$(L^{em})^{\alpha\beta} = \text{Tr}[(\gamma^\beta(\not{k} + m)\gamma^0\gamma^\alpha\gamma^0(\not{k}' + m')]$$
$$= 4[k^\alpha k'^\beta + k'^\alpha k^\beta - g^{\alpha\beta}(k \cdot k' - mm')]. \qquad (7.39)$$

The same trace is obtained if v is used instead of u, since $m \to -m$ and $k \to -k$.

2. Weak Interaction. $M = \bar{u}_l(k')\gamma^\alpha[(1-\gamma^5)/2]u_\nu(k)$. The trace is

$$(L^{weak})^{\alpha\beta} = 2[k^\alpha k'^\beta + k'^\alpha k^\beta - g^{\alpha\beta}(k \cdot k') + i\epsilon^{\alpha\beta\eta\xi}k'_\eta k_\xi]. \qquad (7.40)$$

Here $\epsilon^{\alpha\beta\eta\xi}$ is a completely anti-symmetric tensor, which is 1 for $\alpha\beta\eta\xi = 1230$. Note that the $m_l m_\nu$ term drops out of this expression even if m_ν is not 0.

The same matrix element can serve for $1+2 \to 3+4$ or for $1 \to \bar{2}+3+4$. For the latter case the physical momentum of the outgoing particle 2 is $-k_2$ as shown in Figure 7.2.

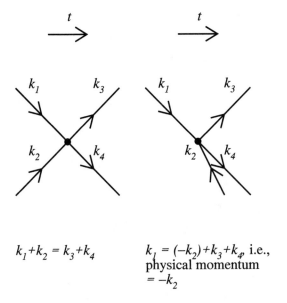

Figure 7.2 Momenta relations in matrix elements.

7.6 $\pi \to l + \bar{\nu}$ DECAY

Is π decay of the same form as nucleon β-decay? (See Figure 7.3.) Could there be a pseudoscalar element (small in nuclear β-decay) that would show up here?

The pion is a pseudoscalar particle and the non-leptonic part of the matrix element is $\langle 0| O_P |\pi\rangle$. Some models for this process are shown in Figure 7.3. There is only one vector available in the term, that of the pion momentum, p_π. Since the matrix element connects the pseudoscalar pion to the scalar vacuum, there can only be A or P terms in the hadronic element. Adding the lepton part, the most general possible matrix element for $\pi \to l + \nu$ is

$$M = \frac{G}{\sqrt{2}} \bar{u}_l [-f_\pi p_{\pi\alpha} \gamma^\alpha \gamma^5 - \sqrt{m_\pi} f_{\pi P} \gamma^5](1-\gamma^5) v_\nu. \quad (7.41)$$

Here f_π and $f_{\pi P}$ are constants. The $1-\gamma^5$ is necessary as the neutrino is left handed. The first term is the axial vector (A) term and the second term is the pseudoscalar (P) term. In the V–A theory, this latter term should be absent. The minus signs and the arbitrary $\sqrt{m_\pi}$ were chosen to agree with definitions in the literature.

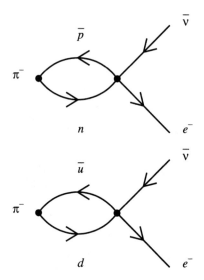

Figure 7.3 Models of pion decay.

Note that $(p_\pi)_\alpha = (p_l)_\alpha + (p_\nu)_\alpha$. Furthermore, from the Dirac equation, $\bar{u}\slashed{p}_l = \bar{u}m_l$ and $\slashed{p}_\nu v_\nu = 0$. Thus

$$M = -\frac{G}{\sqrt{2}}\bar{u}_l[(m_l f_\pi + \sqrt{m_\pi}f_{\pi_P})\gamma^5](1-\gamma^5)v_\nu. \quad (7.42)$$

Also, $\gamma^5(1-\gamma^5) = -(1-\gamma^5)$. Now let

$$d = G(m_l f_\pi + \sqrt{m_\pi}f_{\pi_P}), \quad (7.43)$$

$$M = (1/\sqrt{2})\bar{u}_l[d(1-\gamma^5)]u_\nu. \quad (7.44)$$

The trace of this can now be taken as shown in the previous section. The operator O is $O = 1 - \gamma^5$. The interference term between the 1 and γ^5 term does not contribute. One obtains (see Exercise 7.9)

$$|M|^2 = 4d^2(p_l \cdot p_\nu) = 4d^2[E_l E_\nu - |\vec{p}_l||\vec{p}_\nu|\cos\theta] \quad (7.45)$$

and $\cos\theta = -1$ in the pion rest system.

The transition rate, $\Gamma = 1/\tau$, is given by

$$d\Gamma = \frac{(2\pi)^4}{\Pi_\alpha(2E_\alpha)}|M|^2 \times dP.S._2, \qquad (7.46)$$

where $dP.S._2$ is the two-body phase space which, as noted previously, is given in the pion rest frame by

$$dP.S._2 = \frac{p_l d\Omega_l}{4(2\pi)^6 m_\pi}. \qquad (7.47)$$

The integral over the solid angle just gives 4π. Hence (using $E_\nu = p_\nu = p_l$),

$$\Gamma = \frac{d^2}{2\pi m_\pi^2} p_l^2 E_l \left(1 + \frac{p_l}{E_l}\right). \qquad (7.48)$$

From conservation of energy and momentum it can easily be shown that in the pion rest frame $p_l = m_\pi(1 - m_l^2/m_\pi^2)/2$. Thus,

$$\Gamma = \frac{m_\pi d^2}{8\pi}\left(1 - \frac{m_l^2}{m_\pi^2}\right)^2, \qquad (7.49)$$

where $d = G(m_l f_\pi + \sqrt{m_\pi} f_{\pi P})$.

Consider $R_\pi = \Gamma(\pi \to e\nu)/\Gamma(\pi \to \mu\nu)$. Note that the axial vector term f_π is multiplied by the lepton mass m_l, which is very small for the case of the electron. If there is only an axial vector term, then R_π should be very small, while if there is a pseudoscalar term, R_π should be larger. If there is only an axial vector term, the prediction becomes $R_\pi = 1.2575\times 10^{-4}$. This R_π must still be multiplied by 0.965 to include radiative corrections giving 1.214×10^{-4}. Experimentally it is found that $R_\pi = 1.218 \pm .014 \times 10^{-4}$ in agreement with no pseudoscalar term.[66]

It is easy to see, pictorially, why R_π is small. (See Figure 7.4.) When the π^- decays, the outgoing $\bar\nu$ has spin pointing in its direction of motion (right-handed). V−A makes the e^- left handed with polarization v/c. The pion is spin 0 and conservation of angular momentum forces the e^- to be right-handed. This is possible only to the extent that $v/c < 1$. Since the electron, because of its low mass, moves with v nearly c, its decay mode is strongly inhibited. Experimenters have measured the polarization of the outgoing muon in pion decay. This provides further evidence of the handedness of the neutrino.

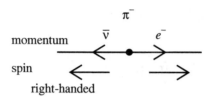

The e^- spin shown is forced by angular momentum conservation, BUT the matrix element wants the e^- to be ⟵ , i.e., to be left-handed with polarization v/c.

Figure 7.4 Lepton polarization in π decay.

The $1-\gamma^5$, which gives the helicity of the neutrino in pion decay, shows that charge conjugation is not conserved in the decay. $C(\pi^+ \to \mu^+ + \nu_L) = \pi^- \to \mu^- + (\bar{\nu})_L$ and P is also needed to give $CP(\pi^+ \to \mu^+ + \nu_L) = \pi^- \to \mu^- + (\bar{\nu})_R$, which is what the $1-\gamma^5$ demands and is what is observed.

The same formalism can be applied to kaon decay. Are strange particle decays also only V and A? Tests can be made for the presence of A and the absence of P. The ratio $R_K = \Gamma(K \to e\nu)/\Gamma(K \to \mu\nu)$ is predicted to be $R_K = 2.5 \times 10^{-5}$ in the absence of P. It is observed to be $R_K = 2.45 \pm 0.11 \times 10^{-5}$.[67] The polarization of the outgoing μ has been measured and found to be consistent with V−A. Thus, strange particle decays may also have V−A form. There is, however, one troubling feature. The absolute rate of the K decays above are about a factor of 14 less than would have predicted from the π decays. This will lead to a new discovery and will be discussed later.

Note that for both the K and π decays, the fact that R agrees so well with theory shows, also, that the weak coupling to muons is the same as the weak coupling to electrons, a silent assumption in the above work. This is known as "μ–e universality."

7.7 FURTHER WEAK INTERACTION RESULTS; SEVERAL NEUTRINOS; CURRENT–CURRENT?

So far it has been shown that in nuclear β-decay the interaction is V−A. The pion interaction tested that A was present and P not present. The kaon decay found the same result in the strangeness-changing two-lepton decays.

Strangeness-changing three-body decays $K^+ \to \pi^0 l^+ \nu_l$ allow one to look for the V and to test against S and T terms. This experiment was first done by a Michigan–Berkeley group with a xenon bubble chamber.[68] The decays were consistent with V and sharply limited the possible contributions of other terms. (A does not contribute to this decay.) Furthermore, comparing the decays with $l = \mu$ to those with $l = e$, compatibility was found with the same coupling to the two leptons, i.e., with μ–e universality for the V coupling as had been found above for the A coupling.

It was suspected that the ν_μ might be unequal to the ν_e because the limit for the decay rate $\mu \to e\gamma$ was very small (B.R. $< 4.9 \times 10^{-11}$).[69] Experiments with neutrino beams showed directly that $\nu_\mu \neq \nu_e$. The pioneering experiment was done at BNL by Lederman, Schwartz, Steinberger, and four coworkers in 1961.[70][71] The three authors specifically noted received the 1988 Nobel Prize for this experiment.

Fifteen GeV protons impinged on a 3 in. thick beryllium target producing many pions. The pions traveled in a 21 m long region during which some of them decayed into $\mu + \nu$. The beam then struck a large 13.5 m thick shielding wall made of steel plates from the armor of an old battleship. This stopped the pions and muons but the neutrinos continued on. (Their cross section is so low they usually go through the entire earth without interacting.) A very few of the neutrinos, however, did interact in the apparatus following the shielding wall. This apparatus was composed of 10 1-ton aluminum plate spark chamber modules. It was triggered by time coincident signals in pairs of scintillators separated by 3/4 in. aluminum plates (to reduce accidental triggerings in a single scintillator). These coincidences were then further required to be in coincidence with the 20 ns beam bursts as determined by a Cherenkov counter in the decay region. This latter requirement reduced the rate due to cosmic rays and to slow neutrons from the proton interactions.

It was observed that only muons were produced, not electrons. $\nu_\mu + N \to N' + \mu$. See Figure 7.5. Thus, muon neutrinos are distinct from electron neutrinos.

Much later tau decay studies led to the conclusion that tau neutrinos are distinct from electron and muon neutrinos. All three neutrino varieties are distinct from one another.

It has been assumed that the weak interaction is essentially a point interaction. It will be seen that this is not true, but the deviations show up only at much higher energies. Evidence for the pointlike coupling at moderate energies came from cross section studies of $\nu_\mu + N \to N' + \mu$, which will be discussed subsequently.

It is also known that ν are distinct from $\bar{\nu}$. If this were not true, then lepton conservation would not be valid and neutrinoless "double β-decay"

Figure 7.5 M. Schwartz standing in front of the Brookhaven National Laboratory experiment confirming that ν_e and ν_μ were distinct particles. *(Photograph courtesy of Brookhaven National Laboratory.)*

could occur, i.e., $N \to N' + e^- + e^-$ with intermediate state but no final state ν's produced. (See the discussion of Majorana neutrinos in Section 7.15). The limits on these decays are many orders of magnitude below the rate expected if ν and $\bar{\nu}$ were not distinct.[72] The expected double β-decay *with* final state neutrinos has been seen in the last few years. Seeing this mode and not seeing the neutrinoless mode adds confidence to the experiments.

These facts lead to the postulate that the interaction can perhaps be written as the product of two currents, a current–current weak interaction. Let $O = \gamma^\alpha(1 - a\gamma^5)/2$. Write

$$j = \langle \overline{\Psi}_p | O | \Psi_n \rangle + \langle \overline{\Psi}_{\nu_\mu} | O | \Psi_\mu \rangle + \langle \overline{\Psi}_{\nu_e} | O | \Psi_e \rangle, \qquad (7.50)$$

$$M \sim \frac{4G}{\sqrt{2}} j_\alpha^\dagger j^\alpha. \qquad (7.51)$$

(The addition of the strangeness-changing term will be discussed subsequently, at which time a more fundamental form of the current involving quarks rather than p and n will be introduced.) It was noted earlier that the Dirac spinors u and v have dimensions $E^{1/2}$. Hence, a current, such as $j_\mu = \bar{u}\gamma^\mu(1/2)(1-\gamma^5)v$, has dimension E. The matrix element for the elementary Fermi theory is $M = (4G/\sqrt{2})j^\dagger j$ as seen previously. This corresponds to two outgoing particles and requires G to have dimension $1/E^2$. G is then quoted by giving the value of GM_p^2, which is dimensionless.

This form has two implications. In the first place all combinations of these couplings must be present, i.e., $pnpn$, $pn\mu\nu_\mu$, $pne\nu_e$, and $\mu\nu_\mu e\nu_e$. Evidence for this last coupling will be examined in the next section.

The second implication is that the normalizations are fixed and the ratios of different processes are predicted. This is then the explanation of μ–e universality. One might, however, expect the strong interactions to renormalize processes in which n and p are involved and this has been seen in the V-1.26A factor in nuclear β-decay.

7.8 Muon Decay

The purely leptonic part of the weak current–current interaction can be examined by considering the decay $\mu^- \to e^- + \bar{\nu}_e + \nu_\mu$. If the current–current picture is correct, then for muon decay, Equations 7.50 and 7.51 yield

$$M = \frac{4G_\mu}{\sqrt{2}} \left(\bar{u}_{\nu_\mu}\gamma^\alpha\left(\frac{1-\gamma^5}{2}\right)u_\mu\right)^\dagger \left(\bar{u}_e\gamma_\alpha\left(\frac{1-\gamma^5}{2}\right)v_{\nu_e}\right)$$
$$= \frac{4G_\mu}{\sqrt{2}}(j_\mu^\dagger)^\alpha(j_e)_\alpha.$$
(7.52)

Note that α is to be summed over. v has been used in the second term as there is a $\bar{\nu}_e$ in the final state. G_μ is used as it may well be different from G for nuclear β-decay. The 1.26 only measures the ratio of possible strong interaction renormalizations of the V and A parts of the np term! Here, the sum over final states and average over initial states gives

$$(1/2)\Sigma|M|^2 = 4G_\mu^2 \text{Tr}[(j_\mu^\dagger)^\alpha(j_\mu)^\beta]\text{Tr}[(j_e^\dagger)_\alpha(j_e)_\beta]$$
$$= 4G_\mu^2(L_\mu^{weak})^{\alpha\beta}(L_e^{weak})_{\alpha\beta},$$
(7.53)

where the last terms are the evaluated traces written out in Section 7.5.

After some algebra this works out to be

$$(1/2)\Sigma|M|^2 = 64G_\mu^2(\mu \cdot \bar{\nu}_e)(\nu_\mu \cdot e) \qquad (7.54)$$
$$\approx 64G_\mu^2(-m_\mu E_{\bar{\nu}_e})(-E_e E_{\nu_\mu}[1 - \cos\theta_{e\nu_\mu}]).$$

It is assumed $p_e \gg m_e$ in the last expression above. Then

$$\Gamma = 1/\tau = \int \frac{(2\pi)^4}{2E_\mu}(1/2)\Sigma|M|^2 dP.S._3, \qquad (7.55)$$

where $dP.S._3$ is the three-body phase space written down earlier:

$$d\Gamma = \frac{(1/2\Sigma|M|^2)}{2m_\mu 8(2\pi)^5 E_e}\left|\frac{p_{\nu_\mu}^3 p_e^2 dp_e d\Omega_e d\Omega_{\nu_\mu}}{p_{\nu_\mu}^2(M_\mu - E_e) + E_{\nu_\mu}\vec{p}_{\nu_\mu}\cdot\vec{p}_e}\right|. \qquad (7.56)$$

After integration over solid angles and some algebra one obtains, letting $x = p_e/(m_\mu/2)$ (see Exercise 7.10)

$$\frac{d\Gamma}{dx} = \frac{8G_\mu^2 m_\mu^5}{2^5(2\pi)^3}x^2\left[1 - \frac{2}{3}x\right]. \qquad (7.57)$$

If the interaction were some general form and not V−A, the square bracket would be replaced with $[2(1 - x + (2/3)\rho(4/3x - 1))]$, where ρ is known as "the Michel ρ parameter" and $\rho = 3/4$ for V−A. As can be seen, ρ is linearly related to the height of the spectrum at the endpoint, $x = 1$, where the square bracket $= (4/9)\rho$. Thus, by carefully measuring the electron spectrum another check of V−A can be obtained. Experimentally ρ is measured to be 3/4 to high accuracy.[73]

Calculation of the leptonic decay of the τ lepton proceeds exactly as the muon calculation. The ρ parameter has been measured for the τ and also found to be 3/4.[74]

It turns out that muon decay cannot measure the interference term between V and A in the muon current, because the interference term drops out in the calculation. However, this term has been measured in the inverse process, $\nu_\mu + e^- \to \mu^- + \nu_e$, and it is found to be consistent with V−A.[75]

Is the lepton quantum number additive or multiplicative? "Additive" means that $\sum L_\mu$ and $\sum L_e$ lepton numbers are conserved separately. "Multiplicative" means that $\sum(L_\mu + L_e)$ is conserved and $(-1)^{L_e}$ is conserved. $\mu^+ \to e^+ + \nu_e + \bar{\nu}_\mu$ is allowed by either law. $\mu^+ \to e^+ + \bar{\nu}_e + \nu_\mu$

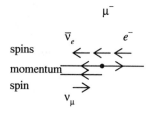

Figure 7.6 Lepton polarization in μ decay.

is allowed by the multiplicative law only. This was examined in an experiment at the Los Alamos Meson Factory.[76] If $\bar{\nu}_e$ were formed, they could be detected via $\bar{\nu}_e + p \to e^+ + n$. The second process was found to be less than 10% of the expected process. Similar results were obtained at CERN from a comparison of $\bar{\nu}_\mu + e^- \to \mu^- + \bar{\nu}_e$ (allowed only under the multiplicative law) with $\nu_\mu + e^- \to \mu^- + \nu_e$, using the polarization of the muon to distinguish the two reactions.[77]

Is ν_μ left-handed as is ν_e? Examine Figure 7.6. In muon decay, high-energy electrons are emitted with both neutrinos recoiling opposite them. From the figure it is then seen that high-energy electrons should go in the direction opposite to the μ spin for μ^- and positrons should go in the direction with the μ spin for μ^+. The pion decay can be used to get polarized muons as previously seen. If the μ^+ is precessed in a magnetic field, a correlation can be obtained between the e^+ direction and time. The results indicate ν_μ is indeed left-handed. Furthermore, this measurement of muon spin precession is a marvelous way of measuring the muon magnetic moment. Muons are trapped in a storage ring and the oscillations measured over many cycles.[78] (This is a development of the experiment for measuring the magnetic moment of the electron by precessing polarized electrons and measuring the final spin direction. This experiment was invented by H.R. Crane of Michigan and carried out with increasing accuracy for many years.[79]) The muon experiment was carried out at CERN. The muon and the electron experiments have yielded extremely accurate values of the magnetic moments (muon and electron "g-factors") of these particles. $g - 2$ is now sufficiently well known for muons to examine sixth-order matrix element terms in the theory, i.e., at the few parts in 10^6 level. At this level the hadronic contributions to $g - 2$ are important. The electron $g - 2$ is known to about 50 times better accuracy. The agreement of theory and experiment to this accuracy is a scientific triumph.

If the preceding differential muon decay rate is integrated, it is found

$$\Gamma = 1/\tau = \frac{G_\mu^2 m_\mu^5}{192\pi^3}\left[1 - \frac{\alpha}{2\pi}\left(\pi^2 - \frac{25}{4}\right)\right], \tag{7.58}$$

where the last term in the square brackets is a first-order radiative correction term, i.e., a term due to radiating soft photons during the decay. From measuring the decay, one finds that $G_\mu M_p^2 \approx 1.02 \times 10^{-5}$, where M_p is the proton mass. This result has almost the same value as in the hadron case. This has far reaching implications which will be discussed in the next section.

7.9 CONSERVED VECTOR CURRENT (CVC)

Why does C_V from O^{14} decay agree so well with G_μ from μ decay?

$$\frac{G_\mu - C_V}{G_\mu} \approx 0.025. \tag{7.59}$$

This includes the radiative correction to the nuclear decay constant as well as the approximately 1/2% radiative correction for muon decay indicated above. When Cabibbo theory will be discussed shortly, it will be seen that even this small difference vanishes. Why is the agreement so good? Don't the strong interactions change $\langle \overline{\Psi}_p | O | \Psi_n \rangle$? Perhaps a conservation law is implied.[80]

Proceed in analogy with the electromagnetic current. There it was found that conservation of the electric current implied conservation of electric charge. That is exactly what is wanted here. Regardless of the clouds of virtual particles around the n or p, the net weak charge is required to be constant. Therefore, assume there is a conserved vector current (CVC) for the weak interaction:

$$\frac{\partial (j^V)^\mu}{\partial x^\mu} = 0, \tag{7.60}$$

$$(J^V)^0 = \int_{(x_0 = t)} dx^3 (j^V)^0 (x) = \text{constant}. \tag{7.61}$$

Note this cannot be applied to both C_V and C_A. It has been seen that C_A/C_V is modified by the strong interactions by 27%. However, it is true that $\partial (j^A)^\mu / \partial x^\mu$ is small, i.e., it is $\propto m_\pi$. This is known as partial conservation of the axial vector current (PCAC). There is a large literature concerning predictions that one can make because of PCAC.

In the limit that the pion mass is taken as zero and the axial vector current is conserved, one can show, using also CVC and the Cabibbo theory to be discussed in the following section, that one can predict $C_A(0)/C_V(0) = |f_\pi g_{\pi N}|/M_N \approx 1.31$ in reasonably good agreement with experiment. Here f_π is the pion coupling constant introduced in Section 7.6, and $g_{\pi N}$ is the pion–nucleon coupling constant. This relation was important historically and is known as the Goldberger–Treiman relation.[81]

7.9 Conserved Vector Current (CVC)

Concentrate now on CVC. Two more facets can be added to it. These are really separate assumptions, but are usually grouped together under CVC.

Suppose that j^V has a definite strong interaction symmetry; it transforms under I-spin as an I-spin vector. Thus suppose it has the symmetry of the isospin raising operator T^+. If this is true, then the same result should be obtained for matrix elements of j^V for any two processes a and b, if the initial state of a has the same I-spin symmetry as the initial state of b, and if the final state of a has the same I-spin symmetry as the final state of b.

The pion has a rare decay mode, $\pi^+ \to \pi^0 + e^+ + \nu$. With the above assumption, the rate of this process should be calculable, given the nuclear β-decay matrix elements. A predicted value is obtained for the branching ratio of 1.05×10^{-8} which agrees well with the experimental value of $1.025 \pm 0.034 \times 10^{-8}$.[82]

Finally add a very bold assumption. Assume that the weak vector current j^V is part of the same current as the electromagnetic current. The meaning of this has to be explained very carefully. The electromagnetic current is spatially a pure vector current. However, as a function of I-spin, it is a mixture of I-spin vector and I-spin scalar. In order to give 1 for a proton and 0 for a neutron, it must be $\propto (1 + \tau_3)/2$. The weak current is a mixture of V and A spatially but is a pure I-spin vector since it is a charged current. This assumption relates the spatial-vector, I-spin vector parts of the two currents. It says that, for this part, the electromagnetic current is a third component (neutral) and the weak current is a charged component of the same current. This implies a deep connection between the electromagnetic and the weak interactions. Is there any evidence for this assumption?

If the weak current is just proportional to the electromagnetic current, then there should be an analogue in the weak current of magnetism in the electromagnetic current, a "weak magnetism."[83] It will be shown in Chapter 8 that for electromagnetic scattering of protons and neutrons, the most general form for $(j^{em})^\mu$ is

$$(j^{em})^\mu \propto \bar{u}(p')[\gamma^\mu F_1^V(q^2) + i\sigma^{\mu\nu}q_\nu \kappa F_2^V(q^2)/2M]u(p),$$

where $\sigma^{\mu\nu} = (i/2)[\gamma^\mu, \gamma^\nu]$, and F_1, F_2 are known as the electric and magnetic form factors. They are functions only of q^2, where q^2 is the square of the momentum of the emitted virtual photon (Figure 7.7). κ is a constant which for the electromagnetic case is the anomalous magnetic moment (= 1.79 for the proton). What should it be for the weak case? The electromagnetic current contains an I-spin scalar and an I-spin vector part. The weak current has only the I-spin vector part, and the same result must be

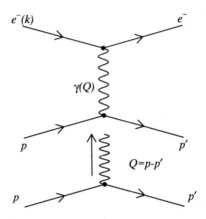

Figure 7.7 $pp\gamma$ vertex.

obtained as with the vector part of the electromagnetic current. To interact with the I-spin vector part of the electromagnetic current, the bra and ket p and n vectors must also form an I-spin vector, so that the result is a scalar as required. The bra part of p and n, i.e., $p^\dagger$ and $n^\dagger$, transform according to the conjugate representation, as do the $\bar{p}$ and $\bar{n}$ (see Section 4.15). The vector combination for them is $p^\dagger p - n^\dagger n$. Hence, for the weak case, κ should be taken as the difference between the anomalous magnetic moments of the proton and neutron ($= 4.7$).

As noted above, conservation of charge says that the integral over all space of j_0 is constant. When Fourier transforms are taken to go from space to momentum space, a restriction on the integral over all space becomes a restriction on the zero frequency part of the transform, i.e., on $F_1(0)$ and $F_2(0)$. In fact it turns out that $F_1(0) = 1$ and $F_2(0) = 1$ for the electomagnetic case.

If the electromagnetic and weak currents are different I-spin states of the same current, then the weak and electromagnetic currents should have the same form factors with the same normalizations. Nuclear β-decay is at low q^2. The q^2 dependence of the first term is hard to see there. However, one can look for the second term, the weak magnetism term, in sufficiently forbidden decays, where it might compete with the first term. The coefficient of the magnetic term for the weak interactions is reasonably large, since $\mu_p - \mu_n = 4.7$.

Consider the I-spin triplet of decays shown in Figure 7.8. The carbon decay can be used for normalization and a search made for the weak magnetism terms in the B^{12} and N^{12} decays. The weak magnetism term should have opposite signs in these two decays. These terms have, in fact, been detected.[84][85]

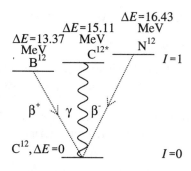

Figure 7.8 B-C-N isospin triplet of nuclei.

As will be seen later the form factor agreement for electromagnetic and weak interactions has also been checked by comparing high-energy electromagnetic eN interactions and high-energy νN interactions. This has been done by examining "inclusive" reactions, comparing $\nu N \to \mu N'$ with $eN \to eN'$. It has also been done using "exclusive reactions." Comparison of $\nu N \to \mu \Delta(1232)$ with $eN \to e\Delta(1232)$ checks PCAC,[86][87] and comparison of weak and electromagnetic diffractive production of the ρ meson checks CVC.[88][89][90]

Decays from B^{12} and N^{12} aligned nuclei can also be used to look for the presence of Second Class Currents. These are V and A currents with G-parity opposite from that of the standard currents. Recall that $G = Ce^{i\pi I_2}$. Under rotations about the I_2 axis, all the charged current terms are I-spin vectors and transform the same. To see differences in how terms transform under G, it is then sufficient to look at the transformation properties under C. Under C it turns out that a Dirac spinor u transforms to $i\gamma^2\gamma^0\overline{u}^T$. Thus, if O is a 4×4 matrix operator, a bilinear form changes as follows:

$$\overline{u}_i O u_j \to \overline{u}_j (\gamma^0 \gamma^2 O^T \gamma^2 \gamma^0) u_i.$$

Hence, for example, $\overline{u}_i \gamma^\mu u_j$ and $\overline{u}_i \sigma^{\mu\nu} u_j$ transform into plus themselves (with i and j reversed), while the second class term $\overline{u}_i u_j$ transforms into minus itself. (Remember the -1 from anti-commuting $\overline{u}_i$ and u_j.) The axial vector terms can be similarly classified.

No evidence is found[91] for second class currents. Besides nuclear decays, observation of $\tau \to \nu \pi \eta$ would be evidence for second class currents (or I-spin violation in the decay), but no evidence is seen for this decay at about the 1% level.[17]

7.10 STRANGENESS-CHANGING WEAK DECAYS

7.10 a. Cabibbo Hypothesis

It was noted previously that the charged K decays into $l\nu$ and into $\pi l\nu$ are compatible with the V–A current–current picture except that the overall rate of these decays is too low by about a factor of 14. The semi-leptonic decays of the hyperons also fit the same pattern. It appears the weak current needs to be written in the awkward form

$$j \propto \langle \bar{p}|\, O\, |n\rangle + \langle \bar{\nu}_\mu|\, O\, |\mu\rangle + \langle \bar{\nu}_e|\, O\, |e\rangle + \sqrt{1/14}\, \langle \bar{p}|\, O\, |\Lambda\rangle, \qquad (7.62)$$

where $O = \gamma^\alpha(1 - a\gamma^5)/2$. Shift now to quark notation and set $a = 1$. Then

$$j = \langle \bar{u}|\, O\, |d\rangle + \langle \bar{\nu}_\mu|\, O\, |\mu\rangle + \langle \bar{\nu}_e|\, O\, |e\rangle + \sqrt{1/14}\, \langle \bar{u}|\, O\, |s\rangle. \qquad (7.63)$$

Note that $\bar{u}$ appears twice. N. Cabibbo[92] hypothesized that perhaps the d that appears in this decay current should be d', a mixture of d and s. The eigenstates for the strong interaction, d and s, may not be the eigenstates for the weak interaction! Let

$$d' = d\cos\theta_c + s\sin\theta_c, \qquad (7.64)$$

$$j = \langle \bar{u}|\, O\, |d'\rangle + \langle \bar{\nu}_\mu|\, O\, |\mu\rangle + \langle \bar{\nu}_e|\, O\, |e\rangle, \qquad (7.65)$$

$$\langle \bar{u}|\, O\, |d'\rangle = \langle \bar{u}|\, O\, |d\rangle \cos\theta_c + \langle \bar{u}|\, O\, |s\rangle \sin\theta_c. \qquad (7.66)$$

The matrix element for this interaction would then be

$$M \sim \frac{4G}{\sqrt{2}}(\langle \bar{u}|\, O\, |d\rangle \cos\theta_c + \langle \bar{u}|\, O\, |s\rangle \sin\theta_c)^\dagger \langle \bar{\nu}_e|\, O\, |e\rangle. \qquad (7.67)$$

Thus, $n \to pe\nu$ corresponds to $d \to u$ and would have a $\cos\theta_c$ in the matrix element, while $\Lambda \to pe\nu$ corresponds to $s \to u$ and would have a $\sin\theta_c$ in the matrix element. The value of this angle, the Cabibbo angle, can be extracted from the ratio of K to π decays into $l\nu$. One obtains $\theta_c = 0.26 \pm 0.001$. So far this is just parametrization. Does this tie into any other experimental facts?

The K or π decays into $l\nu$ only depend on the axial vector (A) part of the current. An independent check for the vector (V) part can be made by looking at the semi-leptonic weak decays. $\theta_c = 0.25 \pm 0.01$ is obtained. This, at least, shows the V−A current–current picture works well, but does not yet test the Cabibbo hypothesis.

If nuclear β-decay is considered, it is noted that there is now a $\cos^2\theta_c$ term in the rate. For the muon decay there is no such term. It was noted before that there is a 2.5% discrepancy between the weak interaction constant from O^{14} β-decay and from μ-decay. The Cabibbo hypothesis now provides an understanding of even this small discrepancy. This tests the vector part of the matrix element. From this comparison $\theta_c = 0.21 \pm 0.025$ is obtained. There are some known small radiative corrections to these first-order determinations. These corrections arise from emission of soft (low-energy) γ rays during the decay. Including these corrections it is found that these different determinations of the Cabibbo angle are in excellent agreement (within $0.05 \pm 0.27\%$).

For strangeness-changing decays, an s quark changes to a u quark. Thus, $\Delta S = \Delta Q$ for the hadrons. This $\Delta S = \Delta Q$ rule allows $K^0 \to \pi^- e^+ \nu$, but prohibits $K^0 \to \pi^+ e^- \nu$. This is observed in practice. In fact, as the K^0 and $\overline{K}^0$ oscillate, oscillations in the charge of the π in the leptonic channel are observed. Oscillation in these leptonic modes was very important in the determination of the $K_L^0 - K_S^0$ mass difference.

For the charm-changing current, $\langle \bar{c}|O|s\rangle$, similar considerations apply. $D^+ \to K^- \pi^+ \pi^+$ (branching ratio= 0.08) is allowed and $D^+ \to K^+ \pi^+ \pi^-$ (branching ratio $< 4 \times 10^{-3}$) is forbidden.

7.10 b. Non-Leptonic Weak Interactions

First, consider non-leptonic weak interactions that do not change strangeness, ($\Delta S = 0$). For these interactions one has the product of two $I = 1$ currents $(\bar{u}d)$, and the isospin of the final state can differ from that of the initial state by $\Delta I = 0, 1, 2$.

For interactions that change strangeness one has the product of an $I = 1$ current $(\bar{u}d)$ and an $I = 1/2$ current $(\bar{u}s)$. Here one can have $\Delta I = 1/2, 3/2$. In practice the $\Delta I = 1/2$ part dominates at about the 95% level.

There is not yet a simple explanation of this, although there are a number of rather complicated plausibility arguments for it. In fact there are other peculiar regularities that occur in non-leptonic hyperon decays (e.g., Σ triangle) which are not explained well in the literature.

A dramatic example of this $\Delta I = 1/2$ rule is seen in the decay of the K_s^0 into two pions. For K_s^0, both $2\pi^0$ and $\pi^+\pi^-$ are allowed and the decay goes quickly. The lifetime is $0.8922 \pm 0.0020 \times 10^{-10}$ s,[17] while for $K^+ \to \pi^+\pi^0$, the decay is not allowed and the decay occurs about a factor of 500 slower than the K_s^0 decay. The K^+ decay is not allowed, because the net pion state must be symmetric. The state is S-wave and therefore the I-spin state must be symmetric, implying even I-spin. Since the two pion state is charged, it cannot be $I = 0$. $I = 2$ means that $\Delta I = 3/2$, as the kaon has $I = 1/2$.

As another example of its use consider the non-leptonic decay of the Λ hyperon. The Λ can decay into $n\pi^0$ or into $p\pi^-$. The final-state particles are an I=1 and an $I = 1/2$ particle. The initial state is an $I = 0$ state. If $\Delta I = 1/2$, then the final state is $I = 1/2, I_3 = -1/2$. From using the Clebsch–Gordan tables it is seen that (labels are I_3)

$$(-1/2) = \sqrt{1/3}(0, -1/2) + \sqrt{2/3}(-1, +1/2)$$
$$= \sqrt{1/3}(\pi^0, n) + \sqrt{2/3}(\pi^-, p).$$

Hence the ratio $\Gamma(\Lambda \to n\pi^0)/[\Gamma(\Lambda \to n\pi^0) + \Gamma(\Lambda \to p\pi^-)] = 1/3$. Because of the slightly different masses of π^-, π^0 and n, p, there are differences in the phase spaces involved that change this prediction to $1/3 \times 1.036 = 0.345$. Experimentally the ratio is 0.358 ± 0.005.[93]

Note that $\Delta I = 3/2$ decays are not suppressed in D decays.

7.11 NEUTRAL CURRENTS

The weak current described so far is a charged current, the initial- and final-state particles having different electric charge, e.g., $\langle \overline{u}| O |d\rangle$. Is there a neutral weak current?

There is good reason to argue against neutral currents in kaon decays. The ratio[17] of rates $\Gamma(K^+ \to \pi^+\nu\overline{\nu})/\Gamma(K^+ \to \pi^0\mu^+\nu) < 3.4 \times 10^{-8}$. The branching ratio for $K_L^0 \to \mu^+\mu^-$ [94] is only $6.3 \pm 1.1 \times 10^{-9}$, which, one might argue, could be due to higher-order interactions. It was argued earlier that a single photon intermediate state could not connect to a spin 0 meson. However, the decay $K^0 \to 2\gamma \to \mu^+\mu^-$ is not prohibited, but is down by α^4. Estimates place it near the present measurement.

Furthermore, the mass difference between K_S^0 and K_L^0 is believed to be due to Feynman diagrams of the form shown in Figure 7.9. These diagrams are second order in the weak interactions. If there were neutral currents, the K_S^0 and K_L^0 might be connected by first-order transitions and the mass difference would be much larger than the one obtained from examining the K^0-$\overline{K}^0$ oscillations.

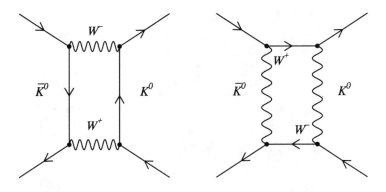

Figure 7.9 Diagrams generating $K_S^0 - K_L^0$ mass difference.

It might therefore seem surprising that neutral currents have in fact been seen. They were discovered in 1973 in an experiment at CERN using a high-energy neutrino beam incident on a large bubble chamber.[95][96] The CERN group measured the ratio NC/CC $= \Gamma(\nu_\mu + N \to \nu_\mu + X)/\Gamma(\nu_\mu + N \to \mu^- + X') \approx 1/4$. Thus, the neutral current interactions are of strength similar to that of the charged current interactions. Figure 7.10 shows neutral current interactions seen at CERN.

Furthermore, neutral currents were even predicted in advance. How can this be? The model was suggested by Glashow, Illiopolis, and Maiani (GIM)[97] in 1970, following an earlier suggestion of Hara, and is known as the GIM mechanism. This group was attempting to understand the Cabibbo mechanism better. They suggested that, perhaps, instead of a triplet of quarks (u, d, s), there were two doublets of quarks (u, d) and (c, s). This suggestion thus required a new quark that the GIM group named charm. This prediction was made before the experimental discovery of charmed particles!

Suppose there are two doublets and the Cabibbo angle mixes the d and s quarks. (This is the most general form of mixing. If the u and c were mixed the same as the d and s, the reactions would be the same as no mixture, corresponding to an orthogonal transformation of the current.) Let

$$d' = d\cos\theta_c + s\sin\theta_c; \quad s' = s\cos\theta_c - d\sin\theta_c. \tag{7.68}$$

Thus, d' and s' are an orthogonal transform of d and s. The hadronic parts of the charged and neutral currents are

$$j = \langle \bar{u}| O |d'\rangle + \langle \bar{c}| O |s'\rangle \text{ (charged)}, \tag{7.69}$$

$$j = \langle \bar{d'} | O' | d'\rangle + \langle \bar{s'}|O'|s'\rangle + \langle \bar{u}| O' |u\rangle + \langle \bar{c}| O' |c\rangle \text{ (neutral)}. \tag{7.70}$$

182   7. Weak Interactions

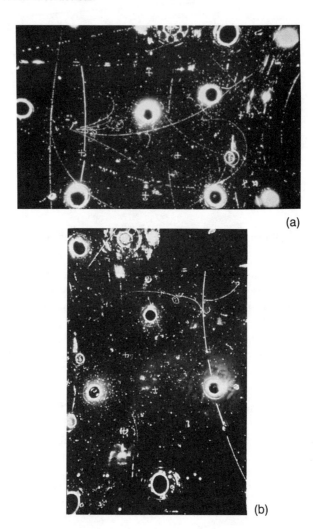

Figure 7.10 First observation of neutral current interactions seen in the Gargamelle heavy liquid bubble chamber at CERN. a) $\nu N \to \nu X$. b) $\nu e \to \nu e$. *(Photograph courtesy of the CERN photographic service.)*

7.11 Neutral Currents

How are the problems concerning kaon decay avoided? The answer is that in the neutral current written above, the strangeness-changing part vanishes! Consider just the quark part. From $\langle \bar{d}' | O' | d' \rangle$ one obtains $\bar{d}d \cos^2 \theta_c + (\bar{d}s + \bar{s}d) \cos \theta_c \sin \theta_c + \bar{s}s \sin^2 \theta_c$. From $\langle \bar{s}' | O' | s' \rangle$ one obtains $\bar{d}d \sin^2 \theta_c - (\bar{d}s + \bar{s}d) \cos \theta_c \sin \theta_c + \bar{s}s \cos^2 \theta_c$. The sum is $\bar{d}d + \bar{s}s$, and the strangeness-changing part has cancelled. There exist no strangeness-changing neutral currents. Similarly there exist no charm-changing neutral currents. This latter prediction was first verified in the mid-1970s in an $\bar{\nu}$ experiment using the Fermilab 15 ft bubble chamber.[98] Consider $\bar{\nu}_\mu + N \to \bar{\nu}_\mu + C + X$. The C (any charmed particle) often decays by $C \to e^+ \nu_e X'$. The event would appear as an event with a final-state e^+ and no μ^+. The limit set was that this was less than 0.87×10^{-3} of the rate for $\bar{\nu}_\mu + N \to \mu^+ + X$. There now exist a number of limits on decays of mesons with b and c quarks via neutral modes. The branching ratios for $b \to \mu^+ \mu^- X$ and $c \to \mu^+ \mu^- X$ are both known to be less than 1%.[99] [100]

If the neutral weak interaction is a current–current coupling, then, as with the charged current, all terms must occur. The $\langle \bar{s} | O' | s \rangle$ term has been seen in bubble chamber neutrino interaction experiments.[101] The $\langle \bar{e} | O' | e \rangle$ term has been seen in a SLAC polarized eD scattering experiment. This beautiful experiment[102] measured the electromagnetic-weak interference (about a 10^{-5} effect) in the scattering of polarized electrons from deuterons.

There is evidence for μ–e universality in the neutral currents. The ν_e neutral current constant was known to be the same as the ν_μ constant within a factor of about 1.4 before LEP started operating. The two constants are now known to be the same to better than 1% as will be discussed in Chapter 13.

Is the operator O' in the neutral current the same as the V–A operator, $\gamma^\alpha(1-\gamma^5)/2$, which appears in the charged current? The surprising answer is that it is a mixture of V and A, but not simply V–A. The fact that it is not V–A and the actual mix provide powerful clues leading toward the development of the Standard Model. This will be discussed in detail in Chapter 10.

The GIM group was even able to estimate an upper limit to the mass of the charmed quark. Because the mass of the s and c quarks are different, the cancellation of these terms is not quite perfect. If the mass of the charmed quark were too large, then in the detailed calculations the kaon difficulties would reappear. They were able to set upper limits on the mass in the 2–2.5 GeV/c^2 mass range. This was truly a remarkable development and an important step in developing the present Standard Model. S. Glashow shared the 1979 Nobel Prize with S. Weinberg and A. Salam for the development of the Standard Model.

7.12 GENERALIZED CABIBBO THEORY: THE CABIBBO–KOBAYASHI–MASKAWA MATRIX

It is now known that there are three families of quark pairs. What kind of general mixing might there be between them for the weak current? This was worked out by Kobayashi and Maskawa[103] before it was certain there even a second family, i.e., before charm was experimentally discovered. The mixing can be given by

$$J^+_{weak} = (\bar{u} \quad \bar{c} \quad \bar{t}) \, O \begin{pmatrix} V_{ud} & V_{us} & V_{ub} \\ V_{cd} & V_{cs} & V_{cb} \\ V_{td} & V_{ts} & V_{tb} \end{pmatrix} \begin{pmatrix} d \\ s \\ b \end{pmatrix}, \qquad (7.71)$$

with $O = \gamma^\mu (1 - \gamma^5)/2$. The diagonalization of each set of quarks occurs through a unitary transformation. Hence, the matrix V is unitary as it is the product of two unitary transforms. Kobayashi and Maskawa found the most general form for the unitary matrix V and it is called the Kobayashi–Maskawa matrix (or, more often, the Cabibbo–Kobayashi–Maskawa (CKM) matrix).

Why did Kobayashi and Maskawa introduce the preceding matrix before the evidence for a second family appeared? The answer is that they were searching for a mechanism for CP violation. Recall from Section 4.6 that a Hamiltonian H is invariant under time reversal if and only if H is real. Thus to get T violation (and hence CP violation if CPT is conserved) a complex phase must be introduced.

Consider now the unitary V matrix in n dimensions corresponding to n generations of quarks ($2n$ quarks). There are n^2 terms in the matrix and each has a real and an imaginary part giving $2n^2$ parameters. The unitary conditions ($V_{ij} V^*_{kj} = \delta_{ik}$) provide n^2 constraints. Furthermore, the phases of each of the $2n$ quarks can be set arbitrarily, although one corresponds only to an ignorable overall phase. Then there are $2n^2 - n^2 - (2n - 1) = n^2 + 1 - 2n$ parameters left.

For two generations of quarks there is only one free parameter, which may be taken as the ratio of the magnitudes of V_{11} and V_{12}. There is no room for an arbitrary complex phase.

For three generations of quarks, there are four parameters. If the V matrix were real, then there would be $n^2 - (n^2 - n(n-1)/2) = n(n-1)/2$ parameters using the constraints $V_{ij} V_{kj} = \delta_{ik}$, and noting that for $i \neq k$, the ik term is the same as the ki term. Thus, a real matrix has three parameters for $n = 3$, and the general unitary matrix has one parameter left to introduce a complex phase, i.e., to introduce CP violation. The parameterization is taken as three real rotations with one phase rotation in the middle. All CP violation measurements up to now are consistent with CP violation being solely due to this phase.

7.12 Generalized Cabibbo Theory

Let $c_{ij} = \cos\theta_{ij}$, and $s_{ij} = \sin\theta_{ij}$, where $i,j = 1,2,3$. Then in one parametrization[17]

$$V = \begin{pmatrix} c_{12}c_{13} & s_{12}c_{13} & s_{13}e^{-i\delta_{13}} \\ -s_{12}c_{23} - c_{12}s_{23}s_{13}e^{i\delta_{13}} & c_{12}c_{23} - s_{12}s_{23}s_{13}e^{i\delta_{13}} & s_{23}c_{13} \\ s_{12}s_{23} - c_{12}c_{23}s_{13}e^{i\delta_{13}} & -c_{12}s_{23} - s_{12}c_{23}s_{13}e^{i\delta_{13}} & c_{23}c_{13} \end{pmatrix}. \tag{7.72}$$

As indicated above, there are now three mixing angles and a phase angle, δ_{13}, not just $\theta_c \approx \theta_{12}$.

What is known about the three weak mixing angles? The evidence indicates that θ_{23} and θ_{13} are both as small or smaller than θ_{12}. Thus the favored decay chain for b quark decay is $b \to c \to s \to u$ and this is observed for b decays by the groups at the Cornell storage ring, CESR, and by the Argus group at DESY.[104]

$|V_{ud}|$ is found by comparing nuclear beta decay to muon decay. $|V_{us}|$ is obtained from analysis of K_{e3} decays (taking account of the isospin violation between K_{e3}^+ and K_{e3}^0) and from hyperon beta decay. (K_{e3}^+ decay is $K \to \pi^0 e^+ \nu_e$, etc.) To find $|V_{cd}|$ one examines neutrino and anti-neutrino production of charm off of valence d quarks, using dimuon production to signal the presence of charm. $|V_{cs}|$ depends on assumptions about the strange quark density in the sea of quark–anti-quark pairs in a nucleon. One then uses D_{e3} analogously to the use of K_{e3} decay for $|V_{us}|$.

The ratio $|V_{ub}/V_{cb}|$ is obtained from the semi-leptonic decay of B mesons produced on the $\Upsilon(4S)$ $b\bar{b}$ resonance by measuring the lepton energy spectrum above the endpoint of the $b \to c l \nu$ spectrum. The magnitude of $|V_{cb}|$ itself can be found if the semi-leptonic bottom hadron partial width is assumed to be the same as that of a b quark decaying with the standard $V - A$ interaction.

As indicated above, the mode $b \to u$ has been seen at a low level. The leptonic decays are seen, but the purely hadronic ones are not yet seen.

The diagonal elements dominate the matrix. Present measurements[17,105] are $|V_{ud}| = 0.9734 \pm 0.0007$, $|V_{cs}| = 0.98 \pm 0.02$, $|V_{td}| \approx 1$. Thus, to an excellent approximation, $s_{12} = |V_{us}|$, $s_{13} = |V_{ub}|$, and $s_{23} = |V_{cb}|$.

The couplings between adjacent generations give $|V_{cd}| = 0.204 \pm 0.017$, $|V_{us}| = 0.226 \pm 0.003$, while $|V_{cb}| = (42 \pm 5) \times 10^{-3}$, $|V_{ts}| \approx 0.04$.

The couplings skipping a generation give $|V_{ub}|/|V_{cb}| = 0.08 \pm 0.02$. One estimate, based on these data, is that $|V_{ub}| = 0.003 \pm 0.001$ including theoretical uncertainties.[106] $|V_{td}| = 0.004 - 0.005$.

Charmed decays follow the preceding pattern. For example, various branching fractions for the charmed D^+ meson are[100]

$$D^+ \to K^-\pi^+\pi^+ (3.9 \pm 1\%), \ c \to s, \ \text{allowed},$$

$$D^+ \to K^+\pi^+\pi^- (<0.2\%), \ c \to \bar{s}, \ \text{forbidden},$$

$$D^+ \to \pi^-\pi^+\pi^+ (<0.31\%), \ c \to d, \ \text{or} \ c \to u, \ \text{suppressed}.$$

Oscillations in strangeness in the neutral kaon system have been discussed previously. One can ask whether an analogous oscillation can occur in neutral states of charmed (D) or bottom (B) mesons. Experimentally no mixing is seen in the $D^0, (\bar{u}c)$ meson, and the ratio $\Gamma(D^0 \to \overline{D}^0 \to \mu^- X)/\Gamma(D^0 \to \mu^+ X) < 0.044$.[107] [108]

For B mesons, mixing was seen by the Argus group and was confirmed at CESR and at LEP.[109] [110] At each of these laboratories, one produces $\overline{B}B$ pairs via $e^+e^- \to \overline{B}BX$. If both B's decay semi-leptonically, and through the favored decay chain $b \to c$, then two reasonably high momentum, oppositely charged leptons occur. If mixing occurs and $B \to \overline{B}$, one obtains events with two same charge high momentum leptons. (See Chapter 13 for experimental results.) The ratio of opposite sign to same sign charge leptons enables a measurement to be made of the amount of mixing. Experimentally, the leptons from the decaying c-quark mesons in the decay chain tend to have lower momentum and, by experimental cuts and theoretical corrections, one can obtain the desired ratio. CP violation is expected to occur in the decays of the B ($b\bar{d}$) and B_s ($b\bar{s}$). This will then be another system in which CP violation can be examined. Unfortunately, millions of decays are required. New accelerators are planned and new large detectors are planned at existing accelerators and $\bar{p}p$ facilities specifically to examine these modes.

The unitarity of the CKM matrix gives constraints. If $UU^\dagger = 1$, the off-diagonal elements of the product must be 0. This gives constraints such as $V_{ud}(V_{ub})^* + V_{cd}(V_{cb})^* + V_{td}(V_{tb})^* = 0$. Plotted in the complex plane, these three terms must form a triangle. From measuring various decay modes, the angles, and hence relative phases, of these terms can be obtained. It will then be possible to examine if CP violation occurs only through the CKM phase, or whether other mechanisms are operative.

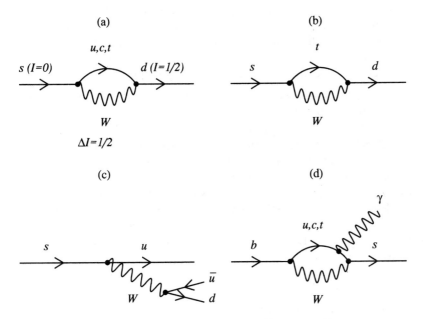

Figure 7.11 Penguin and related Feynman diagrams.

7.13 Penguin Diagrams; Applications to $\Delta I = 1/2$ and to CP Violation

Penguin diagrams are the kind of Feynman diagram illustrated in Figure 7.11a, which shows an s quark going into a d quark. This gives an apparent-strangeness-changing neutral current although in fact only the charged W is involved. The diagrams are named penguins because the original drawing shape was thought to resemble a penguin. In the diagram shown, the $I = 0$ s quark changes into an $I = 1/2$ d quark, which is a change of I of $\Delta I = 1/2$. This diagram was originally proposed as a possible explanation of the $\Delta I = 1/2$ rule in K decays.[111]

It was soon realized that this diagram could also be a possible form of CP violation in K decays according to Figure 7.11b. In the discussion of the CKM matrix it was noted that three families were needed to get a CP violating phase. Unless all three families are present in a given Feynman diagram no CP-violating phase can appear. Here, because of the intermediate t quark, all three families are present and a CP-violating phase can appear. This diagram only contributes to a final $I = 0$ state. In the diagram in Figure 7.11c., which contributes to both $I = 0$ and $I = 2$ final states, such a phase cannot appear. By observing effects involving interference between the two I-spins, as occurs in the $K^0 \to \pi\pi$ decays, the difference between the phases enters and CP-violation effects can be seen.

Recently, clear experimental evidence has been obtained at the Cornell detector Cleo that penguin diagrams do occur and with about the amplitudes calculated.[112] They examined decays of $B^\circ \to K^*(892)^\circ \gamma$ and $B^- \to K^*(892)^- \gamma$, for which the basic process is $b \to s\gamma$. (See Figure 7.11d.) The penguin diagrams are expected to dominate over other Standard Model diagrams by at least an order of magnitude. The branching ratio for $b \to s\gamma$ was estimated to be $2-4 \times 10^{-4}$, and the branching fraction of that decay to $K^*\gamma$ was estimated to be between 5% and 40%. The experimental branching fraction for $B^0 \to K^*(892)^0 \gamma$ and $B^- \to K^*(892)^- \gamma$ was found to be $(4.5 \pm 1.5 \pm 0.9) \times 10^{-5}$, in good agreement with the above estimates.

7.14 INTERMEDIATE BOSONS

Until now the weak interaction has been treated as if it were a direct coupling of the two currents at the same spacetime point. For electromagnetic interactions the current couples to a photon which then couples to the other current. It is natural to ask whether a similar thing happens in the weak interactions. (See Figure 7.12.)

The Standard Model, which will be developed in this text, predicts there should be these analogues of the photon. However, instead of being massless, the theory predicts that they should be quite heavy and actually predicts the masses. These intermediate bosons have been discovered at the CERN $p\bar{p}$ colliding beam facility in 1983.[113] The experimental apparatus is shown in Figure 3.15. C. Rubbia and J. VanderMeer were awarded the 1984 Nobel Prize for their discovery of the bosons and their development of the colliding beam facility and the UA1 detector.

It is found that $M_{W^\pm} \approx 80$ GeV/c^2 and $M_{Z^0} = 91.2$ GeV/c^2. The Z^0 was found by looking at the lepton pair mass spectrum assuming the reaction $\bar{p}p \to Z^0 X$; $Z^0 \to e^+e^-$ or $\mu^+\mu^-$. The $W^\pm$ was found by looking at the single lepton spectrum in the reaction $\bar{p}p \to W^\pm X$; $W^\pm \to e^+\nu_e$ or $\mu^+\nu_\mu$. Because of the large mass of the $W^\pm$, the $W^\pm$ is moving slowly in the laboratory when produced and the leptons have a characteristic momentum spectrum from which the existence and the mass of the $W^\pm$ can be inferred. The Z mass is now measured precisely in the reaction $e^+e^- \to Z^0$ at LEP.

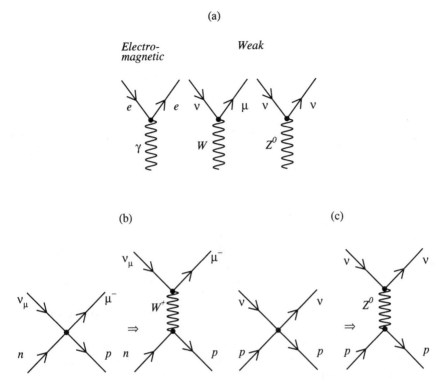

Figure 7.12 Intermediate boson picture of weak interactions.

7.15 Neutrino Oscillations

It has been seen that the strongly interacting quarks of the different generations mix, i.e., that d' is a mixture of d and s. One can ask whether the leptons in the families mix also, specifically if there is a similar mixture in the neutrinos.

If there is such a mixture, and if the neutrinos have mass not equal to zero, then the neutrinos can oscillate between species much as do the neutral kaons or the B system. As indicated previously there is at present no convincing evidence for massive neutrinos. If oscillations are observed, it will serve as a measurement of these masses.

Consider first just electron and muon neutrinos. Let ν_1 and ν_2 be the orthogonal neutrino states with definite mass, i.e., the mass eigenvalue states. Let ν_e and ν_μ be the states produced with electrons and muons, respectively, in the weak interaction. Then

$$\nu_\mu = \nu_1 \cos\theta + \nu_2 \sin\theta; \quad \nu_e = -\nu_1 \sin\theta + \nu_2 \cos\theta. \tag{7.73}$$

The discussion proceeds analogously with the kaon discussion and after a distance L (meters) one has

$$P(\nu_\mu \to \nu_e) = \sin^2(1.27\Delta m^2 L/E) \sin^2 2\theta,$$
$$P(\nu_\mu \to \nu_\mu) = 1 - P(\nu_\mu \to \nu_e), \tag{7.74}$$

where E = energy in MeV and $\Delta m^2 = m_1^2 - m_2^2$ (eV)^2.

Recall now the discussion of the discovery of two neutrinos, the Lederman, Schwarz, and Steinberger experiment. What role does the above formalism play? If the parameters had been such that rapid oscillation between ν_e and ν_μ had taken place, this group would have said (incorrectly) that the muon and electron neutrinos were the same because they would have seen final state electrons at the same rate as muons. The fact that they did not see final-state electrons at all sets limits on the above parameters. Either Δm^2 is small or θ is small or both. This experiment is a prototype, in fact, for a series of experiments that have been performed with a large number of distance scales and a large variety of neutrino energies. One looks for all combinations of neutrino oscillations between ν_e, ν_μ, and ν_τ. None have been seen to date and the limits on Δm^2 are of the order of 1 (eV)^2 unless θ is quite small (which is a realistic possibility).

Another possibility of oscillation is for neutrinos to oscillate with their anti-neutrinos. This is impossible in normal Dirac theory, but can happen in a variant called the Majorana theory of the neutrino.

Let ν_c be the anti-neutrino state that is the charge conjugate of ν, i.e., $\nu_c = C\nu C^{-1}$ and consider the ν-$\bar\nu$ column vector

$$\psi = \begin{pmatrix} \nu \\ \nu_c \end{pmatrix}.$$

The mass term in the Hamiltonian will, in general, be

$$M = \begin{pmatrix} m_{11} & m_{12} \\ m_{12}^* & m_{22} \end{pmatrix}.$$

From CPT invariance, $m_{11} = m_{22}$ and, in normal Dirac theory, $m_{12} = 0$. However, if m_{12} is not zero and is real, then the mass eigenvalue states are $\chi_\pm = (1/\sqrt{2})(\nu \pm \nu_c)$. These states have a mass difference of $2\Delta m$, i.e., $m_\pm = (m \pm m_{12})$. These are the Majorana states. Note that these states go into plus or minus themselves under charge conjugation ($C\chi_\pm C^{-1} = \pm\chi_\pm$).

7.15 Neutrino Oscillations

The majorana states have mixed lepton number ($L = \pm 1$) and transitions with $\Delta L = 2$ are possible. If a weak interaction creates a state with a left-handed neutrino, ν_L, it is a mixture of $(\chi_+)_L$ and $(\chi_-)_L$ that have different masses. The left-handed neutrino can then oscillate into a left-handed anti-neutrino which is a sterile neutrino, i.e., it cannot interact except by oscillating back. Neutrinoless double beta decay also becomes a possibility. In experiments similar to the one described above no evidence is seen for this kind of oscillation, nor, so far, is evidence seen for neutrinoless beta decay. Both kinds of experiments are being actively pursued.

There has been a persistent mystery in the search on earth for neutrinos produced in nuclear processes in the sun. R. Davis and colleagues have measured this process for years[114] and the rate has seemed lower than the calculated rate. The low rate is now confirmed by a Japanese experiment[115] (Kamioka) and by two other experiments (GALLEX[116] and SAGE[117]). Neutrino oscillations and transformations have been suggested as a possible explanation.

A new effect enters here, the elegant Mikheyev–Smirnov–Wolfenstein (MSW) matter enhancement of neutrino oscillations.[118][119][120] When electron neutrinos scatter in matter, they can scatter against electrons through the neutral weak interaction involving Z exchange in the t-channel, or they can scatter through the charged weak current involving $W^\pm$ exchange in the s-channel. This second reaction is unique to electron neutrinos and causes the electron neutrino scattering cross section in matter to be larger than that of the other neutrino families.

Suppose neutrinos are massive and can mix. Consider the time development of a beam of initially electron neutrinos passing through matter. Let a_e be the probability amplitude that a neutrino is an electron neutrino, and a_x be the probability that it is another species. The time evolution is given by

$$i\frac{d}{dt}\begin{pmatrix} a_e \\ a_x \end{pmatrix} = \begin{pmatrix} M_X & M_Y \\ M_Y & M_Z \end{pmatrix} \begin{pmatrix} a_e \\ a_x \end{pmatrix}. \quad (7.75)$$

The matter Hamiltonian is given by

$$M_X = \frac{m_1^2 \cos^2\theta + m_2^2 \sin^2\theta}{2E} + \sqrt{2} G_F N_e,$$

$$M_Z = \frac{m_1^2 \sin^2\theta + m_2^2 \cos^2\theta}{2E},$$

$$M_Y = \frac{m_2^2 - m_1^2}{2E} \cos\theta \sin\theta = \frac{\Delta m^2}{2E} \cos\theta \sin\theta. \quad (7.76)$$

Here m_1 and m_2 are the diagonal mass eigenvalues and N_e is the electron density of the matter. The N_e term in M_X reflects the charged current scattering for the electron neutrino. m_1 refers to the mostly electron neutrino mass eigenstate. If θ is small and if m_1 is also the smallest neutrino mass, then for $N_e = 0$, $M_X < M_Z$. Under these circumstances, the amount of mixing is small. The diagonal states contain of the order of $(M_Y/(M_Z - M_X))^2$ of the "wrong" neutrino. If there is matter, suppose N_e is tuned such that $M_X = M_Z$. For this resonant value a diagonal state is completely mixed:

$$i\frac{d}{dt}(a_e - a_x) = (M_X - M_Y)(a_e - a_x). \tag{7.77}$$

The neutrinos tend toward complete mixing even though the mixing angle, θ is small.

It seems possible that in some parts of the sun, the densities may be in the right range to have this resonant value for mass values with $m_{\nu_x}^2 - m_{\nu_e}^2$ in the few $\times 10^{-5}$ eV^2 range and mixing angles in the 10^{-3} range. If, as the neutrinos pass through the variable density sun, they pass through such a region, then ν_e would be changed to a mixture of ν_e and other neutrinos before leaving the sun. Since present detectors would not detect the other neutrinos, this could account for the solar neutrino deficit.

In this chapter a description of the weak interactions has been given. They are described as interactions of a current with an intermediate, heavy vector boson (Z^0 or $W^{\pm}$). The vector charged current is conserved (CVC) and is related to part of the electromagnetic current, a conclusion that will lead to far reaching consequences to be described later. The current has lepton (μ, e, τ) universality. The quark coupling is also universal once it is recognized that the quark states in the weak current are mixtures of the quark states seen in strong interactions.

7.16 Exercises

7.1 In Chapter 6, the concept of U-spin was introduced. If the u, , d, and s quarks are taken as equivalent, the I-spin operator I^+ can be viewed as changing a d quark to a u quark, and the U-spin operator U^+ can be viewed as changing an s quark to a d quark. I^- and U^- change quarks in the opposite direction from I^+ and U^+, respectively. I^+ can be thought of as the product of a creation operator for a u quark and a destruction operator for a d quark, i.e., $a_u^\dagger a_d$, where the operators are understood to be for the same momentum and spin. (For simplicity, the similar terms for anti-quarks have been omitted.) I^-, U^+, and U^- are handled similarly.

a) Using the anti-commutations relations for creation and destruction operators, show that I^+, and U^- commute.

b) Express the commutator of I^+ and U^+ in terms of creation and destruction operators.

7.2 Verify that the operator $\Lambda_\pm \equiv (\pm \not{p} + m)/(2m)$ indeed is a projection operator for the $\pm$ frequencies of the plane wave solutions of the Dirac equation.

a) Verify that $\Lambda_\pm$ is a projection operator on the Dirac wavefunctions, i.e., show $\Lambda_\pm \Lambda_\pm = \Lambda_\pm$. Hint: In Section 7.1 it was noted that $\gamma^\mu \gamma^\nu + \gamma^\nu \gamma^\mu = 2Ig^{\mu\nu}$. This implies that for two four-vectors a and b, $\not{a}\not{b} + \not{b}\not{a} = 2a \cdot b$, if it is assumed that a_μ, b_ν commute, i.e., $a_\mu b_\nu = b_\nu a_\mu$.

b) Show that if $\Psi = au e^{-ip \cdot x} + bv e^{ip \cdot x}$, then $\Lambda_+ \Psi = au e^{-ip \cdot x}$, and $\Lambda_- \Psi = bv e^{ip \cdot x}$.

7.3 Consider $\Pi_n^+ \equiv .5(1 - \gamma^5 \not{n})$, where n is a unit 4-vector and $\not{n} = \sum \gamma^\mu n_\mu$.

a) Show Π^+ is a projection operator on the Dirac wavefunctions, i.e., show $\Pi^+ \Pi^+ = \Pi^+$.

b) For $n = \hat{z}$ and momentum (p) equal zero, show it projects out the $+\hat{z}$ spin component of the wavefunction.

c) Show that for $n = (|p|/m, \hat{p}(E/m))$ the operator projects out the $+$ helicity part of the wavefunction for any p.

Hint: See the hint for the previous exercise.

7.4 Consider $L = (1 - \gamma^5)/2$.

a) Show that L is a projection operator, i.e., $LL = L$.

b) Suppose one quantizes such that the direction of motion of the particle is the z-direction. Find Lu_r, and show $Lu_1 \to 0$ for E large, and $Lu_2 \to u_2$ in the same limit. Thus, although it does not precisely project out the left-handed part of the wavefunction in general (as does the operator in problem 7.3), in the high-energy limit it does project out the left-handed part of the wavefunction.

7.5 Show that the electromagnetic charge current density $j^\mu = -e\bar{\psi}\gamma^\mu \psi$ satisfies the continuity equation $\partial_\mu j^\mu = 0$.

7.6 For free particle solutions of the Dirac equation

$$\psi = u e^{-ip \cdot x}$$

the four-component spinor u may be written in terms of two two-component spinors

$$u = \begin{pmatrix} \phi \\ \chi \end{pmatrix}.$$

Use the free particle Dirac equation for ψ, $(i\not{\partial} - m)\psi = 0$, and the explicit forms for the γ matrices to show that ϕ and χ satisfy the coupled equations

$$(E-m)\phi = (\vec{\sigma}\cdot\vec{p})\chi; \quad (E+m)\chi = (\vec{\sigma}\cdot\vec{p})\phi.$$

7.7

a) Show that $(\vec{\sigma}\cdot\vec{p})^2 = |\vec{p}|^2$.

b) Show that $\not{p}\not{p} = p^2$.

7.8 Show that to lowest order in nucleon velocities, the nuclear part of each of the S,V,T,A matrix elements reduces to the $\langle|1|\rangle$, or $\langle|\sigma|\rangle$ forms shown in Table 7.3, i.e., the Fermi or Gamow–Teller forms.

7.9

a) Derive Equation 7.45 by starting with Equation 7.44 and using the trace procedure. For this process, since the initial state has spin 0, there is no averaging over initial spins and therefore there is no factor of 1/2 in the trace formula, Equation 7.38. Note that $\text{Tr}\not{a}\not{b} = 4a\cdot b$ and $\text{Tr}(\not{a}\not{b}\gamma^5) = 0$.

b) Show that for $\pi \to \mu + \nu$, in the pion rest frame, $p_\mu = m_\pi(1 - m_\mu^2/m_\pi^2)/2$.

7.10 Show that the differential and integral muon decay rates given in Equations 7.57 and 7.58 indeed follow from Equations 7.54–7.56. For this exercise, ignore the radiative correction term in Equation 7.58.

7.11 For each of the following reaction pairs below indicate whether you expect the ratio of the rates (not lifetimes) to be large, near one, small, or ≈ 0. Hint: The D^0 has quark content $c\bar{u}$.

a) $\sigma_{tot}(\pi^+p)/\sigma_{tot}(\pi^-p)$ near the center of mass energy of the $\Delta(1232)$ resonance.

b) $\pi^+ \to \mu^+\nu/\pi^0 \to \gamma\gamma$.

c) $K^0 \to \mu^+\mu^-/K^0 \to \pi^+\pi^-$.

d) $D^0 \to K^+\pi^-/D^0 \to K^-\pi^+$.

e) $\overline{K}^0 \to \pi^-e^+\nu/\overline{K}^0 \to \pi^+e^-\bar{\nu}$.

7.12 Which interactions (strong, electromagnetic, weak, gravitational) respect and which violate the following symmetries? Is the violation large or small?

a) Parity invariance (P).

b) Charge conjugation invariance (C).

c) Time reversal invariance (T).

d) PC.

e) PCT.

f) Baryon conservation.

g) Isotopic spin invariance.

7.13 Beta decays of the K_L ($K_L \to \pi l \nu_l$) account for 66% of its decay rate. Suppose there is a theory of charm (not quite the Standard Model) in which there is a counterpart to the K_L, the C_L meson, in which the mass is predicted to be 2.2 GeV. Estimate the lifetime of the C_L, assuming that the C_L and K_L have the same beta decay branching ratio and that the same Cabibbo angle obtains for strangeness-changing and charm-changing weak decays. Hint: Only an order of magnitude result is desired. Compare both K_L and C_L decays to the μ decay formula. They are all three-body weak decays and for this order of magnitude calculation ignore the fact that the μ is a fermion and the K_L and C_L are bosons.

7.14 Assume that for a B meson decay ($m_B = 5279$ MeV), the $\bar{u}$, $\bar{d}$, or $\bar{s}$ quark is a spectator doing nothing. Then the decay is $b \to c + W$, followed by $W \to$ lepton $+ \nu$ or $W \to q_1 + \bar{q}_2$. In either case, the Feynman diagram is exactly like the Feynman diagram for muon decay and the muon decay formula (Equation 7.58) can be used, with appropriate mass and other corrections. Use this formula to estimate the lifetime of a B meson and compare it to the experimental lifetime of $(12.9 \pm 0.5) \times 10^{-13}$ s.[17] Remember to include all allowed quarks (with color factors) and leptons in the W decay. Also note that there is a factor $|V_{cb}|^2$ in the $b \to c$ transition to include in the decay formula. In fact there are corrections due to the finite masses of the final state quarks, and partially cancelling QCD corrections, which are not included in the zeroth-order calculation here.

7.15 The charged B meson ($b\bar{u}$) is a pseudoscalar. Consider the decays into lepton plus neutrino. The same formalism holds for this as for charged pion decay (Equation 7.49) except that the masses must be changed appropriately, the rate multiplied by the square of the CKM matrix element, i.e., multiplied by $|V_{ub}|^2$, and f_π (≈ 130 MeV) replaced by f_B, which early lattice gauge calculations estimated at about twice f_π. Using the mass and lifetime given in the previous problem, estimate the branching ratios for $B \to e\nu$, $B \to \mu\nu$, and $B \to \tau\nu$.

8
Elastic and Inelastic Scattering

Consider elastic and inelastic scattering by the electromagnetic and the weak interactions. Extensive information can be obtained from these processes not only on the electromagnetic and weak interactions but also on the strong interactions and the structure of the hadrons. Leptons are believed to be point particles, but nucleons have structure.

The discussion here will proceed in stages. Firstly, elastic electromagnetic scattering of electrons and muons will be described. This is not very practical for experiments, but is the archetypical scattering of two point non-identical particles. It will be seen that looking at the diagram sidewise suddenly gives a very practical result. The discussion will then proceed to add complications in stages. Electromagnetic elastic scattering of leptons on nucleons, then inelastic scattering of leptons on nucleons, and finally inelastic scattering by the weak interaction of neutrinos on nucleons will be studied. The book by F. Close[31] is very good on these subjects. This discussion follows the conventions of J.D. Jackson. Definitions of form factors are from C.H. Llewellyn-Smith.[121]

Recall, from Chapter 5, that the general formula for the cross section for a state α to go into a state β is

$$d\sigma = \frac{(2\pi)^4}{\Pi_\alpha 2E_\alpha} |M_{\beta\alpha}|^2 \times dP.S. \times \frac{E_a E_b}{F_{ab}}. \qquad (8.1)$$

Here $\alpha = a + b$. $dP.S.$ is the phase space factor and $F_{ab}/(E_a E_b)$ is the Møller flux factor:

$$\frac{F_{ab}}{E_a E_b} = \frac{\sqrt{(p_a \cdot p_b)^2 - (m_a m_b)^2}}{E_a E_b}. \qquad (8.2)$$

In the rest frame of b, $F_{ab}/(E_a E_b) = v_a$ and in the center of mass it is

$|\vec{v}_a| + |\vec{v}_b|$.

$$dP.S. \equiv \text{phase space} \equiv \Pi_{\beta_i} \frac{d^3 p_{\beta_i}}{(2\pi)^3 2E_{\beta_i}} \delta^4(p_{\beta_1} + p_{\beta_2} + \cdots + p_{\beta_n} - p). \quad (8.3)$$

For two particles let $E_T = E_1 + E_2$ and $\vec{P}_T = \vec{p}_1 + \vec{p}_2$. Then

$$dP.S._2 = \frac{1}{4(2\pi)^6} \frac{p_1^3 d\Omega_1}{E_T p_1^2 - E_1(\vec{P}_T \cdot \vec{p}_1)} \quad \text{(any frame)}, \quad (8.4)$$

$$dP.S._2 = \frac{p_1 d\Omega_1}{4(2\pi)^6 E_T} \text{(center-of-mass frame; any } m_2\text{)}. \quad (8.5)$$

8.1 Electron–Muon Elastic Scattering

The Feynman diagram for electromagnetic electron–muon scattering is shown in Figure 8.1. k and k' are the initial and final momenta of the electron of mass m. p and p' are the initial and final momenta of the muon of mass M. To avoid some minus signs, it is customary to define $Q^2 = -q^2$, where $q = k - k'$ is the momentum of the virtual photon exchanged. The amplitude for this scattering is given by

$$iM_{e\mu} = -e^2 (j^e)^\beta \left(\frac{-ig_{\beta\alpha}}{Q^2} \right) (j^\mu)^\alpha, \quad (8.6)$$

$$(j^e)^\beta = \overline{\Psi}^e(k') \gamma^\beta \Psi^e(k) \text{ and } (j^\mu)^\alpha = \overline{\Psi}^\mu(p') \gamma^\alpha \Psi^\mu(p). \quad (8.7)$$

Here, $e^2/4\pi = \alpha$, the fine structure constant. The factor e^2 is the product of the two vertex terms and $ig_{\beta\alpha}/Q^2$ is the propagator term. [This is electromagnetic scattering. The electromagnetic coupling involves only γ^β, not $\gamma^\beta(1 - \gamma^5)$, the latter term being a characteristic of the charged weak interaction.]

Next, to calculate the square of the matrix element, assume that the initial particles are unpolarized. Then one can average over initial states and sum over final states. The result of the evaluation of the traces for these currents was noted previously. It is found that

$$|M_{e\mu}|^2 = \frac{(4\pi\alpha)^2}{4Q^4} L^{\beta\gamma} W_{\beta\gamma}, \quad (8.8)$$

$$L^{\beta\gamma} = 4[k^\beta k'^\gamma + k'^\beta k^\gamma - g^{\beta\gamma}(k \cdot k' - m^2)]. \quad (8.9)$$

$W_{\beta\gamma}$ is of the same form as $L^{\beta\gamma}$ with k, k', m replaced by p, p', M, respectively. The factor $1/4$ comes from the averaging over the two spin states in each of the two currents involved.

8. Elastic and Inelastic Scattering

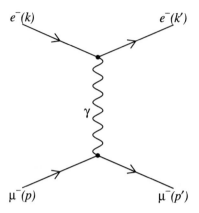

Figure 8.1 Feynman diagram for $e - \mu$ scattering.

The cross section can now be calculated as described before. Work in the laboratory reference frame, defined here as the frame in which the initial muon is at rest. Note that for two bodies a and b in the rest frame of b, $F_{ab} = v_a E_a E_b$. If $v_a \approx c$, then $F_{ab}/(E_a E_b) \approx c = 1$ in the laboratory.

For the initial muon, $p = (M, 0, 0, 0)$. Assume that $|\vec{k}|$ and $|\vec{k}'|$ are large compared to m and M and also that $M \gg m$.

Then $Q^2 \approx 2EE'(1-\cos\theta) = 4EE'\sin^2\theta/2$, where θ is the electron scattering angle in the laboratory, E and E' are the initial and final electron energies, respectively. Let $\nu \equiv E - E'$. By examining energy–momentum conservation at the lower vertex, it is found that $Q^2 = 2M\nu$. Using the delta function of energy conservation, $\nu = E_\mu - M$. Note that when integrating over delta functions, $\int \delta(f(x))dx = (df/dx)^{-1}$ evaluated at the value of x where $f(x) = 0$.

The expressions for the cross section and the two-body phase space expressions written down previously are now used. After some algebra one obtains

$$\frac{d\sigma}{dQ^2 d\nu} = \frac{4\pi\alpha^2}{Q^4}\frac{E'}{E}\cos^2\frac{\theta}{2}\left[1 + \frac{Q^2}{2M^2}\tan^2\frac{\theta}{2}\right]\delta\left(\frac{Q^2}{2M} - \nu\right). \quad (8.10)$$

Next, integrate over ν. Q^2 and ν are not independent. Writing Q^2 in terms of E and ν, the delta function gives a factor of $(1 + 2E/M\sin^2(\theta/2))^{-1}$, which can be calculated to be E'/E. Thus, finally,

$$\frac{d\sigma}{dQ^2} = \frac{4\pi\alpha^2}{Q^4}\left(\frac{E'}{E}\right)^2\cos^2\frac{\theta}{2}\left[1 + \frac{Q^2}{2M^2}\tan^2\frac{\theta}{2}\right]. \quad (8.11)$$

Discussion of this result will be deferred to a later section.

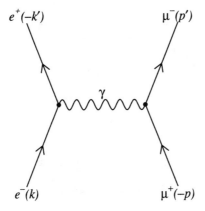

Figure 8.2 Feynman diagram for $e^+e^- \to \mu^+\mu^-$.

8.2 MUON PAIR PRODUCTION IN e^+e^- ANNIHILATION

Consider here the reaction $e^+e^- \to \mu^+\mu^-$. (See Figure 8.2.) The Feynman diagram in the figure is just the Feynman diagram for electron–muon scattering (Figure 8.1) turned on its side. The physical incoming momentum of the e^+ is $-k'$, and that of the outgoing μ^+ is $-p$. With that understanding the matrix element is the same one as in the previous section.

Work in the center of mass where $\vec{k} = -(-(\vec{k'}))$ and $\vec{p} = -(-(\vec{p'}))$. Here $Q^2 = -s$, where s is the Mandlestam variable, the square of the energy in the center of mass. For the Møller flux factor, $F_{ab}/E_a E_b = 2$. One eventually obtains

$$\frac{d\sigma}{d\cos\theta} = \frac{4\pi\alpha^2}{3s}[(3/8)(1+\cos^2\theta)], \qquad (8.12)$$

where θ is the angle between the e^- and μ^- momenta.

$$\sigma_{total} = \frac{4\pi\alpha^2}{3s}. \qquad (8.13)$$

This is also the first-order (Born Approximation) cross section for e^+e^- annihilation into any point spin 1/2 fermion particle–anti-particle pair with unit electric charge. For point quarks just insert the square of the quark electric charge. This formula was already used when the evidence for color was discussed in Section 6.8.

Note that the cross section falls as $1/s$. This has the unfortunate consequence that one needs very high luminosity to examine e^+e^- interactions at high energy (except for the very large cross section at the Z resonance).

200   8. Elastic and Inelastic Scattering

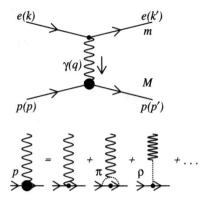

Figure 8.3 Feynman diagram for e-p elastic scattering.

Examination of this and similar reactions has provided excellent evidence for the pointlike nature of muons and electrons at high Q^2. If the deviations from QED are parametrized by a Q^2/Λ^2 term, the limits on Λ exceed 100 GeV. These data nicely complement the extremely accurate low-energy data obtained from the $g-2$ experiments.

8.3   ELECTRON–PROTON ELASTIC SCATTERING

The calculation of e-p elastic scattering contains a new complication. The proton is not a point particle and, as shown in Figure 8.3, the vertex contains contributions from pion clouds, from vector mesons, and from other things. In a different way of dividing up the strong interactions one can think of the contributions as coming from quark and gluon clouds. These contributions can be estimated, but it is not an elementary calculation.

The phenomenological approach will be taken here. Suppose that somehow all of these contributions have been calculated and ask what the most general form is that the resulting vertex function can have. First use Lorentz covariance. The current must be a Lorentz vector with one index. For the proton vertex only p, p', q and the γ matrices are possible vectors. Momentum conservation can be used to eliminate p'. The most general polar vector term is then

$$(j^{proton})^\alpha = \langle \overline{\Psi}(p') | \,[\Gamma_1(Q^2)\gamma^\alpha + \Gamma_2(Q^2)\sigma^{\alpha\beta}g_{\beta\delta}q^\delta$$
$$+ \Gamma_3(Q^2)q^\alpha + \Gamma_4(Q^2)p^\alpha + \Gamma_5(Q^2)\sigma^{\alpha\beta}g_{\beta\delta}p^\delta] \,| \Psi(p) \rangle, \quad (8.14)$$

where Γ_i is a scalar function of Q^2 only and $\sigma^{\alpha\beta} = (i/2)[\gamma^\alpha, \gamma^\beta]$. The terms $\gamma^\alpha\gamma^5$, $\sigma^{\alpha\beta}g_{\beta\delta}q^\delta\gamma^5$, $q^\alpha\gamma^5$, and the similar terms with p^α are axial not polar vector terms.

The fifth term has $\sigma^{\alpha\beta}g_{\beta\delta}p^{\delta}$. Recall that $\gamma^{\alpha}\gamma^{\beta} + \gamma^{\beta}\gamma^{\alpha} = 2Ig^{\alpha\beta}$. Thus, this expression has $i(p^{\delta}g_{\beta\delta}\gamma^{\alpha}\gamma^{\beta} - p^{\alpha})$. The Dirac equation can be written as $(\not{p} - M)u = 0$ or as $\bar{u}(\not{p}' - M) = 0$, where $\not{p} = \gamma^{\beta}p_{\beta}$. Using the Dirac equation the first part of term five becomes $M\gamma^{\alpha}$. Both this part and the second part, p_{α}, are terms that have already appeared among the first four terms. Since term five is just a linear combination of the previous terms, it may be ignored.

There is another restriction that can be used on the form of the vertex. The electromagnetic current is known to be a conserved current, i.e., $\partial j^{\alpha}/\partial x^{\alpha} = 0$. In perturbation calculations the initial and final states are taken as plane waves. $\Psi(p) = e^{-ip\cdot x}u(p)$ and $\overline{\Psi}(p') = e^{ip'\cdot x}\bar{u}(p')$. Since the Γ_i are scalar functions, then j^{α} has an overall factor of $e^{i(p'-p)\cdot x} = e^{iq\cdot x}$, and no other spatial dependence. Thus, $\partial(j^{\text{proton}})^{\alpha}/\partial x^{\alpha} = iq_{\alpha}(j^{\text{proton}})^{\alpha} = 0$.

The plane wave and perturbation theory assumptions are not, in fact, necessary to obtain this last relation. In general, the momentum operator is $P_{\alpha} = P^{\beta}g_{\beta\alpha} = -i\partial/\partial x^{\alpha}$. Thus, for any operator Op, $[P_{\alpha}, Op] = -i\partial_{\alpha}Op$. Then,

$$\langle B| \partial(j^{\text{proton}})^{\alpha}/\partial x^{\alpha} |A\rangle = 0 = i\langle B| [P_{\alpha}, (j^{\text{proton}})^{\alpha}] |A\rangle$$
$$= i(p_{\alpha}(B) - p_{\alpha}(A))\langle B| (j^{\text{proton}})^{\alpha} |A\rangle = iq^{\beta}g_{\beta\alpha}\langle B| (j^{\text{proton}})^{\alpha} |A\rangle.$$

This identity is one of a class known as "Ward–Takahashi identities". Thus:

$$\frac{\partial (j^{\text{proton}})^{\alpha}}{\partial x^{\alpha}} = i\langle \overline{\Psi}(p')| \left[\Gamma_1(Q^2)q_{\alpha}\gamma^{\alpha} + \Gamma_2(Q^2)\sigma^{\alpha\beta}q_{\beta}q_{\alpha} \right.$$
$$\left. +\Gamma_3(Q^2)q^{\alpha}q_{\alpha} + \Gamma_4(Q^2)p^{\alpha}q_{\alpha}\right] |\Psi(p)\rangle. \quad (8.15)$$

The remaining four terms are all independent and each must individually be zero to satisfy current conservation. Consider these terms one by one. The first term has $\gamma^{\alpha}q_{\alpha} = \not{q} = \not{p}' - \not{p}$. Using the Dirac equation as noted above, $\not{p}' - \not{p}$ can be replaced by $M - M = 0$. The first term satisfies current conservation and is allowed.

The second term has $\sigma^{\alpha\beta}q_{\alpha}q_{\beta}$. $\sigma^{\alpha\beta}$ is anti-symmetric in the indices and $q_{\alpha}q_{\beta}$ is symmetric. This term automatically vanishes. For example, consider the 12 + 21 terms. For these terms: $(\gamma^1\gamma^2 - \gamma^2\gamma^1)q_2q_1 + (\gamma^2\gamma^1 - \gamma^1\gamma^2)q_1q_2 = 0$. Thus, the second term satisfies current conservation and is allowed.

The third term has q^2. This quantity is not zero and this term is forbidden by current conservation.

The fourth term has $p \cdot q$. This term is not zero. For example in the rest frame of the initial proton it is $M(E' - E)$. This term is forbidden by current conservation.

The remaining terms are rewritten to conform with standard notation:

$$(j^{\text{proton}})^\alpha = \langle \overline{\Psi}(p')| \left[F_1(Q^2)\gamma^\alpha + \frac{i\kappa F_2(Q^2)}{2M}\sigma^{\beta\alpha}q_\beta \right] |\Psi(p)\rangle, \qquad (8.16)$$

where κ is the anomalous part of the magnetic moment of the proton $= 1.79$. $F_1(Q^2)$ and $F_2(Q^2)$ are form factors, which are one for a point particle, and thus indicate the deviation of the proton from a point particle. As $Q^2 \to 0$, this expression must look like the point charge expression plus the magnetic moment expression. Hence, $F_1(0) = F_2(0) = 1$. It was noted previously that conservation of charge implies that the integral over all space of j_0 is constant and that when the Fourier transform is taken to go from space to momentum space, a restriction on the integral over all space becomes a restriction on the zero-frequency part of the transform, i.e., on $F_1(0)$ and $F_2(0)$.

Next evaluate the cross section. After going through the trace procedure it is found:

$$\frac{d\sigma}{dQ^2} = \frac{4\pi\alpha^2}{Q^4}\left(\frac{E'}{E}\right)^2 \cos^2\frac{\theta}{2} \left[\left\{ F_1^2 + \left(\frac{\kappa Q}{2M}\right)^2 F_2^2 \right\} \right.$$
$$\left. + \frac{Q^2}{2M^2}\{F_1 + \kappa F_2\}^2 \tan^2\frac{\theta}{2} \right]. \qquad (8.17)$$

Note that the quantities in the braces {} were 1 for μ–e scattering. This equation is in the form of the spinless "point scattering" cross section times the quantity in the square brackets and is known as the Rosenbluth form.[122] At fixed Q^2, the form of the angular deviation from point scattering is specified as $A + B \tan^2 \theta/2$. This result depends only on having the process occur through single photon exchange.

Another commonly used notation is

$$G_E \equiv F_1 - \frac{\kappa Q^2}{4M^2}F_2 \text{ and } G_M \equiv F_1 + \kappa F_2. \qquad (8.18)$$

With these form factors there are no interference terms. For $|\vec{q^2}| \ll M_p^2$, G_E and G_M represent the proton's charge and magnetic moment distributions, respectively. Experimentally $G_M(Q^2)/G_M(0) \approx (1 + Q^2/0.7\text{GeV})^{-2}$.

8.4 GEOMETRICAL INTERPRETATION OF FORM FACTORS

There is a simple and useful (but inexact) intuitive interpretation of form factors such as those discussed in the preceding section. Consider an interpretation in terms of a charge distribution in three-dimensional space and further consider $F(Q^2)$ as the Fourier transform of that space distribution:

$$F(Q^2) = \int \rho(r) e^{i \vec{q} \cdot \vec{r}} dr^3. \qquad (8.19)$$

Integrating over the angular variables yields

$$F(Q^2) = \int \rho(r) \frac{\sin qr}{qr} 4\pi r^2 dr. \qquad (8.20)$$

This interpretation cannot be exact because only the three momentum part of q is used here. It is valid only when q^0 can be considered small. In this instance, $Q^2 \approx \vec{q}^2$. When q^0 is not small, the transform must involve time and cannot be simply a spatial distribution.

As an example of the use of this interpretation, consider a charge distribution known as the Yukawa distribution,

$$\rho(r) = \rho_o e^{-\alpha r}/r. \qquad (8.21)$$

Then, letting $|q| = Q$

$$F(Q^2) = 4\pi \rho_o \int (e^{-\alpha r}/Q) \sin Qr \, dr$$

$$= 4\pi \rho_o/(2iQ) \int e^{-\alpha r}(e^{iQr} - e^{-iQr}) dr = 4\pi \rho_o/(\alpha^2 + Q^2).$$

Setting $F(0) = 1$ one obtains

$$F(Q) = \frac{1}{1 + Q^2/\alpha^2} \text{ and } \alpha^2 = 4\pi \rho_o. \qquad (8.22)$$

This is effectively the propagator term for a boson of mass α. The Yukawa form factor corresponds to having the vertex emit an intermediate particle of mass α. The experimental form of G_M is similar but is squared. The proton is a complicated object and cannot be represented by a single vector meson term. Nonetheless, the indications are that a great deal of the form factor is due to effective masses of things of some sort at around 0.7 GeV/c^2 and the range corresponds roughly to this.

The r.m.s. charge radius $\langle r^2 \rangle$ for the Yukawa form factor is given by

$$\langle r^2 \rangle = \int \rho(r) r^2 r^2 dr \bigg/ \int \rho(r) r^2 dr = [\rho_0 6/\alpha^4]/[1/\alpha^2] = 6/\alpha^2. \qquad (8.23)$$

The range corresponding to 0.7 GeV/c^2 is about 0.7 fermi.

One can rewrite $F(Q^2)$ as $F(Q^2) = 1/(1+Q^2\langle r^2 \rangle/6) \approx 1 - Q^2\langle r^2\rangle/6$ for small Q^2. In fact this form can be shown to be a general approximation for any reasonable form factor for small Q^2. (This is done by expanding the exponential in a power series, keeping only the first non-zero term, and using the fact that $F(0) = 1$. (See Exercise 8.6.)

This method has proven very useful for finding estimates of the size of the pion and kaon. These experiments are done by scattering pions (or kaons) off of atomic electrons. Because the electron is very light, only small Q^2 can be reached with even large initial pion momentum. The formalism differs a bit from the preceding section since the pion is a boson, but it is essentially similar. The first order terms have been measured. 0.56 ± 0.04 fermi and 0.53 ± 0.05 fermi for the r.m.s. charge radii of the pion[123] and kaon[124] respectively have been obtained.

8.5 Electron–Proton Inelastic Scattering

Figure 8.4 shows the Feynman diagram for inelastic e–p scattering. The initial proton mass is M and the final effective mass of the hadron system is M'. The argument for the restrictions on the form factors proceeds exactly as in the discussion of the preceding section for elastic scattering and has the same conclusions. Conventionally, the form factors are named differently. The result is

$$\frac{d\sigma}{dQ^2 d\nu} = \frac{4\pi\alpha^2}{Q^4} \left(\frac{E'}{E}\right) \cos^2\frac{\theta}{2} \left[W_2 + 2W_1 \tan^2\frac{\theta}{2}\right]. \qquad (8.24)$$

Here there is a factor E'/E to the first power. The other factor of E'/E in e–p elastic scattering came from the conservation of energy delta function. For e–p inelastic scattering, conservation of energy is understood, but because M' is variable, no extra restriction is placed on k' and p'. Unlike the elastic scattering case, Q^2 and ν are independent here.

It is desirable now to develop the restrictions on the form factors in another way, convenient when neutrino scattering is discussed. The summed and averaged matrix element squared can be written as

$$|M_{ep-inel}|^2 = \frac{(4\pi\alpha)^2}{4Q^4} L_{\alpha\beta} W^{\alpha\beta}. \qquad (8.25)$$

$L_{\alpha\beta}$ is the electromagnetic trace used before. For $W^{\alpha\beta}$ the most general form possible is sought.

8.5 Electron–Proton Inelastic Scattering

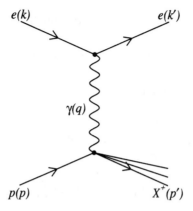

Figure 8.4 Feynman diagram for $e-p$ inelastic scattering.

Because of the fact that the final mass M' is not fixed, there may be more possibilities than for the elastic case. As vectors choose p, q. (Since traces have already been taken, the gamma functions are no longer available.) The form factors could be functions of the two scalars Q^2 and M'. However, it is useful to consider them functions of the equivalent variables Q^2 and $\nu = E - E'$. The most general form with two indices is then

$$W^{\alpha\beta} = 8[-W_1 M^2 g^{\alpha\beta} + W_2 p^\alpha p^\beta + i(W_3/2)\epsilon^{\alpha\beta\gamma\delta} p_\gamma q_\delta + W_4 q^\alpha q^\beta$$

$$+ (W_5/2)(p^\alpha q^\beta + p^\beta q^\alpha) + (W_6/2)(p^\alpha q^\beta - p^\beta q^\alpha)]. \quad (8.26)$$

Here $\epsilon^{\alpha\beta\gamma\delta}$ is the completely anti-symmetric tensor in four indices with $\epsilon^{1230} = 1$. This term can be shown to give rise to parity violation (i.e., introduces pseudovectors) and is forbidden for the parity conserving electromagnetic interaction. Hence $W_3 = 0$.

Note that since $L_{\alpha\beta}$ is symmetric in its indices, only the symmetric part of $W^{\alpha\beta}$ will contribute to the unpolarized particle cross section. Thus, the anti-symmetric W_6 term will not contribute.

Again use current conservation. Since W is composed of products of the current, the analogous expression to that of the preceding section is $q_\alpha W^{\alpha\beta} = 0$.

$$[-W_1 M^2 q^\beta + W_2(p \cdot q) p^\beta + W_4 q^2 q^\beta + W_5((p \cdot q) q^\beta + p^\beta q^2)/2] = 0. \quad (8.27)$$

Since q^β and p^β are independent four-vectors, the coefficients of each have to separately be zero. This gives two further restrictions:

$$-W_1 M^2 + W_4 q^2 + W_5 (p \cdot q)/2 = 0 \text{ and } W_2(p \cdot q) + W_5 q^2/2 = 0. \quad (8.28)$$

W_4 and W_5 can be eliminated to obtain

$$W^{\alpha\beta} = 8M^2 \left\{ W_1(\nu, Q^2) \left(-g^{\alpha\beta} - \frac{q_\alpha q_\beta}{Q^2} \right) \right.$$
$$\left. + W_2(\nu, Q^2)/M^2 \left[\left(p^\alpha + \frac{(p \cdot q)}{Q^2} q^\alpha \right) \left(p^\beta + \frac{(p \cdot q)}{Q^2} q^\beta \right) \right] \right\}. \quad (8.29)$$

From this expression, the previously shown cross section can be obtained. Comparing this cross section with the electron–muon scattering cross section it is seen that

$$W_1^{e\mu} = \frac{Q^2}{4M^2} \delta\left(\frac{Q^2}{2M} - \nu \right) \text{ and } W_2^{e\mu} = \delta\left(\frac{Q^2}{2M} - \nu \right). \quad (8.30)$$

8.6 Inelastic Charged Current ν–p Scattering

Turn next to the weak interaction and consider the process $\nu + p \to \mu^- + X$. The Feynman diagram for this process is shown in Figure 8.5. It is very similar to electromagnetic scattering with the photon replaced by a W^+ boson. The weak amplitude for this process is given by

$$iM^{\nu p} = \frac{4G}{\sqrt{2}} \langle \mu | j_\alpha | \nu \rangle \frac{i M_W^2 (-g^{\alpha\beta} + q^\alpha q^\beta / M_W^2)}{(M_W^2 + Q^2)} \langle X | j_\alpha | p \rangle. \quad (8.31)$$

The current is the V–A charged weak current:

$$\langle a | j_\alpha | b \rangle = \overline{\Psi}^a \gamma^\alpha \left(\frac{1 - \gamma^5}{2} \right) \Psi^b. \quad (8.32)$$

The central term is a propagator term for a vector W-boson. For calculations ignoring closed loop diagrams ("tree-level"), the $q^\alpha q^\beta$ term gives terms in the cross section proportional to m_E^2/M_W^2, where m_E is the mass of an external particle. These terms are, therefore, negligible unless a t-quark, a W, or other heavy particle is involved in the final state.

8.6 Inelastic Charged Current   207

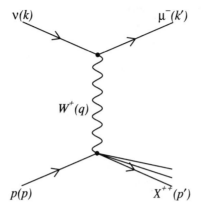

Figure 8.5 Feynman diagram for ν-p charged current scattering.

The $q^\alpha q^\beta$ term in the propagator actually depends on the gauge chosen. The choice above corresponds to the "unitary gauge." In this gauge the particle spectrum is the standard one, not true in all gauges as "ghost particles" can occur in the Lagrangian that are not physically present. For an example of this see Section 11.10

Recall that for the leptonic charged weak current, the summed and averaged trace is

$$(L^{weak})^{\alpha\beta} = 2[k^\alpha k'^\beta + k'^\alpha k^\beta - g^{\alpha\beta}(k \cdot k') + i\epsilon^{\alpha\beta\eta\xi} k'^\eta k^\xi]. \qquad (8.33)$$

For the hadronic vertex the most general term possible is desired. The procedure is similar to that used for the derivation of inelastic electromagnetic e–p scattering. The most general trace is

$$W_{\alpha\beta} = 2[-W_1 M^2 g^{\alpha\beta} + W_2 p^\alpha p^\beta + iW_3 \epsilon^{\alpha\beta\gamma\delta} p_\gamma q_\delta/2$$
$$+ W_4 q^\alpha q^\beta + W_5(p^\alpha q^\beta + p^\beta q^\alpha)/2 + W_6(p^\alpha q^\beta - p^\beta q^\alpha)/2]. \qquad (8.34)$$

This time there is neither parity conservation nor current conservation. (Remember only the vector part of the charged weak current is conserved.) Nonetheless, several of the coefficients do not contribute or are small.

When the cross section is calculated, the coefficients of W_4 and W_5 are proportional to m_μ^2 and can be ignored since, for the usual case, E, E', etc., are all much greater than m_μ. These terms would contribute for ν_τ interactions.

208   8. Elastic and Inelastic Scattering

The W_6 term does not contribute to the cross section. From the anti-symmetry of the term, the only possible contribution would come from the $L_{\alpha\beta}$ term involving ϵ and this gives zero since $q = k - k'$, which means that two of the coefficients of ϵ are the same (e.g., $\epsilon^{\alpha\beta\gamma\delta} k_\alpha k_\beta \cdots$). Because of the anti-symmetry of the ϵ tensor the term vanishes.

W_1, W_2, W_3 still remain. This last term is a parity-violating new term, which is not present in the electromagnetic interaction. After calculation, the cross section is found to be

$$\frac{d\sigma}{dQ^2 d\nu} = \frac{G^2 E'}{2\pi E} \left(\frac{M_W^2}{M_W^2 + Q^2}\right)^2 \cos^2\frac{\theta}{2}$$
$$\times \left[W_2 + 2W_1 \tan^2\frac{\theta}{2} \mp \frac{W_3}{M}(E + E')\tan^2\frac{\theta}{2}\right], \quad (8.35)$$

where the minus sign is chosen for neutrino scattering and the plus sign for anti-neutrino scattering.

An example of a neutrino interaction is given in Figure 3.12.

8.7   FORM FACTORS IN THE QUARK–PARTON PICTURE

A summary of the relations found for inelastic lepton-nucleon scattering is given below. For inelastic electromagnetic scattering,

$$\frac{d\sigma^{em}}{dQ^2 d\nu} = \frac{4\pi\alpha^2}{Q^4}\left(\frac{E'}{E}\right)\cos^2\frac{\theta}{2}\left[W_2^{em} + 2W_1^{em}\tan^2\frac{\theta}{2}\right]. \quad (8.36)$$

For inelastic weak charged current (CC) ν-nucleon,

$$\frac{d\sigma^{CC}}{dQ^2 d\nu} = \frac{G^2 E'}{2\pi E}\left(\frac{M_W^2}{M_W^2 + Q^2}\right)^2$$
$$\times \cos^2\frac{\theta}{2}\left[W_2^{CC} + 2W_1^{CC}\tan^2\frac{\theta}{2} \mp \frac{W_3^{CC}}{M}(E+E')\tan^2\frac{\theta}{2}\right], \quad (8.37)$$

where the minus sign is chosen for neutrino scattering and the plus sign for anti-neutrino scattering.

As will be shown, there is extensive experimental data on these form factors. The fact that they depend only on the two variables Q^2 and ν, gives strong support for the local character of the lepton current for the charged weak interactions, i.e., that the lepton fields $\Psi_l(\vec{r}, t)$ and $\Psi_{\nu_e}(\vec{r}, t)$ are always at the same spacetime point in the lepton current operator $j_\beta^\pm$.

8.7 Form Factors in the Quark–Parton Picture

For elastic electromagnetic scattering of point unit-charged particles,

$$W_1 = \frac{Q^2}{4M^2}\delta\left(\frac{Q^2}{2M} - \nu\right), \tag{8.38}$$

$$W_2 = \delta\left(\frac{Q^2}{2M} - \nu\right). \tag{8.39}$$

It was indicated at the beginning of the chapter that a lot can be learned about the structure of the nucleons from these experiments. The nucleons are very complicated structures. How can these form factors be related to fundamental properties, e.g., the distributions of the quarks and other things in a nucleon?

The answer is surprisingly simple and beautiful. The nucleon has valence quarks as well as a sea of transient clouds of quark–anti-quark pairs and gluons being continually created and destroyed. Collectively these quark, anti-quark, and gluon constituents of the nucleon are called "partons." It is now believed that, at high momentum transfers, the strong interactions become weaker, and perturbation theory can be used. Imagine moving at a very high velocity past a nucleon. Watching the nucleon from this moving coordinate system all of the processes of creation and destruction are slowed down by time dilation and an instantaneous snapshot of the system can be seen. If the momentum transfers are high enough, this can be viewed as a collection of almost free particles since the binding is much less than q. Furthermore, at high momentum transfers the strong interactions are weak and the lepton scattering can be considered to occur on a single one of these almost free partons. Suppose a single parton has a fraction "x" of the total proton longitudinal momentum and transverse momentum negligible compared to this. The scattering is then like the scattering from a single point particle of momentum px, where p is the momentum of the nucleon.

What effective mass should be used for this parton? See Figure 8.6. The final-state momentum of the parton satisfies

$$(xp + q)^2 = M^2_{\text{parton}} \approx 0. \tag{8.40}$$

Here q is the absorbed four-momentum. It has been assumed that the momentum and the Q^2 are large enough that the mass above is negligible in this equation. The quantity $(xp)^2 = x^2 M^2_{\text{nucleon}}$ is also then negligible and $x \approx -q^2/(2(p \cdot q))$. In the laboratory system this initial-state nucleon

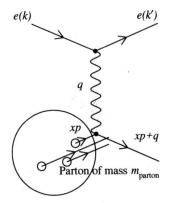

Figure 8.6 Scattering of a lepton from a parton in a nucleon.

has $p = (M_{\text{nucleon}}, 0, 0, 0)$ and $(p \cdot q) = M_{\text{nucleon}}\nu = M_{\text{nucleon}}(E - E')$. Thus,

$$x \approx \frac{-q^2}{2M_{\text{nucleon}}\nu} = \frac{Q^2}{2M_{\text{nucleon}}\nu}. \tag{8.41}$$

If the preceding argument is applied to a free quark (not in a nucleon), the same calculation would apply except that $x = 1$ and $M_{\text{nucleon}} \to M_{\text{parton}}$. Hence, for a free parton $Q^2 = -q^2 = 2M_{\text{parton}}\nu$. Comparing this with the previous expression it is seen that,

$$M_{\text{parton}} = xM_{\text{nucleon}}. \tag{8.42}$$

Assume next that the point form factors can be used for each parton. Furthermore, assume for now that only quarks and anti-quarks are present. The quark charge squared must be included. For the electromagnetic case this is $(2/3)^2$ for u quarks, $(1/3)^2$ for d quarks, etc. For example, for a u quark and the electromagnetic case one would have

$$W_1^{em} = (2/3)^2 \frac{Q^2}{4(Mx)^2} \delta\left(\frac{Q^2}{2Mx} - \nu\right), \tag{8.43}$$

$$W_2^{em} = (2/3)^2 \delta\left(\frac{Q^2}{2Mx} - \nu\right). \tag{8.44}$$

Let $f_i(x)$ be the probability of finding a parton of type i with momentum fraction x in a proton. (Note that protons are now specifically being discussed, rather than nucleons, which could be p or n.) The $f_i(x)$ are known as "structure functions." Generically $\overline{f}_i$ will refer to anti-particles. Let

8.7 Form Factors in the Quark–Parton Picture

$u(x) \equiv x f_u(x)$, $d(x) \equiv x f_d(x)$, $s(x) \equiv x f_s(x)$, etc. It was noted previously that when integrating over delta functions $\int \delta(g(x))dx = (dg/dx)^{-1}$. For $\int \delta(Q^2/(2Mx) - \nu)dx$ one has $1/[d/dx(Q^2/2Mx)] = 2Mx^2/Q^2 = x/\nu$. The following relations refer only to quarks since gluons do not have electric or weak charge:

$$W_2^{em} = \int \Sigma_i q_i^2 (f_i(x) + \overline{f}_i(x)) \delta\left(\frac{Q^2}{2Mx} - \nu\right) dx$$

$$= \frac{1}{\nu} \Sigma_i q_i^2 x (f_i(x) + \overline{f}_i(x))|_{x=Q^2/2M\nu},$$

$$W_1^{em} = \frac{1}{2M} \Sigma_i q_i^2 (f_i(x) + \overline{f}_i(x))|_{x=Q^2/2M\nu}. \quad (8.45)$$

For this latter relation $Q^2/4M^2x^2 = (Q^2 4M^2\nu^2)/(4M^2Q^4) = \nu^2/Q^2$, and $\nu^2/Q^2 \times x/\nu = \nu x/Q^2 = 1/(2M)$ was used.

For the weak interaction form factors W_1^{CC} and W_2^{CC}, similar expressions are obtained. The q_i^2 term is taken as 0 or 2 if the CKM mixing angles are set to 0. This matches the point quark cross sections. (Exercise 8.7 covers this factor and the other points listed in this present paragraph.) For the term W_1 in the weak interactions, in principle, $1/2M$ should be replaced by $(1/2M + x/2yE)$. However, the second term is small unless Q^2 is quite small. [The ratio of the second term to the first is just $2(Mx)^2/Q^2$.] The W_3^{CC} term, which comes from V–A interference, is also found in Exercise 8.7 using the quark picture and is

$$W_3^{CC} = \frac{2}{\nu} \Sigma_i (\overline{f}_i(x) - f_i(x)). \quad (8.46)$$

Thus $W_3^{CC} = \frac{2}{\nu}(f_{\overline{u}} + f_{\overline{d}} + \cdots - f_u - f_d - \cdots)$.

The mixing angles will often be ignored in the following, setting $\sin\theta_i = 0$. The strange, charmed, etc., quarks are only in the particle–anti-particle sea and are suppressed there over the values one would guess from $SU(3)$ because of their large mass. For the strange particles this suppression is of the order of a factor of 2 or more. If it is desired to include the strange quarks, that will be indicated explicitly. Comparisons between the weak and electromagnetic form factors are much simpler when these few percent corrections are ignored. Ignoring for now the mixing angles and c, b, t quarks, then $\nu d \to u$, $\nu u = 0$, $\nu s = 0$, $\nu \overline{u} \to \overline{d}$, $\nu \overline{d} = 0$, $\nu \overline{s} = 0$.

Note that the form factors are only functions of x, not two variables (Q^2, ν). This is an important prediction of this free-particle scattering model and is called "scale invariance." It is a sign of hard point particle scattering. The actual distributions[125] are mainly scale invariant, but, because this is not quite free-particle scattering, and because of the charm threshold, there are significant violations that tend to decrease as Q^2

increases. The gluon contributions, which violate the free-particle scattering assumptions, change slowly, as $\ln Q^2$. The scale invariance violations in the weak interactions were first seen in the 1970s in a Fermilab bubble chamber experiment,[126] but they were examined most beautifully in a series of CERN bubble chamber[127] and CERN and Fermilab counter experiments.[17] The violations of scale invariance are in agreement with theoretical estimates.

As was indicated previously, because of the higher mass of the c quark, some violation of the approximate scaling behavior (scale invariance behavior) appears until Q^2 is quite large. One often multiplies the charm part of the cross section by the factor $[1 - m_C^2/(2MEx)]$ for the ν-quark or $\bar{\nu}$-anti-quark part of the interaction and by the factor $[1 + m_C^2/(2MEx(1-y))]$ for the ν-anti-quark or $\bar{\nu}$-quark part of the interaction. Some of the effects of the charm quark mass threshold (m_C) are taken into account in this manner and scaling appears at a lower Q^2. In this correction, m_C is found to be $1.6 \pm 0.20 \pm 0.12$ GeV experimentally.

Because of approximate scale invariance, it often is desirable to change variables from (Q^2, ν) to (x, y), where

$$x \equiv \frac{1}{\omega} \equiv \frac{Q^2}{2M\nu}, \qquad (8.47)$$

$$y \equiv \frac{\nu}{E} = \frac{E - E'}{E}. \qquad (8.48)$$

These are known as the "Bjorken" x and y variables. As seen previously, the momentum fraction x is in approximate agreement with this new x.

At high energies both x and y range from close to 0 to close to 1. Introduce new form factors:

$$F_1 = MW_1; \quad F_2 = \nu W_2; \quad F_3 = -\nu W_3. \qquad (8.49)$$

Then,

$$\frac{d\sigma^{em}}{dxdy} = \frac{8\pi\alpha^2 ME}{Q^4}\left[F_2^{em}\left(1 - y - \frac{Mxy}{2E}\right) + F_1^{em} y^2 x\right] \qquad (8.50)$$

$$\frac{d\sigma^{CC}}{dxdy} = \frac{G^2 ME}{\pi}\left(\frac{M_W^2}{M_W^2 + Q^2}\right)^2 \left[F_2^{CC}\left(1 - y - \frac{Mxy}{2E}\right)\right.$$
$$\left. + F_1^{CC} y^2 x \pm y(1 - y/2)x F_3^{CC}\right]. \qquad (8.51)$$

The plus sign is for neutrinos and the minus sign is for anti-neutrinos.

8.7 Form Factors in the Quark–Parton Picture

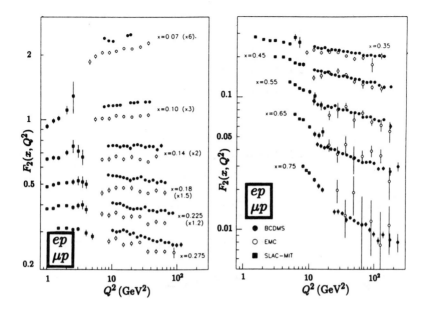

Figure 8.7. The proton structure function $F_2^p(x, Q^2)$ measured by the SLAC–MIT, EMC, and BCDMS experiments on hydrogen versus Q^2 for fixed bins of x. The error bars show only statistical errors. The SLAC–MIT and EMC data are interpolated to the BCDMS bins of x. $R = \sigma_L/\sigma_T = 0.21$ is assumed for the SLAC–MIT data, $R = 0$ for the EMC data, and $R = R_{QCD}$ for the BCDMS data.[17]

Experimentally, x distribution shapes similar to the ones shown in Figures 8.7–8.13 are found. The low x part of the distribution ($x \lesssim 0.15$) is due to the sea and the high x part is due to the valence quarks. As the energy increases, the low x part increases, corresponding to an increase of sea quarks at higher energies.

With the availability of data from HERA, it has become possible to explore the low x region down to about $x = 10^{-4}$. The predictions of important LHC cross sections involve x values in the 10^{-5} region. Reducing the distance of extrapolation from measured values considerably increases the accuracy of these predictions. In addition, these low x values are important in the comparison of various quark model sum rules with data. The sum rules, a few of which will be discussed later in this section, involve integrals over x of various combinations of quark functions.

214   8. Elastic and Inelastic Scattering

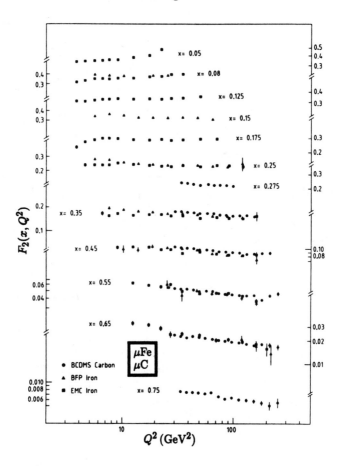

Figure 8.8 The nucleon structure function F_2 measured in electromagnetic scattering of muons on iron (BFP,EMC) and carbon (BCDMS) targets, versus Q^2 for fixed bins in x. For x of 0.05, 0.125, 0.175, 0.275, 0.45, 0.65 use the right-hand scale; for all other bins of x use the left-hand scale. Only statistical errors are shown.[17]

In terms of the quark functions,

$$F_1 = \Sigma_i \frac{1}{2} q_i^2 (f_i(x) + \overline{f}_i(x)),$$

8.7 Form Factors in the Quark–Parton Picture

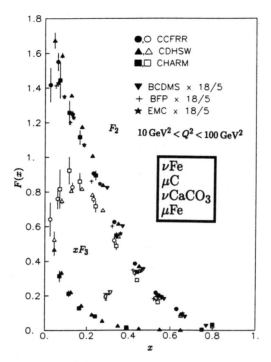

Figure 8.9 The structure functions F_2, xF_3, and $\overline{q}^\nu$ measured in different experiments on isoscalar targets as functions of Bjorken x. The CCFRR, CDHSW, BFP, and EMC data were taken with iron targets; the CHARM data with a marble ($CaCO_3$) target; and the BCDMS with a carbon target. Only statistical errors are shown.[17]

$$F_2 = \Sigma_i x q_i^2 (f_i(x) + \overline{f}_i(x)),$$
$$F_3 = \Sigma_i q_i^2 (f_i(x) - \overline{f}_i(x)). \quad (8.52)$$

Substituting for F_1, F_2, F_3 in terms of the f_i leads to the forms

$$\frac{d\sigma^{em}}{dx\,dy} = \frac{4\pi\alpha^2 MEx}{Q^4} \sum_i q_i^2 (f_i(x) + \overline{f}_i(x)) \left[(1-y)^2 + 1 - \frac{Mxy}{E}\right], \quad (8.53)$$

216   8. Elastic and Inelastic Scattering

Structure Functions

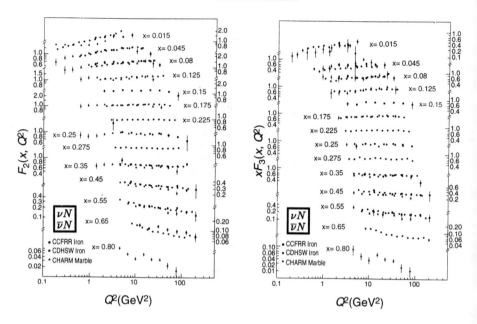

Figure 8.10 The nucleon structure functions F_2 and xF_3 measured in charged-current neutrino and anti-neutrino scattering on iron (CCFRR, CDHSW) and marble (CHARM) targets, versus Q^2, for fixed bins of x. Closed symbols are read on the right-hand scale, open symbols (appearing for alternate x values) on the left-hand scale. Only statistical errors are shown.[17]

$$\frac{d\sigma^{CC}}{dxdy} = \frac{2G^2 MEx}{\pi} \left(\frac{M_W^2}{M_W^2 + Q^2}\right)^2 \left[\sum_{\text{allowed } i} f_i(x) + (1-y)^2 \sum_{\text{allowed } i} \overline{f}_i(x) - \frac{Mxy}{2E} \sum_{\text{allowed } i} (f_i(x) + \overline{f}_i(x))\right]. \quad (8.54)$$

For the CC cross section $f_i(x)$ and $\overline{f}_i(x)$ refer to allowed quarks and anti-quarks respectively. Because the CC current changes charges, there are restrictions on which quarks may occur. For neutrino scattering, d, s, $\overline{u}$, $\overline{c}$ are allowed from the first two generations and for anti-neutrino scattering,

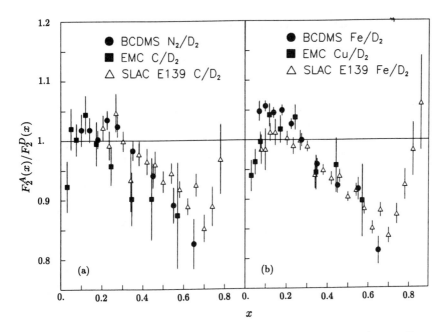

Figure 8.11 EMC effect. The ratio of the structure functions $F_2^A(x)/F_2^D(x)$ for nuclear targets A compared to deuterium D, measured in deep inelastic electron (SLAC-E139) and muon (BCDMS, EMC) scattering. Left graph medium-weight targets (A = N, C); right graph heavier-weight targets (A = Fe, Cu). Only statistical errors are shown.[17]

u, c, $\bar{d}$, $\bar{s}$ are allowed. For anti-neutrino scattering f_i and $\bar{f}_i$ reverse places in the preceding equation.

The definition of f_i is the same in electromagnetic and charged current weak scattering. As a check, the transition between electromagnetic and weak scattering can be seen readily. For electromagnetism, the vertex and propagator factors are $e^4/Q^4 = (4\pi\alpha)^2/Q^4$. For the charged weak interaction at low Q^2, this factor is $[4G/(\sqrt{2}M_W^2)]^2$. In the electromagnetic interaction, the electric current has left- and right-handed components, each of which contribute, one giving the factor "1" and the other giving the factor "$(1-y)^2$." Thus, the $4\pi\alpha^2 MEx/Q^4$ in the electromagnetic cross section becomes $(1/4\pi)(8G^2MEx/M_W^2) = 2G^2MEx/\pi$, which is the factor in front of the CC cross section above.

At high energy the term $Mxy/(2E)$ is small. If that term is small, and if the momentum transfer is small compared to the mass of the intermediate

218   8. Elastic and Inelastic Scattering

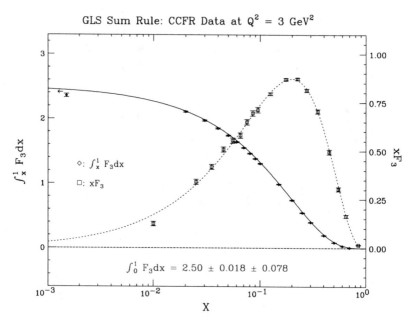

Figure 8.12 CCFR results for $xF_3^{[128]}$ at $Q^2 = 3$ Gev/c^2.

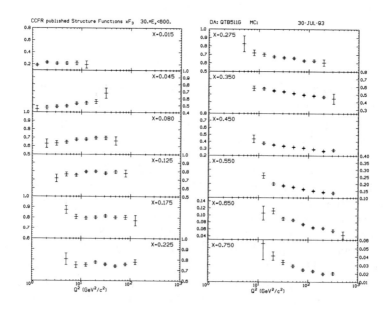

Figure 8.13 CCFR results for $xF_3^{[128]}$ as a function of Q^2.

8.7 Form Factors in the Quark–Parton Picture

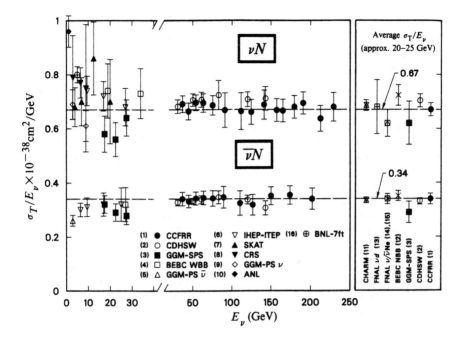

The error bars include both statistical and systematic errors.

Figure 8.14 σ_T/E_ν for ν_μ and $\bar{\nu}_\mu$ charged current total cross sections as a function of E_ν.[17]

boson, then when the weak charged current differential cross section is integrated over x and y, it is evident from the equations above, that the total cross section is proportional to the laboratory energy of the neutrino, E.

To the extent that the mass of the intermediate boson is large compared to the momentum transfer, the interaction is a point interaction. The fact that experimentally the cross section is found to be proportional to E is then experimental evidence of the point nature of the weak interaction at medium energies. See Figure 8.14.

Comparing the expressions for W_1 and W_2 in terms of quark probabilities and the expressions $F_1 = MW_1$ and $F_2 = \nu W_2$, then, in this picture, $F_2 = 2xF_1$. This is known as the Cällen–Gross relation. It is true for spin 1/2 quarks only and therefore can serve as a spin test.

The predominantly $(1-y)^2$ dependence of anti-neutrino scattering (see Figure 8.15) depends on the left handed V−A coupling. Examining this mode is then a test of V−A at high energies. This test has been done for both muon and electron neutrinos. The muon polarization has also been measured in deep inelastic scattering and provides further confirmation of the V−A form at high energies.

By examining Figure 8.16, the origin of this behavior can be seen. Remember that the weak interaction involves couplings of left-handed particles (and right-handed anti-particles). A point interaction is essentially an S-wave scattering. From the figure, it is seen that for νq both helicities are -1 and $J_z = 0$. Hence, an isotropic distribution is obtained. For $\nu \bar{q}$, one has $J_z = -1$. This leads to a $(1 + \cos\theta)/2 \to 1 - y$ in the amplitude (noting $1 - y \propto E'/E$) or a $(1-y)^2$ in the cross section.

There is still another standard notation found in the literature and for purposes of completeness it will be listed here. Define σ_R, σ_L, σ_S, σ_T as the cross sections on nucleons for right-handed, left-handed, scalar (helicity 0), and transverse bosons or photons. Since these are cross sections for virtual particles ("off mass shell"), there is some ambiguity in how they are defined and the parameter κ used subsequently has been defined in two different ways. L. Hand[129] defines $\kappa_{\text{Hand}} = \nu - Q^2/2M$, while F. Gilman[130] defines $\kappa_{\text{Gilman}} = |\vec{q}|$. For the electromagnetic interactions, for either definition, one then defines

$$W_1 = \frac{\kappa}{4\pi^2\alpha}\sigma_T \text{ and } W_2 = \frac{\kappa}{4\pi^2\alpha}(\sigma_T + \sigma_S)\frac{Q^2}{Q^2+\nu^2}, \quad (8.55)$$

while for the weak interactions, one has

$$W_1 = \frac{\kappa}{\pi G\sqrt{2}}(\sigma_R + \sigma_L), \; W_2 = \frac{\kappa}{\pi G\sqrt{2}}\frac{Q^2}{Q^2+\nu^2}(\sigma_R + \sigma_L + 2\sigma_S), \quad (8.56)$$

$$W_3 = \frac{\kappa}{\pi G\sqrt{2}}\frac{2M}{\sqrt{Q^2+\nu^2}}(\sigma_R - \sigma_L). \quad (8.57)$$

Consider the Cällen–Gross relation in terms of $\sigma_{R,L,S}$:

$$W_1/W_2 = (\nu F_1)/(MF_2) = \frac{Q^2+\nu^2}{Q^2}(\sigma_R + \sigma_L)/(\sigma_R + \sigma_L + \sigma_s).$$

If $Q^2 \ll \nu^2$, then $2xF_1/F_2 \to (\sigma_R + \sigma_L)/(\sigma_R + \sigma_L + 2\sigma_S)$. Thus, in the quark parton model $\sigma_S = 0$ for $Q^2 \ll \nu^2$.

8.7 Form Factors in the Quark–Parton Picture

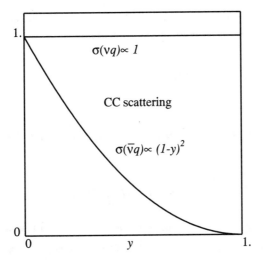

Figure 8.15 Expected y-distributions for νq and $\bar{\nu} q$.

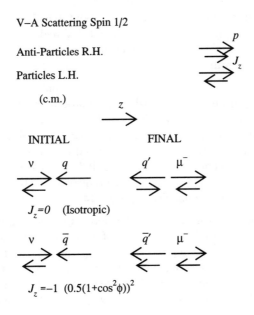

Figure 8.16 y distributions in the quark model.

Consider again the more standard F_1, F_2, F_3 notation. If one integrates the neutrino cross section over all x, one obtains

$$\frac{d\sigma}{dy} \propto [(1-y-y^2/2) \pm B(y-y^2/2) + R'(1-y)], \tag{8.58}$$

$$B = \int xF_3 dx \Big/ \int 2xF_1 dx = \int (u+d-\bar{u}-\bar{d})dx \Big/ \int (u+d+\bar{u}+\bar{d})dx, \tag{8.59}$$

$$R' = \int (F_2 - 2xF_1)dx \Big/ \int 2xF_1 dx. \tag{8.60}$$

The plus sign is chosen for neutrinos and the minus sign for anti-neutrinos.

R' measures the violation of the Cällen–Gross relation. It is found that R' is of the order of 5% or less at high energies. Thus, the approximate validity of the Cällen–Gross relation is confirmed providing further evidence for spin 1/2 quarks.

Ignoring R', then for pure νq or $\bar{\nu}\bar{q}$, B equals $+1$, which implies $d\sigma/dy \propto 1$. For $\nu\bar{q}$ or $\bar{\nu}q$, B equals -1 and $d\sigma/dy \propto (1-y)^2$. See Figure 8.15. Experimentally,[131] it is found that B varies in magnitude from about 0.9 at neutrino energies of about 1 GeV to about 0.6 for neutrino energies above 100 GeV. A typical value in the 20–50 GeV region is about 0.75. The difference from one is seen to be a measure of the anti-quark and, hence, the sea contribution in the nucleon.

Suppose scattering is considered on a nucleus that has an equal number of neutrons and protons. To first order, the quark distributions in a neutron are the same as in a proton, but with u and d interchanged. Thus for nuclei with equal numbers of protons and neutrons, $f_u = f_d$ and the contributions of u and d are the same. Integrate the preceding cross section expression over y from 0 to 1. Then, for B equal 1, it is found that the ratio, $R = \sigma(\bar{\nu}N)/\sigma(\nu N) = 1/3$. The actual measured ratio is about 1/2 reflecting the anti-quark contributions. See Figure 8.14.

A remarkable formalism has been developed in which the structure functions F_1, F_2, and F_3 are only functions of x and are expressed in terms of the probability of the various partons having that x, i.e., that fraction of the momentum of the nucleon. These results are expected to hold at high enough Q^2 that the strong interactions are weakened enough to use perturbation theory. There are a great many experimental consequences of this simple picture.

Consider the ratio of $(F_2)_{em}/(F_2)_{wkCC}$ for nuclei with equal numbers of neutrons and protons. Let $f = f_u = f_d$ and $\bar{f} = f_{\bar{u}} = f_{\bar{d}}$. Since $F_2 = \sum q_i^2 x f_i(x)$, $(F_2)_{em} = (2/3)^2 x(f_u + f_{\bar{u}}) + (1/3)^2 x(f_d + f_{\bar{d}}) = (5/9)x(f+\bar{f})$.

8.7 Form Factors in the Quark–Parton Picture

For the weak interactions, as noted previously, the effective q^2 is taken as 2. For neutrino charged current interactions, only the d or the $\bar{u}$ can contribute as the change in charge of the hadron must be $+1$. Thus, $(F_2)_{wkCC} = 2x(f_d + f_{\bar{u}}) = 2x(f + \bar{f})$. The ratio of $(F_2)_{em}/(F_2)_{wkCC} = 5/18$. This is indeed seen in the data.[132] Including the effect of the strange quarks, the relation becomes

$$\frac{(F_2)_{em}}{(F_2)_{wkCC}} = \frac{5}{18}\left(1 - \frac{3}{5}\frac{s(x)+\bar{s}(x)}{q(x)+\bar{q}(x)}\right). \tag{8.61}$$

Here s and $\bar{s}$ refer to strange quarks and anti-quarks; q and $\bar{q}$ refer to all quarks and anti-quarks. At low x, the correction is about a 10% effect.

Neutrino experiments also have been useful in examining the coupling to charm quarks. One can ask how many strange particles one should see in connection with charm production by neutrinos.[133] For $\nu N \to \mu^- XD$, followed by D decay, one can have $\nu d \to \mu^- c$ (valence quark but suppressed by $\sin^2\theta_c$). Here one sees the strange particle associated with D decay. One can also have $\nu s \to \mu^- c$ (suppressed as the interaction occurs on a sea quark). Here one sees two strange particles, the first associated with D decay and the second from the left over $\bar{s}$ quark in the s–$\bar{s}$ sea pair. The expressions for the two mechanisms are similar in magnitude and both terms contribute to the cross section.

For $\bar{\nu} N \to \mu^+ XD$, only the analogue of the second term exists, $\bar{\nu} s \to \mu^+ \bar{c}$, and more strange particles per event should be seen for charm production by anti-neutrinos than by neutrinos. This is verified by the data.

The x and y dependence in charm production also is qualitatively predictable. Neutrino production of charm involves production from valence and sea quarks while anti-neutrino production of charm involves production from sea anti-quarks only. For V–A currents, both neutrino and anti-neutrino charm production should have an approximately flat y distribution, and this is indeed observed. For anti-neutrinos this is quite different from the $(1-y)^2$ distribution observed from the bulk of the CC events. The anti-neutrino events are predominantly produced at low x as would be expected from the production from sea quarks. Thus, every indication shows that at high energies the charm current couples in the same manner as the other currents. (At lower energies the decay structure picture of the charmed particles provides detailed information confirming the current structure.)

It has been seen that in the electromagnetic and weak scattering, the effective charge of the quarks is different. This helps to separate out the contributions of different quarks. In addition, the comparison between neutrino and anti-neutrino scattering and the comparison between scattering on neutrons and scattering on protons can be used to separate out the contribution of various quarks and anti-quarks.

In heavy nuclei it is found experimentally that nuclear effects make the quark distributions different than just the simple sum of the quark distributions of the individual nucleons[134] (EMC effect). This has provided important information for nuclear physics.

Next consider the fraction of momentum in a proton carried by the quarks.[135] It is necessary that $\int_0^1 xp(\Sigma_i f_i)dx$ = the total momentum of the proton. If it is supposed some of that momentum comes from gluons and if it is recalled that $u = xf_u$, etc.,

$$\int_0^1 (u + \bar{u} + d + \bar{d} + s + \bar{s})dx = 1 - \epsilon_g, \qquad (8.62)$$

where ϵ_g is the fraction of momentum carried by gluons. Gluons are not directly seen by a photon probe since gluons are not electrically charged. Integrating over the data one obtains

$$\int F_2^{ep}(x)dx = \frac{4}{9}\epsilon_u + \frac{1}{9}\epsilon_d = 0.18 \text{ and } \int F_2^{en}(x)dx = \frac{1}{9}\epsilon_u + \frac{4}{9}\epsilon_d = 0.12. \qquad (8.63)$$

Here, ϵ_u and ϵ_d are the fractions of nucleon momentum in the proton carried by u and d quarks, respectively. The neutron has been regarded as having the same wavefunction as the proton with d and u reversed. Thus, ignoring the small strange quark and all anti-quark contributions, $\epsilon_g = 1 - \epsilon_u - \epsilon_d = 1 - 0.36 - 0.18 = 0.46$. About half of the momentum in the proton is carried by gluons.

Several other sum rules have been derived by methods similar to the above.

The Gross – Llewellyn-Smith (GLS) sum rule measures the number of valence quarks in the nucleon with small QCD corrections. Including up to three loop QCD corrections, the sum rule is

$$\frac{1}{2}\int_0^1 \left[F_3^{\nu P}(x) + F_3^{\bar{\nu} P}\right] dx = 3\left[1 - \frac{\alpha_s(Q^2)}{\pi} - \frac{43}{12}\left(\frac{\alpha_s(Q^2)}{\pi}\right)^2 - 18.975\left(\frac{\alpha_s(Q^2)}{\pi}\right)^3\right]. \qquad (8.64)$$

This prediction is $2.55 \pm .05$ and experimentally[136] the result, including the data shown in Figure 8.11 and other data, is $2.50 \pm 0.018 \pm 0.07$ in excellent agreement with the theory. Note that even the highest-order corrections were significant within the experimental errors.

The Gottfried sum rule is

$$\int_0^1 [F_2^{ep}(x) - F_2^{en}(x)]\frac{dx}{x} = (e_u^2 - e_d^2)\int_0^1 (f_u^P(valence) - f_d^P(valence))dx$$

$$= \frac{1}{3}.$$

Experimental results from the NMC group[137] have given 0.242 ± 0.04 confirming older SLAC results of 0.28. This sum rule seems to be violated. The sum rule is really $\frac{1}{3}[1 - \int dx(\overline{f}_d(x) - \overline{f}_u(x))]$. These results may indicate that the $\bar{d}$ sea is not the same as the $\bar{u}$ sea in the proton. This may come from natural charge asymmetries in that there is a positive charge in a proton. Pion clouds leading to a nucleon resonance or interactions of valence quarks (more u than d) with the sea (this is known as "cannibal quarks") may be the cause of the sum rule results.

A similar sum rule for charged weak currents is known as the Adler sum rule:

$$\frac{1}{2}\int \frac{dx}{x}[F_2^{\bar{\nu}P} - F_2^{\nu P}] = \int dx\left[(f_u - \overline{f}_u) - (f_d - \overline{f}_d)\right] = 1. \quad (8.65)$$

Another sum rule examines the fraction of proton spin carried by quarks. It was mentioned in Section 6.2 that an experiment in 1987 seemed to find that only about 1/6 of the spin is carried by quarks at high Q^2. There is still theoretical and experimental controversy concerning the meaning of this result.

In this chapter elastic and inelastic electromagnetic and weak scattering have been examined. It has been seen that at high energies the structure of a nucleon has a simple picture in terms of the quark model. This picture leads to a number of experimentally useful consequences. It allows the measurement of the momentum distribution of the various quarks in the nucleons by means of scattering of charged leptons and neutrinos on nucleons.

8.8 EXERCISES

8.1 Consider the process $q\bar{q} \to W^+$.

    a) For what quark pairs i, j (d, u, s, c, b) is this possible? (Ignore the t quark here.)

226   8. Elastic and Inelastic Scattering

b) Show by diagrams (no equations) that in the $q\bar{q}$ center of mass the direction of spin of the W^+ is determined.

c) Show that, therefore, in the reaction $\bar{q}q' \to W^+ \to e^+ + \nu$ the positron is emitted preferentially in the direction of the initial anti-quark momentum, i.e., there is an asymmetry in the decay.

8.2 Suppose the various quark and lepton scattering processes are parametrized as $\sigma_i = c_i/\hat{s}$, where $\hat{s}$ corresponds to the s of the interacting quarks or leptons.

a) For $e^+e^- \to \mu^+\mu^-$ write down $c_{ee \to \mu\mu}$.

b) For $q\bar{q} \to q\bar{q}$ at high Q^2, estimate $c_{q\bar{q} \to q\bar{q}}$. (Note this involves the strong interaction coupling constant α_s.)

c) For $\bar{p}p$ scattering imagine a quark in the $\bar{p}$ with $\bar{x}$ scattering from a quark in the p with x. Find $\hat{s}$ in terms of the s (center-of-mass energy squared) of the $\bar{p}p$ system. (Assume the energy is large compared to the masses.)

d) Consider u and d quarks in the proton [probability distribution $f_u(x)$ and $f_d(x)$] scattering from $\bar{u}$ and $\bar{d}$ quarks in the anti-proton. Consider valence quarks only. Write an expression for the $\bar{p} + p \to 2$ jets + beamjets cross section.

8.3 Suppose only valence quarks are considered and suppose $\int u \, dx = 2 \int d \, dx$ for a proton. Here $\int u \, dx$ is the total momentum fraction in u quarks in a proton. What would be the ratio of the form factor $\int F_2 dx$ for deep inelastic scattering of e on p to that for e on n?

8.4 Examine some of the considerations in building a neutrino beam.

a) At present accelerators, the principal beams have been built using neutrinos from π and K decays. For the major mode of each, estimate the fraction of the original high-energy laboratory meson momentum that a neutrino emitted at $90°$ in the meson rest frame will retain. For charged pions, almost 100% decay into $\mu + \nu$. For charged kaons, 63.5% decay into $\mu + \nu$.

b) Suppose a reasonably parallel and monoenergetic beam of π and K secondaries is made and maintained over a considerable decay distance, but one small compared to the π and K laboratory decay lengths. Estimate roughly the ratio of neutrinos from π's to neutrinos from K's (for ν and $\bar{\nu}$ separately) that will be seen at a detector located after the decay region. Assume that for fixed momentum in the beam, the ratio of $K^+/\pi^+ = 0.15$ and the ratio of $K^-/\pi^- = 0.05$. The lifetime × velocity of light for the pion is $c\tau_\pi = 7.804$ m. The same quantity for the kaon is $c\tau_K = 3.709$ m.

c) From a), the average energies of these two classes of neutrinos will be quite different. If the charged particle beam was close to parallel, what can be said about the angular distributions of the two classes of neutrinos? Assuming the original beam was small in the transverse direction and that the decay distance was smaller than the shielding distance required afterward to stop the many muons from the decays before the detector, would a correlation between transverse position of a neutrino interaction and neutrino energy be expected?

8.5 A beam of ν or $\bar{\nu}$ interacts with a nucleus with approximately equal numbers of protons and neutrons. Consider the production of $K^\pm$ of high energy in the forward direction. These are mostly K's, one of whose quark components is the struck quark. What can be said about the expected ratio of K^+/K^- for a ν beam? A $\bar{\nu}$ beam? What can be said about the $\nu/\bar{\nu}$ ratios of production of the various fast forward K's? The strange part of the quark sea is believed to be suppressed over the $SU(3)$ expectation because of the mass of the strange quark. Can measurements of any of these ratios give some evidence concerning the amount of this suppression?

8.6 Derive the relation discussed in Section 8.4, i.e., that for small Q^2, the form factor, $F(Q^2)$, can be written as $F(Q^2) \approx 1 - Q^2 \langle r^2 \rangle / 6$.

8.7 An alternate way of obtaining the quark form for the weak form factors W_1 and W_2, which also works for W_3–W_5, is to start from the general hadronic trace, $W^{\alpha\beta}$, given in Equation 8.34. Assume that the "hadron" is a single point quark. Put $W^{\alpha\beta}$ into the same form as $L_{\alpha\beta}$, but with p, p' replacing k, k'. By identifying terms reproduce the relations given for W_1–W_3, and find the appropriate forms for W_4 and W_5. Since $W^{\alpha\beta}$ is meant to be the point quark trace here, M should be replaced by the quark "mass," Mx. Also, since this represents quark elastic scattering there is an implicit $\delta(Q^2/(2Mx) - \nu)$, which brings in an extra x/ν. Include terms proportional to the final lepton mass, which are normally dropped.

8.8 The charged weak deep inelastic scattering cross section given in Equation 8.35, ignores lepton mass terms. Using $W^{\alpha\beta}$ and $L^{\alpha\beta}$, find these terms and, by comparison with Equation 8.35, find the modified equation including these terms. For ν_τ interactions these lepton terms are not negligible. This equation is needed for analysis of some of the searches for neutrino oscillations.

9
The Strong Interaction: Quantum Chromodynamics and Gluons

It is believed that the strong force is mediated by a set of vector bosons called gluons. The gluons are massless, and form an $SU(3)$ octet under the color symmetry. The strong interaction coupling constant, α_s, varies as a function of momentum transfer, Q^2. It is large at low Q^2 ("infrared slavery"), and small at high Q^2 ("asymptotic freedom").

9.1 Evidence for Gluons

Evidence for gluons will be outlined here. Some of this evidence will be discussed in more detail in later sections of this chapter. It was noted previously that half of the momentum in a proton is carried by something other than quarks. This is indirect evidence for gluons. More direct evidence follows from looking at the reaction $e^+e^- \to q\bar{q}$. At high energies, much of the time these events appear as two jets, one formed from the materialization of the quark and the other formed from the anti-quark. However, a fraction of the time three jets are seen. This is believed to be due to the process $q \to q + $ gluon. Calculations of this process using gluons, including the angular correlations and energy distributions of the three jets, give results agreeing with experiment if vector (i.e., spin 1) gluons are used. The results are not in agreement for scalar gluons.

Decays of heavy $J = 1$ $q\bar{q}$ resonances give more evidence for spin 1 gluons. These decays are believed to be dominated by a three-gluon decay mode. Two gluons are forbidden by the same arguments used in Section 4.7a to show that a $J = 1$ two photon state is forbidden. The decay structure is then quite different for spin 0 and spin 1 massless gluons. Experimentally, the data are in good agreement with spin 1 and in strong disagreement with spin 0.

Furthermore, if gluons had even spin, they should couple to particles and anti-particles in the same way and the qq states should be degenerate with the $q\bar{q}$ states in contradiction with observation.

The experimental evidence for asymptotic freedom is quite strong. At low energies strong binding characterized by many bound states and resonances is found. At high energies, as seen most dramatically in deep inelastic scattering, the partons act as quasi-free objects. This is shown by the approximate scaling observed. A number of groups at PEP, PETRA, Tristan, and LEP have determined experimentally the values of the quark–gluon strong coupling constant α_s. The results seem to lie in the region of 0.15–0.20 at the lower PETRA energies and definitely lower, 0.11–0.12, at the higher LEP energies.

Color symmetry is believed to remain unbroken since evidence for free colored particles has not been seen. If color were freed, a whole new spectrum of very strongly interacting particles would appear. The arguments for the color structure of gluons are discussed in the following section.

In the present Standard Model, the bare masses of the gluons are zero. One assumes that, in analogy with QED, the strong interactions are a gauge theory with local gauge invariance. This then requires the gluons to be massless. The analogous result for photons in QED will be discussed in Chapter 11. If the bare masses were not zero, a great number of Higgs particles (see Chapter 11) would be required. In such models one finds that asymptotic freedom is not obtained or that the spectrum of hadrons in phenomenologically unacceptable. The physical mass of a gluon is ill-defined, since gluons are confined; free gluons are not seen. The mass can be defined in various operational senses. Thus, of necessity, the experimental evidence for gluons being massless is soft. The generally good fits to the various "onium" (charmonium, $\cdots$) spectra put some restrictions on the range of the force. If the spectrum of high $p_\perp$ pions in nucleon–nucleon scattering can be shown to approach $p_\perp^{-4}$, then this would be strong evidence for massless gluons. (This is the Rutherford dependence which, as seen several times earlier, just comes from a massless propagator.) However, even if this dependence were seen, it would only imply that the masses of gluons would be small compared to the $p_\perp$ examined. Using a particular operational definition, one estimate, from data, is that gluons can be considered to have an effective mass of about 0.37 GeV.[138]

A gluon phenomenon that may exist is the production of "glueballs," which are particles made up entirely of gluons. These could occur in the several GeV region. They would be flavor singlets, I-spin singlets, and would have charge zero. There are several possible candidates for glueballs, particularly in the region around 1440 MeV, but none is considered certain at present.

9.2 COLOR FORCES

Gluons must carry color. If gluons were colorless, then there would be no reason to favor colorless states and there would be, e.g., eight colored pions, degenerate in mass and other properties with the known colorless pion. If the symmetry group were not $SU(3)$, but $SO(3)$, for example, then

9. The Strong Interaction

the colorless states would not always be favored as it will be shown they are for $SU(3)$. The theory with gluons, in many ways analogous to QED, is known as Quantum Chromodynamics (QCD).

Consider a toy model.[139] Imagine that the single gluon exchange diagram (shown in Figure 9.1a) can be taken as the dominant interaction and that the gluon symmetry group is indeed $SU(3)$. Consider here only the color dependence. For both baryon and meson states, it will be shown that a colorless combination is the lowest-energy state.

Label the eight gluons by their color indices. Consider red (R), blue (B), and yellow (Y) as the primary colors. The octet is formed from pairs of color–anti-color. For six pairs the grouping is obvious: $R\overline{B}$, $R\overline{Y}$, $B\overline{R}$, $B\overline{Y}$, $Y\overline{R}$, $Y\overline{B}$. For the other two one must take account of the fact that the totally symmetric combination, $R\overline{R} + B\overline{B} + Y\overline{Y}$, is a singlet and is not in the octet representation. The last two states must, therefore, be constructed to be orthogonal to this one. (Note that $\langle R\overline{R} \mid B\overline{B}\rangle = 0$, etc.) The states, $(1/\sqrt{2})(R\overline{R} - B\overline{B})$, $(1/\sqrt{2})(R\overline{R} - Y\overline{Y})$, $(1/\sqrt{2})(B\overline{B} - Y\overline{Y})$, are all orthogonal to the singlet state, but only two of the three states are independent. Choose the state $(1/\sqrt{2})(B\overline{B} - Y\overline{Y})$ as one of our two states and then construct the last state from the two other states in such a way that it is orthogonal to $(1/\sqrt{2})(B\overline{B} - Y\overline{Y})$. Since it is a combination of the above states it still will be orthogonal to the singlet state. By inspection it can be seen that the normalized sum of the first two combinations, $(1/\sqrt{6})(2R\overline{R} - B\overline{B} - Y\overline{Y})$, is the proper state. A different set could have been chosen as the basis of the set of orthogonal octet states, but the physical results would be the same.

Start by considering baryon states of two or three quarks. Consider the three-gluon exchange graphs shown in Figure 9.1b. The color factors shown at the vertices are obtained by looking at the gluon color combinations derived previously. For example, the first diagram has a vertex between a gluon and a red quark. The only gluon in the octet with an $R\overline{R}$ in it is the last, $(1/\sqrt{6})(2R\overline{R} - B\overline{B} - Y\overline{Y})$, and the coefficient of the $R\overline{R}$ term is $2/\sqrt{6}$. The product of the factors at each vertex is $(2/\sqrt{6})^2 = 2/3$. The other diagrams follow similarly.

Diagram one in Figure 9.1b shows $RR \to RR$ and has a coefficient of $2/3$. It is symmetric in the exchange of the colors of the two quarks. From diagrams two plus three it is seen that $RB \to -(1/3)RB + (1)BR$. The diagrams for BR instead of RB are exactly similar, of course. Thus, taking the symmetric or anti-symmetric combinations one has (color factors only)

$$RB + BR \to (-1/3)(RB + BR) + (1)(BR + RB) = +(2/3)(RB + BR),$$

$$RB - BR \to (-1/3)(RB - BR) + (1)(BR - RB) = -(4/3)(RB - BR).$$

These results can be summarized by noting that the coefficient can be written as $V_{QQ} = -(1/3) + P$, where $P = \pm 1$ depending on whether the

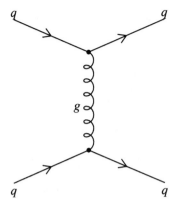

Figure 9.1a Dominant interaction for toy model.

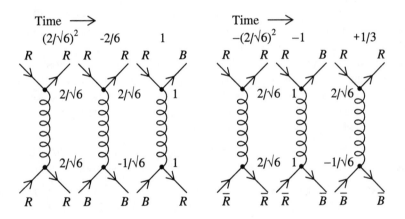

Figure 9.1b qq interactions. Figure 9.1c $q\bar{q}$ interactions.

Figure 9.1 Feynman diagrams for colored gluon exchange. Note that there is an intrinsic − sign for gluon exchange between q and $\bar{q}$.

quark combination is symmetric or anti-symmetric under color exchange. Although only interactions involving at least one red quark have been examined here, the same result must hold for all colors because of the overall color symmetry. There is nothing special about the name "red."

Assume now, for the model, that the energy of a state depends linearly on the expectation value of this quantity V_{QQ}. In addition include a term

Table 9.1 Color Factor Energies for Multi-Quark States

Representation	Symmetry	$\langle E \rangle$
$[QQ]_{3^*}$	AS	$4C/3$
$[QQ]_6$	S	$10C/3$
$[QQQ]_1$	AS	0
$[QQQ]_{10}$	S	$6C$

for the mass of the quarks. Then, $\langle E \rangle = C[\langle V_{QQ} \rangle + (4/3)(\langle N_Q \rangle + \langle N_{\overline{Q}} \rangle)]$. The factor $4/3$ is a convenient normalization as will be seen.

Consider states of two quarks as well as the usual three quarks. For two quarks, $[QQ]$, the possible states can be shown to be 3^* and 6, where the 3^* representation is anti-symmetric and the 6 representation is symmetric. For three quarks, consider the $[QQQ]$ singlet (1) and decuplet (10) representations as examples. Consideration of the mixed symmetry states does not change the conclusions, but they will not be considered here. Note that for the three-quark state there are three pairs of quark–quark couplings, which then introduce a factor of 3 into V_{QQ}. Applying the above formulas for $\langle E \rangle$ and V_{QQ}, Table 9.1 is obtained.

$\langle E \rangle$ is zero for the color singlet and is greater than zero for all of the other states. If $C \to \infty$, then only the color singlet state is allowed.

Next consider the meson states, i.e., the $Q\overline{Q}$ states. Consider the set of three color exchange diagrams shown in Figure 9.1c. For vector exchange between a quark and an anti-quark, it turns out that there is an intrinsic minus sign that reverses the net coefficients for each diagram as shown. This minus sign introduces a difference between the coupling of QQ and $Q\overline{Q}$ states. If the gluon spin were even, this difference would not occur. As noted in the previous section, the observation of $Q\overline{Q}$ states and not the analogous QQ states provides evidence against even-spin gluons.

Consider the singlet $Q\overline{Q}$ state. From the first two diagrams in Figure 9.1c (and from one exactly similar to the second but with $B \leftrightarrow Y$), it is seen that (color factors only) $R\overline{R} \to (-(2/3)R\overline{R} - B\overline{B} - Y\overline{Y})$. Because of the overall color symmetry, the same formula must be true for the other quark–anti-quark initial states with R, B, Y appropriately permuted. Thus, for the symmetric state, $R\overline{R} + B\overline{B} + Y\overline{Y}$, one must have a net coefficient of $-2/3 - 1 - 1 = -8/3$ for V_{QQ}. Combined with the mass term, $(4/3)(\langle N_Q \rangle + \langle N_{\overline{Q}} \rangle)$, this gives $\langle E \rangle = 0$.

Table 9.2 Color Factor Energies for Quark–Anti-Quark States

Representation	Symmetry	$\langle E \rangle$
$[Q\overline{Q}]_1$	S	0
$[Q\overline{Q}]_8$	AS	$9C/3$

For the anti-symmetric state a coefficient of $+1/3$ is obtained as seen from the third diagram in Figure 9.1c. ($R\overline{B} \to 1/3R\overline{B}$ and therefore $R\overline{B} - B\overline{R} \to 1/3[R\overline{B} - B\overline{R}]$.) Table 9.2 summarizes the results.

Again only the singlet has $\langle E \rangle = 0$.

Thus, at least in this toy model, $SU(3)$ color seems to give quark bound states corresponding to the observed nucleons and mesons.

9.3 ARE QUARKS FRACTIONALLY CHARGED?

After discussing color, it is now possible to address the question of whether quarks are fractionally charged. The evidence presented so far certainly indicates the charge of the quarks averaged over color is fractional. The R value and other successes of the colored model of quarks listed in Chapter 6 and the observed ratio of F_2^{CC}/F_2^{em} discussed in Chapter 8 require $(1/3)\sum_{i=r,b,y} q^2(u_i) = (2/3)^2$, etc.

However, one might imagine a model in which the quarks were integrally charged, but the charge of the red, blue, and yellow quarks displaced with respect to each other in a way that the preceding averages were satisfied. Such a model was constructed by M.Y. Han and Y. Nambu.[140] In this model, the electromagnetic current consists of a color singlet (J^0) term (the usual term), but, in addition, has a color octet term (J^8). In the absence of open color, most experiments give the same results for this model as for a fractionally charged quark model.

If color were freed at some energy, and if quarks were integrally charged, the F_2 value would be expected to rise dramatically above threshold. This occurs because, above threshold, the octet part of the electromagnetic current becomes useful. Thus, $(2/3)^2 u(x)$ becomes $(1/3)(1^2 + 1^2 + 0^2)u(x) = (2/3)u(x)$ and similarly for $d(x)$. $F_2(x)$ would increase by a factor of $5/3$. This would be a dramatic signal, which is not seen. However, color is not believed to be freed, so this cannot be used as an argument against the integral charge model.

In one photon processes, $\langle X | J | Y \rangle$, between the usual color singlet final states, only the singlet term contributes. Two photon processes offer a hope of distinguishing the two models. In $\langle X | J^2 | Y \rangle$, the octet contributes because $(J^8)^2$ has a singlet component. However, in most processes such as $e^+e^- \to e^+e^- +$ two jets, there is a dynamical suppression for normal matrix elements. This occurs because the $(J^8)^2$ term involves colored intermediate states. If these are effectively very heavy, then there is an energy denominator from the propagator term that damps out their contribution.

This argument does not apply for π^0, η, η' decays because these decays occur via anomalous triangle diagrams. In these very singular diagrams, the contribution comes from the very-high-mass region and the suppression does not occur. π^0 decay does not distinguish the two models, but η and η' decay do.

The decay $\eta' \to \gamma\gamma$, and the production reaction $e^+e^- \to e^+e^-\eta'(958)$, which goes through a two-photon process, have been used to measure quark charge. The decay width of the $\gamma\gamma$ decay for η' has been measured[141] to be 5.8±1.1 KeV. With standard assumptions (small pseudoscalar octet–singlet mixing angle and equal singlet and octet decay constants), a fractionally charged model predicts 6 KeV and an integrally charged model predicts 26 KeV, the latter in gross disagreement with the data. The assumptions that go into this calculation have been the subject of considerable theoretical discussion,[142] but are now believed to be valid. M.S. Chanowitz has defined a parameter, ξ^2, which is one for fractional and four for integer charges. The above experimental results give $\xi^2 = 1.15 \pm 0.25$.

Some variants of the Han–Nambu model can satisfy even these tests. However, they require either a new photonlike particle[143] or a light Higgs particle.[144] The photonlike particle would require deviation from QED at high Q^2, now ruled out by experiment, and the limits on light Higgs particles from LEP exclude the latter model. Thus, the evidence strongly favors the fractional quark model, although the integral charge model has been surprisingly resilient.

9.4 ASYMPTOTIC FREEDOM; $\alpha_s(Q^2)$

Start by considering QED. At the beginning of the course it was noted that the effective vertex coupling, e, is the product of the bare coupling and correction terms induced by higher-order diagrams such as those shown in Figure 9.2a. These higher-order electron–positron loops act to screen the (infinite?) bare charge of the electron just as a charge is screened by the polarization in a dielectric material.[145] Since these correction terms allow screening in the vacuum similar to dielectric polarization, they are known as "vacuum polarization" terms. At large Q^2 these terms can be summed

9.4 Asymptotic Freedom

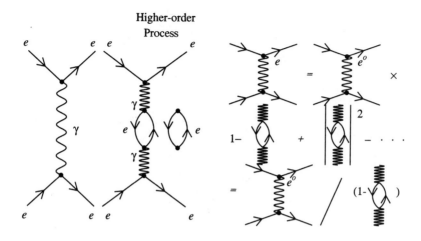

Figure 9.2 Feynman diagrams for QED. Schematic picture of summing interactions to get effective vertex coupling e.

to all orders. Each individual term is proportional to the original element, but has extra factors of $\ln Q^2$. Thus, one can define an effective "running" coupling constant, $\alpha(Q^2)$. One obtains

$$\alpha(Q^2) = \alpha \left[1 + \frac{\alpha}{3\pi} \ln \frac{Q^2}{m^2} + \cdots + \right]. \tag{9.1}$$

For $Q^2/m^2 \to \infty$, one can show that:

$$\alpha(Q^2) = \alpha \bigg/ \left[1 - \frac{\alpha}{3\pi} \ln \left(\frac{Q^2}{m^2} + 1 \right) \right]. \tag{9.2}$$

Diagrammatically one can get a qualitative feeling for this sum from the diagrams in Figure 9.2b.

Note that $\alpha(Q^2)$ increases as Q^2 increases. At high Q^2 the two particles get closer together and penetrate some of the screening cloud of electron–positron pairs. They then feel more the stronger, original charge. When $\alpha(Q^2)$ gets sufficiently large, then perturbation theory breaks down. However, this only occurs at extremely high energy.

Next consider QCD. For this theory, $\alpha_s = g^2/4\pi$ plays an analogous role to α in QED. The diagrams shown in Figure 9.3a play an analogous role in QCD to the role played by the diagrams shown in Figure 9.2a in QED. They again have the same effect. If this were the whole story, then α_s would increase as Q^2 increases just as α does.

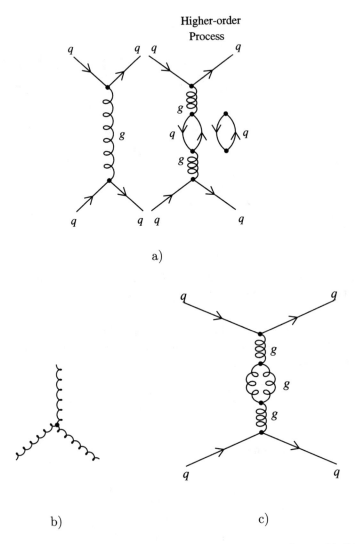

Figure 9.3 Feynman diagrams for QCD. a) QCD quark couplings. b) QCD three-gluon vertex. c) Higher-order process unique to QCD.

However, a new effect enters for QCD that works in the opposite direction. The gluon field carries color, i.e., gluons possess a color charge. This is quite different from photons that are neutral particles, i.e., do not carry the electromagnetic charge. Gluons can interact with themselves as shown in Figure 9.3b. There is a paramagnetic spin–spin attraction between the gluons. Thus, a gluon can transport color away from the original quark leaving the quark less charged. Effectively, the color source is smeared out and made less pointlike. See Figure 9.3c. (Although this argument is very

9.4 Asymptotic Freedom

qualitative, and these diagrams go into each other when Lorentz frames are changed, the argument can be made rigorous for a non-abelian theory.)

Thus the quark–anti-quark loops provide screening and the gluon pair loops provide anti-screening. Which wins? It turns out that it depends on how many different kinds of gluons (8) there are and how many different kinds of quarks there are, since there is a separate loop for each kind of quark and each kind of gluon. Count each quark $(u, d, s, \cdots)$ as a separate flavor (f) of quark. Because of color there are $3f$ different kinds of quarks. If $f \leq 16$, then anti-shielding wins; otherwise shielding wins. It is believed that there are 6 flavors (three generations) and there is then net anti-shielding in agreement with the experimental evidence of asymptotic freedom and infrared slavery. (If f is greater than 16, but only because there are some ultra heavy new quarks, then there would be temporary freedom in which the coupling constant would decrease as Q^2 increases, until the new mass threshold is passed.)

By looking at higher-order diagrams one can obtain a formula for $\alpha_s(Q^2)$ relative to $\alpha_s(\mu^2)$, where μ is an arbitrary mass. It is found that as $Q^2 \to \infty$

$$\alpha_s(Q^2) = \alpha_s(\mu^2) \left[1 + \frac{\alpha_s(\mu^2)}{12\pi} (2f - 33) \ln \frac{Q^2}{\mu^2} + \cdots + \right]. \tag{9.3}$$

The diagrams can be summed in the leading log approximation in the same manner as the QED diagrams (see Figure 9.2e) to obtain for $Q^2 >> \mu^2$:

$$\alpha_s(Q^2) = \alpha_s(\mu^2) \bigg/ \left[1 + \frac{\alpha_s(\mu^2)}{12\pi} (33 - 2f) \ln \frac{Q^2}{\mu^2} \right]. \tag{9.4}$$

Let Λ be defined by $\ln \Lambda^2 = \ln \mu^2 - (12\pi)/((33 - 2f)\alpha_s(\mu^2))$. Then for $Q^2 >> \Lambda^2$:

$$\alpha_s(Q^2) = \frac{12\pi}{(33 - 2f) \ln \frac{Q^2}{\Lambda^2}}. \tag{9.5}$$

If Λ is taken as 500 MeV, then for $Q^2 \approx (1 \text{ GeV})^2$, it is found that $f_{\text{effective}} = 3$ and for $Q^2 \approx (10 \text{ GeV})^2$, it is found that $f_{\text{effective}} = 5$. This is expected since at 1 GeV, the heavy quark degrees of freedom are not yet active. α_s is about 1 at $Q^2 \approx (1 \text{ GeV})^2$, and 0.15–0.2 at PETRA and PEP energies. This value is still much greater than $\alpha(\text{QED}) = 0.0074$. α_s decreases very slowly (logarithmically) and, because it is large, QCD perturbation theory converges much more slowly than QED perturbation theory.

In the reaction $e\bar{e} \to$ hadrons there are events with two, three, four, and more jets. The hadronic decays of the Z also form a collection of hadronic jets produced at high energy. These can be used to make QCD investigations. The definition of what constitutes a jet experimentally is somewhat arbitrary. A useful algorithm has proven to be the "JADE" algorithm.[146] One takes the effective mass of pairs of tracks or clusters of hits for the event. The pair with the lowest mass is combined into a single cluster. Each cluster is taken as massless in the calculation of effective mass. The procedure is then repeated until the smallest scaled invariant mass squared is greater than some cutoff. This is then compared with Monte Carlo calculations using the same cutoff. The three jet events include a gluon jet and hence measure the strength of α_s. From looking at the ratio of events with three jets to events with two jets, the strong interaction coupling can be obtained. Using this result together with several other methods, the combined LEP[147] result is

$$\alpha_s(\sqrt{s} = M_Z) = 0.119 \pm 0.006,$$

$$\Lambda_{\overline{MS}} = 192^{+69}_{-55} \text{ MeV}.$$

The theoretical errors are dominated by renormalization scale uncertainties. Using the three jet ratio only, the LEP result[148] is $\alpha_s(\sqrt{s} = M_Z) = 0.019 \pm 0.010$.

The running of the coupling constant can be seen by looking at the ratio of three jet to two jet events in e^+e^- collisions as a function of center of mass energy. This ratio is seen in Figure 9.4. The ratio clearly falls with energy indicating the decrease of α_s with energy as expected according to asymptotic freedom.

QCD is a non-abelian theory. The gluon carries the color charge, while the photon has no electromagnetic charge. If the diagrams corresponding to $e\bar{e} \to$ four jets (Figure 9.5) are examined, it is seen that the diagrams fall into two classes, leading to $q\bar{q}gg$ and $q\bar{q}q\bar{q}$, respectively. The two diagrams in the first class transform into one another by appropriate gauge transforms. One might imagine a theory in which only the second class of diagrams were present. The predictions for the four-jet events are quite different in QCD and in this abelian theory. Define an angle (Nachtmann–Reiter angle) as the angle between the momentum vector differences of jets 1,2 and 3,4, where the jets are in order of decreasing total momentum. Figure 9.6 shows a comparison of the Nachtmann–Reiter angle with L3 Z decay data,[150,151] for QCD, and for the "abelian" theory with only the second class of diagrams. The data clearly require the first class of diagrams. (However, in the usual gauge the first class is dominated by the double bremsstrahlung not the three-gluon coupling. An improved analysis method for four-jet events by the ALEPH group removed this difficulty.[152])

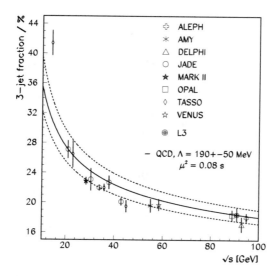

Figure 9.4 Variation of three-jet fraction in $e^+e^- \to$ jets as a function of center of mass energy, $\sqrt{s}$.[149]

The gluon–gluon interaction qualitatively helps explain confinement. For e^+e^-, the lines of force of the electric field correspond to a dipole as shown in Figure 9.7a. For two quarks, the attractive interaction between gluons tends to compress the lines of force together. Although not quite proven, it appears that the lines of force are confined into a tube of radius approximately the range of the force as is shown in Figure 9.7b. The energy in the field is given by an integral of the square of the field over the volume as with QED. However, for the tube, this leads to an energy proportional to the separation of the particles. If the confining potential is linear in the separation, the force is constant rather than decreasing with distance as in QED. Hence, the quarks do not become free. When the energy is sufficient, a pair of new quarks can be created, but the quarks remain confined.

Consider the anti-shielding concept further. For color this corresponds to an effective color dielectric constant, κ, less than 1. Consider the limit, $\kappa \to 0$. In this limit, there is a strong repulsion between a color charge and the normal vacuum. If the vacuum is considered as a material, it can be shown that, in this limit, a hole or bubble is formed around the charge. The particle stays in this bubble and, within it, is approximately free. The particle wavefunction goes to zero at the boundaries of the bubble.

This is the basis of the "bag models" of hadrons. In these models, the hadron is treated as a bag, within which quarks are free. (It can be shown that this corresponds to treating hadrons as a type of solution known as

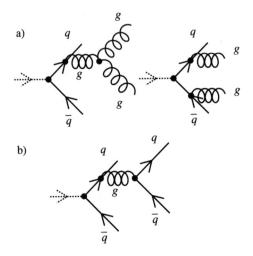

Figure 9.5 Two classes of couplings leading to four-jet events.

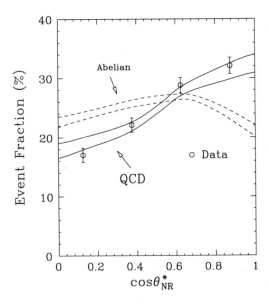

Figure 9.6 Test of QCD using four-jet events. Event fraction versus Nachtmann–Reiter angle. This angle is the angle between the momentum vector differences of jets 1,2 and 3,4 where the jets are ordered by momentum.[151]

(a) (b)

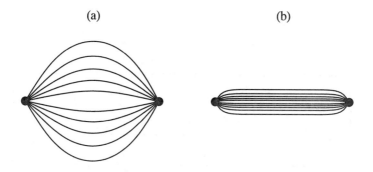

Figure 9.7 Comparison of lines of force for QED and QCD. a) QED. b) QCD.

a "soliton" solution.) The mass of a hadron phenomenologically can be written as[1]

$$m = m_0 + N\frac{\xi}{R} + \frac{4\pi}{3}R^3 p + 4\pi R^2 s, \qquad (9.6)$$

where ξ, p, and s are parameters. Because the medium exerts "pressure" on the bubble, there is a volume energy, $(4\pi/3)R^3 p$. "Surface tension" accounts for the area term, $4\pi R^2 s$, and "thermodynamical energy" of the gas inside the bubble accounts for the kinetic energy term, $N\xi/R$. From the uncertainty principle it is seen that for zero mass quarks, the kinetic energy is proportional to $1/R$. The constants are not all independent. R is determined by minimizing m. This leads to

$$4\pi R^3(2s + Rp) = N\xi. \qquad (9.7)$$

Two popular bag models correspond to zero surface energy, $\xi = 2.0428$ (MIT bag)[153] and zero volume energy, $\xi = 1$ (SLAC bag).[154] When gluons are included, the constant ξ changes to $\xi_p = \xi + O(s)$.

For zero quark mass, comparison with experiment is given in Table 9.3. Introducing quark mass terms can correct the poor estimate for C_A/C_V, the axial vector to vector weak coupling constant ratio. The model also makes some predictions involving the ratio of ρ, Δ, π, and nucleon masses, valid to about 20%.

It has been seen that the effective QCD coupling constant decreases and the effective QED coupling constant increases with energy. Is there some high energy at which these constants and perhaps also the weak interaction coupling constant become equal? It is believed this does happen at an

Table 9.3 Comparison of Bag Model Predictions with Experiment

Observable	Experiment	MIT bag	SLAC bag
$R_p = R_{\text{proton}}$	$3.86/m_N$	$4.25/m_N$	$3.21/m_N$
μ_p	$2.79/(2m_N)$	$2.36/(2m_N)$	$2.14/(2m_N)$
μ_n	$-0.685\mu_p$	$-2/3\mu_p$	$-2/3\mu_p$
C_A/C_V(nucleon)	1.26	1.09	5/9

energy called the "unification energy." This is shown in Figure 9.8. There, $\alpha_s = g_3^2/4\pi, g = g_2$, and $g_1 = (\sqrt{5/3})g'$. g is the charged current weak coupling constant to the W. The electromagnetic coupling constant is given by $gg'/\sqrt{g^2 + (g')^2}$ (see Chapter 10). The $\sqrt{5/3}$ factor depends on assumptions about further unification and is chosen here to correspond to $SU(5)$. (See Chapter 12.) There seems to be a much larger symmetry above the unification energy which is broken as one goes down in energy to the more accustomed energy scales. This will be discussed further later.

9.5 Q^2 Evolution of Structure Functions

Suppose one could look inside of a quark at shorter and shorter distance scales, i.e., at higher and higher Q^2.[135] One would see more and more structure as shown in Figure 9.9a. At high Q^2 the quark dissociates into a quark and a gluon as shown in Figure 9.9b. With an electron probe, one interacts with a quark that is degraded in energy, at $x' < x$. Furthermore, at high Q^2 there are more sea quark pairs with which to interact. Thus, at larger Q^2 the structure function is expected to be more concentrated at small x and less at large x. The crossover point is about $x \sim 0.2 - 0.3$. Experimental results are shown in Figures 8.7, 8.8, 8.10, and 8.13.

Consider a proton. Let $f_u(x, Q^2)$ be the probability that a u quark be within dx of x when probed at Q^2. Go to a Lorentz frame moving very fast with respect to the proton ("infinite momentum frame"). It was argued in Chapter 8, that, in this frame, because of time dilation, a snapshot is taken of the instantaneous state of the components of the proton. Suppose there is no scale factor, i.e., suppose that the fractional change in f_u is proportional to the fractional change in Q^2. Then,

9.5 Q^2 Evolution of Structure Functions

$$\frac{df_u(x,Q^2)}{d\ln Q^2} = \alpha_s(Q^2) \int_{y=x}^{y=1} \left\{ f_u(y,Q^2) \left[P_{QQ}\left(\frac{x}{y}\right) \right] \right.$$

$$\left. + f_g(y,Q^2) \left[P_{QG}\left(\frac{x}{y}\right) \right] \right\} dy. \quad (9.8)$$

Here $\alpha_s P_{QQ}(z)$ is the probability of a quark splitting into $q+g$ and the quark having a fraction z of the original momentum. $\alpha_s f_u(y,Q^2) P_{QQ}(x/y)$ is the probability of a quark having momentum fraction y to begin with and then splitting down to x. The second term in the equation is a similar term for an initial gluon giving a quark by way of a quark–anti-quark loop. This is known as an Altarelli–Parisi equation. As is evident one can write a similar equation for gluons, etc. From looking at Feynman diagrams one can show to first order

$$P_{QQ}(z) = \frac{4(1+z^2)}{3(1-z)_+} + 2\delta(1-z), \quad (9.9)$$

$$P_{QG} = \frac{1}{2}(z^2 + (1-z)^2), \quad (9.10)$$

$$P_{GQ} = \frac{4[1+(1-z)^2]}{3z}, \quad (9.11)$$

$$P_{GG} = 6\left(\frac{1-z}{z} + \frac{z}{(1-z)_+} + z(1-z)\right). \quad (9.12)$$

Here $1/(1-z)_+$ is defined by:

$$\int_0^1 \frac{h(z)}{(1-z)_+} dz \equiv \int_0^1 \frac{h(z) - h(1)}{1-z} dz. \quad (9.13)$$

Thus $(1-z)_+ = 1-z$ for $z < 1$ and is infinite for $z = 1$.

The main point of this is that one can calculate, in terms of simple functions, the evolution of the structure functions. It gets very complicated when taking higher-order diagrams into account, but these effects can be estimated and comparisons made with experiments. As Q^2 increases, the amount of splitting and fraction of low x quarks and gluons increases, increasing the effective size of the sea.

In this chapter $SU(3)$ color symmetry, asymptotic freedom, and the Q^2 evolution of structure functions have been discussed as an introduction to the strong interactions. It has been indicated how one can hope to make calculations with QCD to compare with experiments.

244 9. The Strong Interaction

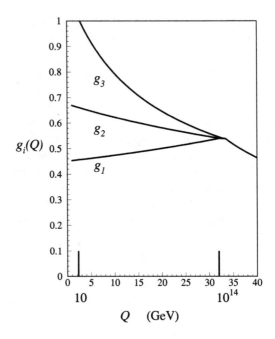

Figure 9.8 Effective coupling constants versus energy.

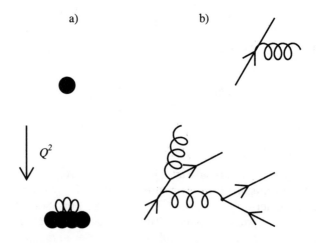

Figure 9.9 Evolution of quark structure at high Q^2.

9.6 Exercise

9.1 The lowest upsilon state ($b\bar{b}$ state) is colorless, has spin-parity = 1^- and $l = 0$. Show that decays via one or two gluons are forbidden; the lowest allowed decay will involve three gluons. (For other spins, the two gluon mode is not necessarily forbidden.) Hint: for the two gluon state first consider $C(\bar{b}b)$, referring back to the e^+e^- discussion in Section 4.11. Note that instead of writing gluon fields as g_k, $k = 1, \cdots, 8$, one can write G_{ij}, where i and j index from one to three, i.e., index over the three colors. In this representation, to get eight independent fields, the trace of G is taken to be zero. As a generalization of charge conjugation for photons, $CG_{ij} = -G_{ji}$. A two-gluon, colorless state must have no free color indices and hence must be of the form $G_{ij}G_{ji}$. Find C for this state and compare with that of the upsilon.

10
The Standard Model

10.1 $SU(2) \otimes U(1)$ Weinberg–Salam Model; Neutral Currents

There are now two neutral currents, j^{NC} and j^{em}, where NC stands for the weak neutral current, and em for the electromagnetic current. It is known that the charged weak current, j^{CC} is related to $j^{em}(I=1)$ since they refer to the same I-spin multiplet. This was the conserved vector current hypothesis. Is there a relation between these two neutral currents?

First define weak isospin. For weak isospin consider not only the quark doublets, but also the lepton doublets. A distinction will be made between right- and left-handed leptons and quarks. Let L be left-handed, $(1-\gamma^5)/2$, and R be right-handed, $(1+\gamma^5)/2$. Consider as left-handed doublets under I^{weak}:

$$\begin{array}{cccccc} u & c & t & \nu_e & \nu_\mu & \nu_\tau \\ d & s & b & e^- & \mu^- & \tau^- \end{array} \tag{10.1}$$

This is an $SU(2)$ group. Since it refers to left-handed objects, it is labeled $SU(2)_L$. Next recall that $Q = I_3 + (B+S)/2 = I_3 + Y/2$, where Y is called hypercharge. $Y = (B+S) = 2(Q - I_3)$.

Define $Y^{weak} \equiv 2(Q - (I_3)^{weak})$. Y^{weak} has the same value for each member of a doublet. For example, for the doublet containing ν_e and e^-, $Y^{weak} = -1$. For the quark doublet containing u and d, $Y^{weak} = 1/3$. Y^{weak} corresponds to a $U(1)$ symmetry and the combined I^{weak}, Y^{weak} set of transforms corresponds to $SU(2)_L \otimes U(1)$. I^{weak} is not equal to I. I^{weak} is $1/2$ for the left-handed s quark and I is 0. I^{weak} is $1/2$ for left-handed leptons and I is 0. Let χ_L be a left-handed two-component spinor in weak isospin space.

Take the right-handed particles (and left-handed anti-particles) as weak singlets ($I^{weak}=0$) having, therefore, $Y^{weak} = 2Q$.

The electromagnetic current for the u and d quarks and for the ν_e and e^- leptons can then be written

$$(j^{em})^\alpha(u,d) = \left[\bar{\chi}_L^{ud}\left((I_3)^{weak} + \frac{1}{6}\right)\gamma^\alpha \chi_L^{ud} + \frac{2}{3}\bar{u}_R \gamma^\alpha u_R - \frac{1}{3}\bar{d}_R \gamma^\alpha d_R\right], \tag{10.2}$$

10.1 $SU(2) \otimes U(1)$ Weinberg–Salam Model

$$(j^{em})^\alpha(\nu_e, e^-) = -\bar{e}\gamma^\alpha e = -\bar{e}_L\gamma^\alpha e_L - \bar{e}_R\gamma^\alpha e_R$$

$$= \left[\bar{\chi}_L\left((I_3)^{weak} - \frac{1}{2}\right)\gamma^\alpha\chi_L - \bar{e}_R\gamma^\alpha e_R\right]. \quad (10.3)$$

In both cases this can be written as

$$(j^{em})^\alpha = [\bar{\chi}_L((I_3)^{weak} + Y^{weak}/2)\gamma^\alpha\chi_L + (Y^{weak}/2)\bar{\chi}_R\gamma^\alpha\chi_R] \quad (10.4)$$
$$= [\bar{\chi}((I_3)^{weak} + Y^{weak}/2)\gamma^\alpha\chi].$$

Note that χ_R here is a weak isospin singlet.

The weak charged current couples to $(I^+)^{weak}$ and the electromagnetic current couples to a mixture of $(I_3)^{weak}$ and Y^{weak}. Suppose it is imagined that there are two neutral currents, mixtures of weak and electromagnetic, such that one couples to Y^{weak} and has a boson B^0 associated with it and the other couples to $(I_3)^{weak}$ and has a boson W^3 associated with it. The physical photon and Z are, then, mixtures of the W^3 and B^0. Thus, it is suggested that the weak and electromagnetic forces are inextricably mixed, part of one grand electromagnetic–weak force! The interactions above and those in the previous chapters can be considered as pieces of the Lagrangian (see Appendix A) of fundamental particles. Try, as a theoretical assumption, a Lagrangian for the interaction part for both the electromagnetic and the weak forces (neutral and charged) as follows:

$$\mathcal{L} = g(j^\alpha \cdot W_\alpha) + \frac{g'}{2}(j^{Y^{weak}})^\alpha B_\alpha, \quad (10.5)$$

where the dot product is over the I-spin indices. Here j is the weak I-spin current,

$$(j^i)^\alpha = \bar{\chi}_L(I_i)^{weak}\gamma^\alpha\chi_L, = \bar{\chi}(I_i)^{weak}\gamma^\alpha\chi(\text{carrying } I^{weak}) \quad (10.6)$$

and $j^{Y^{weak}}$ is the weak hypercharge current,

$$(j^{Y^{weak}})^\alpha = \bar{\chi}_L Y^{weak}\gamma^\alpha\chi_L + \bar{\chi}_R Y^{weak}\gamma^\alpha\chi_R, = \bar{\chi}Y^{weak}\chi(\text{carrying}Y^{weak}). \quad (10.7)$$

Note that the electromagnetic current $j^{em} = j^3 + j^{Y^{weak}}/2$. As noted above, in these equations, there are the bosons $W^\pm, W^3, B^0$, while the physical bosons are $W^\pm, Z, A$, where A stands for the photon. Set

$$W^\pm_\alpha \equiv \frac{1}{\sqrt{2}}(W^1_\alpha \pm iW^2_\alpha); \text{ and } j^\pm_\alpha \equiv (j^1_\alpha \pm ij^2_\alpha), \qquad (10.8)$$

$$W^3_\alpha \equiv \frac{gZ_\alpha + g'A_\alpha}{\sqrt{g^2 + (g')^2}}, \qquad (10.9)$$

$$B^0_\alpha \equiv \frac{-g'Z_\alpha + gA_\alpha}{\sqrt{g^2 + (g')^2}}. \qquad (10.10)$$

The first equations are the usual transform to get to $+,-$ from 1,2. They are seen in the discussion of angular momentum raising and lowering operators in Appendix C. The last two equations define Z, A as orthogonal transforms of W, B. With this new notation,

$$\mathcal{L} = g((j_1)^\alpha W^1_\alpha + (j_2)^\alpha W^2_\alpha) + (j_3)^\alpha \frac{1}{\sqrt{g^2 + (g')^2}} (g^2 Z_\alpha + gg' A_\alpha$$

$$+(g')^2 Z_\alpha - gg' A_\alpha) + \frac{1}{\sqrt{g^2 + (g')^2}} (j^{em})^\alpha (-(g')^2 Z_\alpha + gg' A_\alpha), \qquad (10.11)$$

$$\mathcal{L} = \frac{g}{\sqrt{2}}((j^-)^\alpha W^+_\alpha + (j^+)^\alpha W^-_\alpha)$$

$$+\sqrt{g^2 + (g')^2}(j^3)^\alpha Z_\alpha + \frac{1}{\sqrt{g^2 + (g')^2}} (j^{em})^\alpha (-(g')^2 Z_\alpha + gg' A_\alpha). \qquad (10.12)$$

Next set:

$$g'/g = \tan\theta_W \implies \frac{g'}{\sqrt{g^2 + (g')^2}} = \sin\theta_W \text{ and } \frac{g}{\sqrt{g^2 + (g')^2}} = \cos\theta_W. \qquad (10.13)$$

Then,

$$\mathcal{L} = \frac{g}{\sqrt{2}}((j^-)^\alpha W^+_\alpha + (j^+)^\alpha W^-_\alpha) + \frac{g}{\cos\theta_W}(j^3)^\alpha Z_\alpha$$

$$+(j^{em})^\alpha(-g\sin\theta_W \tan\theta_W Z_\alpha + g\sin\theta_W A_\alpha), \qquad (10.14)$$

$$\mathcal{L} = \frac{g}{\sqrt{2}}((j^-)^\alpha W^+_\alpha + (j^+)^\alpha W^-_\alpha)$$

$$+ \frac{g}{\cos\theta_W}((j^3)^\alpha - (j^{em})^\alpha \sin^2\theta_W)Z_\alpha + g\sin\theta_W (j^{em})^\alpha A_\alpha. \qquad (10.15)$$

In the last equation the first term is the weak charged current, the second term is the weak neutral current, and the last term is the electromagnetic current. θ_W is called the Weinberg angle.

10.1 $SU(2) \otimes U(1)$ Weinberg–Salam Model

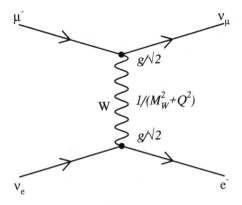

Figure 10.1 Feynman diagram: effective weak coupling constant.

A number of things can be concluded from this representation. One must have $g \sin\theta_W = e$, since the strength of the electromagnetic current is known to be e.

What about the $(1 - \gamma^5)$? This is in the preceding equations correctly for the weak charged current. $I^{weak} = 0$ for right-handed particles; $(j_i)^\alpha = \overline{\chi}_L I^{weak^i} \gamma^\alpha \chi_L$. Thus, the weak charged current is left-handed as required by experiment.

The neutral current in the preceding theory is not a pure left-handed current because of the presence of j^{em}. It is $V - CA$, where the constant C is not one. The force does not contain any component coupling left-handed to right-handed neutrinos. CVC and PCAC come about because there is a gauge theory of electroweak interactions and the strong interactions respect that symmetry.

The mass of the charged W boson can be obtained from the preceding Lagrangian. By comparing with the effective β-decay matrix element for muon decay (Equation 7.52) it is seen that $4G/\sqrt{2} = g^2/2M_{W^\pm}^2$. See Figure 10.1. Note also that for muon decay $Q^2 \ll M_{W^\pm}^2$. Hence,

$$M_{W^\pm}^2 = \frac{g^2 \sqrt{2}}{8G} = \frac{e^2 \sqrt{2}}{8G \sin^2\theta_W} \implies M_{W^\pm} = \frac{37.28}{\sin\theta_W} \text{GeV}. \quad (10.16)$$

Radiative corrections increase this estimate by about 4 GeV to 80.2±1.1GeV given the measured value of $\sin^2\theta_W$. The measurement of the mass of the W boson (80.22 ± 0.26 GeV) is in good agreement with this value.

The W magnetic moment[1] in this model is $\vec{M} = (e/m)\vec{s}$. If parametrized in the conventional way, $\vec{M} = (1+\kappa)(e/2m)\vec{S}$, then the anomalous magnetic moment, $\kappa = 1$.

The mass of the Z is not as simple. It depends on the model. In the model of Weinberg and Salam, there is one Higgs scalar doublet (to be discussed in the next chapter) and

$$M_Z = M_{W^\pm}/\cos\theta_W = \frac{74.56}{\sin 2\theta_W} \text{ GeV} + \sim 3 \text{ GeV}(\text{ radiative corrections }). \quad (10.17)$$

Using the measured value for $\sin^2\theta_W$, the Z mass is predicted to be 92 ± 0.7 GeV.[155] The measured Z mass is 91.187 ± 0.007 GeV.

What is the experimental evidence for this model? The $W^\pm$ has been found with the predicted mass. The Z has been found with the mass predicted by the simple one-Higgs model. One can start with the most general sets of coupling constants for the various lepton–lepton, and lepton–quark couplings. From the many experiments now available it is possible to show that the Standard Model couplings previously given are essentially the only solution. These experiments include, among others, neutrino scattering experiments, the observation of γ–Z interference, and the set of measurements of $e^+e^- \to Z$ described in Chapter 13. The helicity and isospin structure are firmly established. The set of measurements at the Z resonance has obtained accurate values for the NC vector and axial vector couplings for the various leptons and quarks (except for the t). A value of $\sin^2\theta_W$ is obtained from NC/CC ratios in neutrino interactions. This is just a measurement of the relative strength of the NC coupling compared to the CC coupling. The prediction of the mass of the W is essentially the same thing, not really a prediction. However, given those measurements, the precise form of the NC coupling is then a prediction. A very beautiful experiment at SLAC[156] measured the electromagnetic–weak interference (about a 10^{-5} effect) in the scattering of polarized electrons from deuterons. This was in fine agreement with the preceding model and in disagreement with pure V–A force. Further measurements of $\sin^2\theta_W$ come from $ee \to \mu\mu$ asymmetries (electromagnetic–weak interference terms), νe elastic scattering, νN elastic scattering, atomic parity violation terms, etc. All of these give consistent values of $\sin^2\theta_W$ if the mass of the t quark is under about 200 GeV as will be discussed later. This can be looked at as giving an upper limit on the t mass. An average of many low-energy results yields $\sin^2\theta_W = 0.2284 \pm 0.0043$. As will be seen in Chapter 13, the values of the vector and axial vector parts of the NC coupling are now known with high precision from Z decays, and are an accurate confirmation of this model. Using the LEP results, the result becomes $\sin^2\theta_W = 0.23172 \pm 0.00024$.[49]

Thus by the mid 1970s the weak and electromagnetic forces had merged into a single combined law. This development is one of the glories of twentieth century physics and ranks with the merging by Faraday and Ampere of the electric and magnetic forces, and with the merging by Maxwell of electromagnetic theory and light.

10.2 Weak Neutral Current Neutrino Interactions

In this section the neutrino scattering results of Chapter 8 will be extended to weak neutral current interactions. First consider the relative overall normalizations of NC and CC interactions. For $Q^2 \ll M_W^2$, a CC neutrino matrix element will involve $(g/\sqrt{2})^2/M_W^2$, i.e., the product of the two vertex couplings and $1/M_W^2$ from the propagator term. A neutral current neutrino interaction will have $(g/\cos\theta_W)^2/M_Z^2$. However, as noted above, in the simplest case, $M_Z^2 = M_W^2/\cos^2\theta_W$, and $(g/\cos\theta_W)^2/M_Z^2 = g^2/M_W^2$, a factor of 2 different from the CC case. Next separate out the left- and right-hand part of the NC couplings. Let g_L be the left-handed coupling of the Z to the fermion and g_R be the right-handed coupling. Write the part of the Lagrangian coupling the Z to fermions (f) as

$$\mathcal{L}_{NC} = \frac{g}{\cos\theta_W}\left(g_L \bar{f}\gamma^\alpha \frac{(1-\gamma^5)}{2}f + g_R \bar{f}\gamma^\alpha \frac{(1+\gamma^5)}{2}f\right) Z_\alpha. \quad (10.18)$$

For a given fermion, one obtains from the Lagrangian, Equation 10.15, and from the definitions of the currents,

$$g_L = I_3^{weak} - q\sin^2\theta_W, \quad g_R = -q\sin^2\theta_W. \quad (10.19)$$

Here q is the electric charge of the fermion in units such that the charge on the electron is -1. One sometimes uses the vector (g_V) and axial vector (g_A) parts of the coupling, which are given by

$$g_V = g_L + g_R = I_3^{weak} - 2q\sin^2\theta_W; \quad g_A = g_L - g_R = I_3^{weak}. \quad (10.20)$$

Examine muon neutrino–electron elastic scattering. For the left-handed current, the argument follows that given in Chapter 8 for CC neutrino interactions with the preceding change of normalization. For the right-handed part, the anti-neutrino CC scattering argument applies. The leptons have no structure, so the x-distributions are of the form $\delta(x-1)$. Use the form of F_1, F_2, and F_3 for the CC interactions given in Chapter 8 [e.g., $F_2(x) = 2x \sum_i f_i(x)$, etc.] One needs the neutrino couplings to the Z and the electron couplings to the Z. Recognizing $g_L = 1/2$, $g_R = 0$ for the neutrino coupling to the Z and integrating over x, one obtains

$$\left.\frac{d\sigma}{dy}\right|_{\nu e(NC)} = \frac{2G^2 mE}{\pi}(g_L^2 + g_R^2(1-y)^2), \quad (10.21)$$

where g_L and g_R refer to the electron couplings to the Z. For anti-neutrino electron scattering g_L and g_R switch places. For electron neutrino–electron scattering there is an additional W exchange diagram and the results are somewhat more complicated. (See Exercise 10.3.)

Deep inelastic neutrino neutral current scattering from hadrons can be derived starting from the neutrino electron considerations above. m in the νe scattering is replaced by $xM_{nucleon}$, the effective quark mass, as was done in Chapter 8 for the CC current. Since for deep inelastic scattering, Q^2 is not always small, the propagator term $M_Z^2/(Q^2+M_Z^2)$ must be added and one does not integrate over x. One obtains

$$\frac{d\sigma}{dxdy}\bigg|_{\nu N(NC)} = \frac{2G^2 ME}{\pi}\left(\frac{M_Z^2}{Q^2+M_Z^2}\right)^2 \left(\sum_i [(g_{Li}^2 + (1-y)^2(g_{Ri}^2))xf_i(x)]\right.$$
$$\left. + \sum_i [((1-y)^2 g_{Li}^2 + (g_{Ri}^2))x\overline{f}_i(x)]\right). \quad (10.22)$$

For the specific case of neutrino scattering from protons and including only u and d quarks, i.e., ignoring s and all anti-quarks,

$$\frac{d\sigma}{dxdy}\bigg|_{\nu p(NC)} = \frac{2G^2 ME}{\pi}\left(\frac{M_Z^2}{Q^2+M_Z^2}\right)^2 \left(\left[\frac{1}{2}-\frac{2}{3}\sin^2\theta_W\right]^2 u(x)\right.$$
$$+\left[-\frac{1}{2}+\frac{1}{3}\sin^2\theta_W\right]^2 d(x) + (1-y)^2\left[\left(\frac{2}{3}\sin^2\theta_W\right)^2 u(x)\right.$$
$$\left.\left. + \left(\frac{1}{3}\sin^2\theta_W^2\right)^2 d(x)\right]\right). \quad (10.23)$$

Recall that u and d are defined as the appropriate $xf(x)$ for the proton. Hence, the neutrino–neutron cross section is given by replacing $u(x)$ with $d(x)$ and vice versa. For anti-neutrinos, the $(1-y)^2$ factor switches from the right-handed to the left-handed terms (or in the more general case with anti-quarks, between the anti-quark and the quark terms). Figure 7.10 shows examples of neutral current neutrino interactions.

10.3 TRIANGLE ANOMALIES

A class of Feynman diagrams such as that illustrated in Figure 10.2 is very singular. The integral over the closed fermion loop violates current conservation. This introduces non-renormalizability in higher orders due to the loss of gauge invariance. The diagram has been useful, for example, in the

10.3 Triangle Anomalies

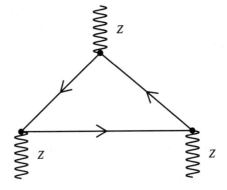

Figure 10.2 Triangle diagram with three external bosons.

Table 10.1 Triangle Anomaly Contributions

Particle	I_3^{weak}	$-4I_3^{weak} q \sin^2 \theta_W$	$4q^2 \sin^4 \theta_W I_3^{weak}$
ν_e	$-1/8$	0	0
e^-	$+1/8$	$+(1)\sin^2 \theta_w$	$-2\sin^4 \theta_w$
u	$N_c(-1/8)$	$N_c(-2/3)\sin^2 \theta_w$	$N_c(+8/9)\sin^4 \theta_w$
d	$N_c(+1/8)$	$N_c(+1/3)\sin^2 \theta_w$	$N_c(-2/9)\sin^4 \theta_w$
TOTAL	0	0	0

discussion of π^0 decay as noted in Chapter 6, but often it gives unwanted infinities. One sums over the diagrams for all possible loop particles, quarks, and leptons.

Similar singularities occur in box (four-vertex) and pentagon (five-vertex) diagrams. Diagrams with larger numbers of vertices can be described in terms of subdiagrams with three, four, or five vertices.

In the Standard Model, the anomalous parts of some classes of triangle singularity diagrams cancel. If (some of) the vertices are weak vertices, it is the parts with an odd number of γ^5 matrices that give singular contributions. Thus VVA and AAA are singular, while VVV and VAA are not.

The cancellation of the leading parts of these diagrams (which are the anomalous parts) can easily be shown for the neutral current VVA term. Note from the preceding that $J_V^{NC} \propto I_3^{weak} - 2q\sin^2 \theta_W$ and $J_A^{NC} \propto I_3$.

Summing diagrams for the VVA term then involves $\sum I_3^{weak} (I_3^{weak} - 2q\sin^2\theta_W)^2$, where the sum extends over the quarks and leptons. Table 10.1 shows that this sum is indeed zero for each generation. ($N_c=3=$ number of colors.) For AAA only the first column would occur. Note that fractionally charged quarks and color are both necessary to obtain zero.

The Standard Model presented thus far is tremendously successful in the range of data it explains. The weak and electromagnetic interactions are intertwined into a combined interaction. However, the picture is not yet complete. In the next chapter the problems of preventing infinities from creeping into the experimental predictions will be discussed and the present Standard Model completed.

10.4 Exercises

10.1 From Equation 10.19 work out g_L and g_R for the neutral weak interaction couplings of the Z to the various fermions.

10.2 Using the results of Problem 10.1, calculate the branching ratios for the $Z \to \mu^+\mu^-$, e^+e^-, $\tau^+\tau^-$, and $\bar{q}q$. Also find the branching ratio for the various neutrino pair decay modes. Only the ratios of various processes are needed (and use only lowest order). Remember to include the color factor for the quark decay modes. Hint: find g_L and g_R and note that the decay rate is proportional to $g_L^2 + g_R^2$.

10.3 Using the results of Problem 10.1, work out the appropriate g_L and g_R in the cross section formula given in the text (Equation 10.19) for $\nu_\mu e \to \nu_\mu e$, $\bar{\nu}_\mu e \to \bar{\nu}_\mu e$, $\nu_e e \to \nu_e e$, and $\bar{\nu}_e e \to \bar{\nu}_e e$. Note that these latter two processes also have CC contributions, which can be included using the effective g_L, g_R for the CC part. These are $g_L = 1$, $g_R = 0$ for the third process and $g_L = 0$, $g_R = 1$ for the last process. Note that, as indicated in the text, g_L and g_R refer to the electron currents. Also note that when going from ν to $\bar{\nu}$, L and R essentially switch places.

10.4 For neutrino interactions, estimate the ratio of neutral to charged current events expected in hydrogen for incident neutrinos; for incident anti-neutrinos. Assume $\int f_u(x)dx \sim 2 \int f_d(x)dx$ and consider only valence terms. Suppose the hydrogen is replaced by a material having equal numbers of neutrons and protons in its nucleus. What do the ratios become? Assume $Q^2 \ll M_W^2$, M_Z^2.

11
Spontaneous Symmetry Breaking: The Higgs Mechanism

As beautiful as the theory described in the preceding chapter is, there is a potentially fatal flaw. If one goes beyond the lowest order in the perturbation, the theory diverges and the high-energy behavior of matrix elements is bad. This bad high-energy behavior will be examined later. For now, note that the process $e^+e^- \to W^+W^-$ will be shown to grow quadratically with energy in the center of mass. This growth is so fast that it violates unitarity (conservation of probability) by 250 GeV in the center of mass. The same problem occurs for the similar second-order process $\nu_e \bar{\nu}_e \to \nu_\mu \bar{\nu}_\mu$. See Figure 11.1. Something must happen to prevent those catastrophes.

In the next sections it will be shown that much of the problem comes from the longitudinal polarization component of massive vector bosons. Extra neutral currents or extra heavy leptons can help the problem by getting some of the diagrams to cancel at high energies, but one cannot get rid of all of the bad behavior in this manner.

The only mechanism now known that can make the theory finite is a combination of a broader form of gauge theory and a new idea called "spontaneous symmetry breaking." In this theory each of the bosons above is associated with a local gauge transformation. In order to give all of these particles but the photon mass requires new particles, "Higgs particles," which have not yet been seen. With these new particles the theory is then renormalizable. This proof was worked out by G. t'Hooft, while a student of M. Veltman.

At present the lower mass limit on Higgs bosons is about 60 GeV coming from the examination of Z decays at LEP. With the present $\bar{p}p$ colliders, and with LEP it is hoped to be able to look up to nearly 100 GeV. With the next generation of hadron colliders, it is hoped to be able to search up to around 1000 GeV.

Although very beautiful, this is still not a confirmed theory. Many theorists now suspect that a finite elementary particle Higgs with finite mass may not be found and a similar but more complicated mechanism, possibly involving composite Higgs, or a dynamical mechanism acting in a manner similar to the Higgs, may be responsible for the renormalizability.

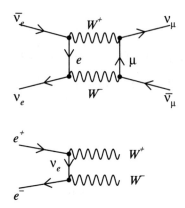

Figure 11.1 Some badly behaved processes in the Standard Model with no Higgs boson.

In the next few sections the Higgs mechanism will be examined. A "toy" model will be used, which, nevertheless, has most of the real physics contained in it. The treatment below is patterned after that given by H. Fritzsch and P. Minkowski.[157]

11.1 Gauge Theory in QED for the Dirac Equation

The ideas in this section have already been discussed for the Schrödinger equation. The application to the Dirac equation is very similar. Consider the transformation

$$\Psi(x) \to e^{-i\alpha}\Psi(x). \tag{11.1}$$

The Lagrangian for the Dirac equation is:

$$\mathcal{L} = \overline{\Psi}(x)(i\partial_\mu \gamma^\mu - m)\Psi(x) \tag{11.2}$$

($\partial_\mu = \partial/\partial x^\mu$). As usual, the Greek indices are spacetime indices going from 0 to 3. This Lagrangian is invariant under the transformation if α is constant. As seen in Section 4.9, this symmetry implies that there is a conserved current that is $j^\mu(x) = \overline{\Psi}(x)\gamma^\mu\Psi(x)$ with $\partial_\mu j^\mu = 0$.

Now let $\alpha = \alpha(x)$. The Lagrangian is not invariant under this transformation. However, if, in the Lagrangian, the derivative is replaced by a "covariant derivative," $\mathcal{D}_\mu \equiv \partial_\mu - ieA_\mu$, where A_μ is called a "gauge field," then the theory is invariant if and only if the gauge field transforms as

$A_\mu(x) \to A_\mu + (1/e)\partial_\mu \alpha(x)$. This form of adding the electromagnetic current is known as the "minimal electromagnetic interaction." One can add a term depending on $F^{\mu\nu}$, where $F_{\mu\nu} = \partial_\mu A_\nu - \partial_\nu A_\mu$; this corresponds to an anomalous magnetic moment term, $\Psi^\dagger \gamma^0 \gamma^\mu \gamma^\nu \Psi' F_{\mu\nu}$. The minimal interaction conserves the third component of I-spin, I_3. The preceding anomalous term is gauge invariant. Ψ and Ψ' must transform the same way under charge rotation, but I_3 may or may not be conserved depending on the Ψ and Ψ' chosen. It is assumed here, that in the fundamental Lagrangian, such anomalous terms are not present, because of renormalizability.

If it is desired to add a gauge invariant kinetic energy term involving A_μ to the Lagrangian, it can be shown that it must be in the form $(-1/4)F_{\mu\nu}F^{\mu\nu}$. Thus, the gauge transformation and the assumption of minimal electromagnetic coupling fixes the form of the photon interaction. In fact, it guarantees that the mass of the photon is zero, since a mass term $m^2 A_\mu A^\mu$ is not gauge invariant.

11.2 Non-Abelian Gauge Transformations

Consider the example of I-spin.[158] The Dirac Lagrangian for the nucleon system is

$$\mathcal{L} = \overline{N}(i\partial_\mu \gamma^\mu - m)N, \tag{11.3}$$

where $N = \begin{pmatrix} p \\ n \end{pmatrix}$.

Consider the transformation $N \to e^{-i(\vec{\alpha}(x)\cdot\vec{\tau})/2}N$. Here $\vec{\tau}$ are the Pauli spin matrices in I-spin space. Furthermore, let $T_i = \int \overline{N}(\gamma^0 \tau_i/2)N dx^3$ and $[T_i, T_j] = (1/4)\int \overline{N}\gamma^0[\tau_i, \tau_j]N d^3x$, for $i,j = 1,2,3$. From the commutation properties of the Pauli matrices, it is easily seen that $[T_i, T_j] = i\epsilon_{ijk}T_k$, where ϵ_{ijk} is the totally anti-symmetric tensor with $\epsilon_{123} = 1$. Next proceed in analogy with the QED process and ask how to make the Lagrangian invariant under this transformation. It is found that a triplet of gauge fields must be introduced:

$$\mathcal{D}_\mu = \partial_\mu + \frac{ig}{2}(\vec{\tau}\cdot(\vec{W})_\mu), \tag{11.4}$$

$$W^i_\mu \to W^i_\mu + \epsilon_{ijk}\alpha^j W^k_\mu + (1/g)\partial_\mu \alpha^i. \tag{11.5}$$

When the kinetic energy term for W is added, the Lagrangian becomes

$$\mathcal{L} = \overline{N}(i\gamma^\mu \mathcal{D}_\mu - m)N - (1/4)(G_i)_{\mu\nu}(G_i)^{\mu\nu}, \tag{11.6}$$

where $(G_i)_{\mu\nu}$ are field strength tensors:

$$(G_i)_{\mu\nu} = \partial_\mu W^i_\nu - \partial_\nu W^i_\mu - g\epsilon_{ijk}W^j_\mu W^k_\nu. \tag{11.7}$$

For a more general group the same technique is used but the I-spin matrix goes to the corresponding group matrix and the $\epsilon_{ijk} \to f_{ijk}$, the "structure constants" of the group, which are defined by $[T^i, T^j] = if_{ijk}T^k$.

The preceding Lagrangian has trilinear and quartic terms in the W^i_μ, i.e., the W^i_μ are "charged." Two W's will interact with each other. This was not true for A_μ.

For both this example and the previous one with A^μ, there are massive fermions and massless vector bosons.

11.3 Spontaneous Symmetry Breaking

Next spontaneous symmetry breaking will be discussed. This is a separate idea from gauge invariance. In the next section it will be shown how a combination of the two ideas leads to remarkable results.

Consider a real scalar field interacting with itself. A Lagrangian for this field is

$$\mathcal{L} = (1/2)(\partial_\alpha \phi \partial^\alpha \phi) - (1/2)\mu^2|\phi|^2 - (1/4)\lambda|\phi|^4. \tag{11.8}$$

The quadratic term is a mass term (mass = μ), for $\mu^2 > 0$ and the quartic term is the interaction term. Note that the Lagrangian is invariant under the transformation R, where $\phi_R = -\phi$.

A potential can be defined:

$$V(\phi) = (1/2)\mu^2|\phi|^2 + (1/4)\lambda|\phi|^4. \tag{11.9}$$

This potential is sketched in Figure 11.2a,b. For $\mu^2 > 0$, there is a curve with a single minimum centered at the origin. However, for $\mu^2 < 0$, there is a double minimum. The potential is minimum for $\phi = \pm\sqrt{-\mu^2/\lambda}$. By the vacuum state is meant the ground state and for the value of ϕ at minimum one has

$$\text{vacuum} = v = \langle 0|\phi|0\rangle. \tag{11.10}$$

Thus, there are two symmetric ground states for $\mu^2 < 0$. Suppose the system settles in one of these, i.e., suppose it spontaneously breaks the symmetry. This state is not invariant under R since one of the two minima

11.3 Spontaneous Symmetry Breaking

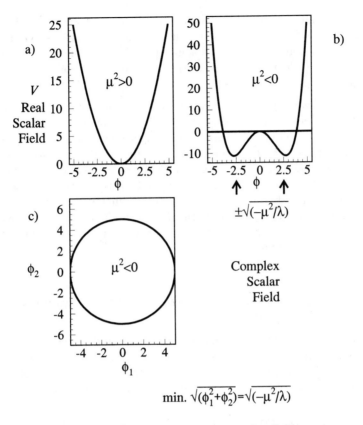

Figure 11.2 Potential for interacting scalar fields.

has been chosen. Define $\phi' = \phi - v$, where v is the vacuum value above. The Lagrangian becomes

$$\mathcal{L} = (1/2)(\partial_\alpha \phi' \partial^\alpha \phi') + \mu^2 (\phi')^2 - \lambda v (\phi')^3 - (1/4)\lambda(\phi')^4 + \text{constant}. \quad (11.11)$$

ϕ' is the excitation with respect to the the field vacuum. Remembering that $\mu^2 < 0$, there is now a term corresponding to a positive mass of $\sqrt{2}|\mu|$, as can be seen by comparing with the original Lagrangian.

Next apply this same concept to a complex scalar field $\phi = (1/\sqrt{2})(\phi_1 + i\phi_2)$ with the same Lagrangian and potential as above. For $\mu^2 < 0$, the region where the potential is minimum is a circle in ϕ_1, ϕ_2 space. (See Figure 11.2c.) Suppose a point is chosen on the circle on the ϕ_1 axis (or phases rotated so one is on the ϕ_1 axis). Then proceed as above to define ϕ'. $\phi'_1 = \phi_1 - v$ and $\phi'_2 = \phi_2$. Proceeding as above, it is found that ϕ'_1 corresponds to a particle of mass $\sqrt{2}|\mu|$, and ϕ'_2 corresponds to a particle of mass 0.

This is an example of the Goldstone theorem, which states that a massless boson ("Goldstone boson") appears if a continuous symmetry of a physical system described by a group G is spontaneously broken. The number of Goldstone bosons is equal to the number of broken generators of the group.

Spontaneous symmetry breaking occurs all the time. For example, even in the absence of external magnetic fields, a piece of iron settles into a state with a number of domains, no longer completely spatially invariant. Imagine a dinner at a round table where the wine glasses are centered between pairs of diners. This is a symmetric situation and one doesn't know whether to use the right or the left glass. However, as soon as one person at the table makes a choice, the symmetry is broken and the glass for each person to use is determined. It is no longer right–left symmetric. Even though a Lagrangian has a particular symmetry, a ground state may have a lesser symmetry.

11.4 COMBINING GAUGE INVARIANCE AND SPONTANEOUS SYMMETRY BREAKING

Now put the concepts of gauge invariance and spontaneous symmetry breaking together. Consider the complex ϕ field. Let $\mu^2 < 0$ and consider the gauge transformation $\phi \to e^{-i\alpha(x)}\phi$ with gauge field A_ν. The appropriate Lagrangian is ($\phi^\star$ = complex conjugate of ϕ)

$$\mathcal{L} = \mathcal{D}_\beta^\star \phi^\star \mathcal{D}^\beta \phi - \mu^2 \phi^\star \phi - \lambda(\phi^\star \phi)^2 - (1/4)F_{\beta\nu}F^{\beta\nu}, \qquad (11.12)$$

where $F_{\beta\nu} = \partial_\nu A_\beta - \partial_\beta A_\nu$; $\mathcal{D}_\beta = \partial_\beta - igA_\beta$; and $A_\beta \to A_\beta + (1/g)\partial_\beta \alpha(x)$.

Let $\phi_1 \equiv \phi_1' + \langle 0|\phi_1|0\rangle \equiv \phi_1' + v$; $v = \sqrt{-\mu^2/\lambda}$; substitute this expression for ϕ_1' into the preceding Lagrangian. New terms involving A together with ϕ or v are

$$(1/2)g^2 v^2 A_\nu A^\nu - gvA_\nu \partial^\nu \phi_2. \qquad (11.13)$$

The first term is a mass term for A_ν. This field has acquired a mass!

Consider the infinitesimal transform $\phi \to (1 - i\alpha)\phi$. In terms of ϕ_1 and ϕ_2, this is $\phi_1 \to \phi_1 + \alpha\phi_2$ and $\phi_2 \to \phi_2 - \alpha\phi_1$. Thus, $\phi_1' \to \phi_1' + \alpha\phi_2$ and $\phi_2 \to \phi_2 - \alpha v - \alpha\phi_1'$. Note that this last transformation is an inhomogeneous gauge transform. Since $\alpha(x)$ can be chosen arbitrarily, one always can choose a gauge such that $\phi_2 = 0$. In this gauge, the second term in 11.13 vanishes.

One started with a massive scalar field (one state), a massless Goldstone boson (one state) and a massless vector boson (two polarization states). After this transform there is a massive vector meson, A^μ, with three states of polarization and a massive scalar boson, ϕ'_1, which has one state. Thus, the Goldstone boson has been eaten up by the gauge field, which has become massive.

This combination of ideas has provided a specific form for the interaction, and has provided mass for the gauge bosons. Most importantly, it can be shown that this theory is renormalizable. There is, finally, a theory of interactions that is not plagued by higher-order infinities.

It was necessary to introduce complex scalar fields to get masses. At least one scalar field itself gets a mass and is left over (ϕ'_1 in the preceding example). This is the Higgs boson. Variants of the theory can have more than one Higgs, and can have charged Higgs. In some variants, (e.g., "technicolor"), the Higgs is not a fundamental field, but is a composite or a dynamical object.

In technicolor, new heavy quarks exist whose bound states give the intermediate vector bosons. Extended technicolor also has fermion states composite. Extended technicolor has difficulty avoiding strangeness-changing neutral currents and also has difficulty satisfying the precision data coming from LEP. No evidence has yet shown up for the compositeness of quarks or leptons. M. Veltman finds the absence of $\mu \to \gamma e$ strong evidence against having the μ and e made from the same building blocks. Experimentally the branching ratio for this decay is $< 4.9 \times 10^{-11}$ at 90% C.L.[69]

Unfortunately, there are still serious problems with the standard theory. The mass of the Higgs is not predicted. However, if it is beyond about 1000 GeV, then it does not prevent a violation of unitarity in perturbative calculations of the WW scattering above. The 1000 GeV limit here is $1/\sqrt{G}$ with a few numerical factors.

The Higgs self-interaction is not asymptotically free. The coupling increases with energy eventually becoming infinite at some energy E^*. It appears that even this theory is still an "effective theory" and not yet the final fundamental theory. As the mass of the Higgs increases, E^* decreases. At the limit $M_H = E^*$, the theory becomes inconsistant. At one loop level this occurs at 650 GeV.

Another serious problem is that with the Higgs, after the spontaneous symmetry breaking, there is a vacuum with extensive Higgs fields that, apparently, would cause an unacceptably large cosmological constant, according to present theories of the origin of the Universe. The cosmological constant is a constant that appears in General Relativity. It corresponds to a net attractive force of one part of space on another, independent of the masses, tending to curve space and close the Universe. At present, however, one does not know how to combine quantum mechanics with gravity in a straightforward fashion, and the role of the cosmological constant is not well understood. It may be that there are ways around this problem or ways in which the constant would be forced to small values or zero, or not even be present.

262   11. Spontaneous Symmetry Breaking

Furthermore, unless there is extensive fine tuning, the vacuum value v, which is shown in the next section to be 246 GeV, would get one loop quantum corrections tending to bring it up to the highest mass scale in the theory. In spite of its problems, this theory avoids many of the previous higher order infinites and is the present "state of the art."

11.5 THE STANDARD MODEL: GAUGE THEORY AND SYMMETRY BREAKING APPLIED TO THE WEINBERG–SALAM MODEL

Recall from the last chapter that

$$\mathcal{L} = g(j^\alpha \cdot W_\alpha) + \frac{g'}{2}(j^{Y^{weak}})^\alpha B_\alpha$$

expresses the interactions between fermions and the gauge bosons. Now add to that interaction Lagrangian terms for the kinetic energy of the fermions and the scalar meson terms following the procedure of the last sections:

$$\mathcal{L} = \left[\bar{X}_L \gamma^\beta \left(i\partial_\beta - g\frac{1}{2}\tau \cdot W_\beta - g'\frac{Y}{2}B_\beta \right) X_L + \bar{X}_R \gamma^\beta \left(i\partial_\beta - g'\frac{Y}{2}B_\mu \right) X_R \right.$$
(11.14)

$$-\frac{1}{4}W_{\beta\nu} \cdot W^{\beta\nu} - \frac{1}{4}B_{\beta\nu}B^{\beta\nu}$$

$$\left. + \frac{1}{2}\left| \left(i\partial_\beta - g\frac{1}{2}\tau \cdot W_\beta - g'\frac{Y}{2}B_\beta \right) \phi \right|^2 - V(\phi) \right],$$

where

$$V(\phi) = \frac{1}{2}\mu^2 \phi^\dagger \phi + \frac{1}{4}\lambda(\phi^\dagger \phi)^2 \quad (11.15)$$

$$W^i_{\beta\nu} = \partial_\beta W^i_\nu - \partial_\nu W^i_\beta - g\epsilon_{ijk}W^j_\beta W^k_\nu; \quad B_{\beta\nu} = \partial_\beta B_\nu - \partial_\nu B_\beta. \quad (11.16)$$

Here X_L denotes a left-handed fermion (lepton or quark) doublet, and X_R denotes a right-handed fermion singlet. The first two terms are the gauge invariant kinetic energy terms for the fermions and their interactions with the gauge bosons. The next two terms are the kinetic energy terms for the gauge bosons and the last two terms are the Higgs masses and couplings.

There are many choices for ϕ, each corresponding to a separate theory. To keep the gauge invariance intact, ϕ_i must belong to $SU(2)_{Iwk} \otimes U(1)_Y$ multiplets. Furthermore, the breaking must not break the entire symmetry but must leave a $U(1)_{em}$ symmetry so that the photon remains massless.

11.5 The Standard Model

The most economical choice is that made by Weinberg in 1967 which completes the Standard Model. He chose a single weak isospin doublet with weak hypercharge $Y = 1$:

$$\phi = \begin{pmatrix} \phi^+ \\ \phi^0 \end{pmatrix},$$

with

$$\phi^+ \equiv (\phi_1 + i\phi_2)/\sqrt{2}; \quad \phi^0 \equiv (\phi_3 + i\phi_4)/\sqrt{2}.$$

To generate gauge boson masses the vacuum expectation value is chosen as

$$\phi = \frac{1}{\sqrt{2}} \begin{pmatrix} 0 \\ v \end{pmatrix}.$$

ϕ^0 with $I^{weak} = 1/2$, $I_3^{weak} = -1/2$, and $Y^{weak} = 1$ has, thus, been chosen as the nonzero vacuum part, i.e., as the direction of symmetry breaking. This breaks both the $SU(2)_{I^{weak}}$ and the $U(1)_{Y^{weak}}$ symmetries. However, ϕ^0 is neutral. As long as only neutral scalars are allowed to acquire vacuum expectation values, a massless photon will be preserved. Thus, the $U(1)_{em}$ symmetry with generator $Q = I_3^{weak} + Y^{weak}/2$ stays unbroken. Because $Q = 0$, then for any $\alpha(x)$

$$\phi^0 \rightarrow (\phi^0)' = e^{i\alpha(x)Q}\phi^0 = \phi^0.$$

Three of the four fields are eaten up giving mass to the $W^\pm$ and the Z. The photon remains massless. If the absolute value squared term in $\mathcal{L}$ is expanded, one obtains

$$\left(\frac{1}{2}vg\right)^2 W_\beta^+(W^-)^\beta + \frac{1}{8}v^2 \begin{pmatrix} W_\beta^3 & B_\beta \end{pmatrix} \begin{pmatrix} g^2 & -gg' \\ -gg' & (g')^2 \end{pmatrix} \begin{pmatrix} (W^3)^\beta \\ B^\beta \end{pmatrix}.$$

The first term yields

$$M_W = \frac{1}{2}vg.$$

This term is independent of the form of the ϕ as seen in the last chapter. Since this must equal the result obtained in the last chapter, namely, $M_{W^\pm} = 37.28/\sin\theta_W\,\text{GeV}$, then one must have

$$v = 246 \text{ GeV}.$$

The remaining term is not diagonal in the W, B basis. When it is diagonalized, one finds that the eigenvectors are the Z, A basis given in the last chapter and the eigenvalues correspond to

$$M_A = 0; \quad M_Z = \frac{1}{2}v\sqrt{g^2 + (g')^2}.$$

This latter mass corresponds to the mass of the Z, $M_Z = M_W/\cos\theta_W$, given in the last chapter. It does depend on the specific form of the ϕ chosen.

There are still massless fermions. To give mass to the fermions still another gauge invariant term must be added to the Lagrangian. First, the form will be given for electrons and then, afterward, generalized:

$$\Delta(e \text{ only})\mathcal{L} = -G_e\left[(\bar{\nu}_e \ \bar{e})_L \begin{pmatrix}\phi^+ \\ \phi^0\end{pmatrix}e_R + \bar{e}_R(\phi^- \ \bar{\phi}^0)\begin{pmatrix}\nu_e \\ e\end{pmatrix}_L\right].$$

Introduce now the spontaneous symmetry breaking recognizing that ϕ_3 is the only field remaining and call the new scalar H. H is the ϕ_3 field with respect to the vacuum value, i.e., is the physical Higgs particle:

$$\phi^0 = \frac{1}{\sqrt{2}}\begin{pmatrix}0 \\ v + H(x)\end{pmatrix}.$$

Choose G_e such that $m_e = G_e v/\sqrt{2}$. $\Delta\mathcal{L}$ then becomes

$$\Delta\mathcal{L} = -m_e\bar{e}e - \frac{m_e}{v}\bar{e}eH.$$

For the quarks a similar addition is used. Here one must also introduce the field $\phi_c \equiv -i\tau_2\phi^*$. ϕ_c transforms identically to ϕ under $SU(2)_{Iweak}$ (see the similar argument in Section 4.15). However, Y is reversed. For ϕ, $Y^{weak} = 1$, and for ϕ_c, $Y^{weak} = -1$. The appropriate final form for $\Delta\mathcal{L}$ is

$$\Delta\mathcal{L} = -(G_1^i \bar{L}\phi R + G_2^i \bar{L}\phi_c R) + \text{hermitian conjugate}.$$

As for the electron case, the quark mass terms for generation i become

$$-m_d^i \bar{d}_i d_i \left(1 + \frac{H}{v}\right) - m_u^i \bar{u}_i u_i \left(1 + \frac{H}{v}\right).$$

The masses depend on arbitrary couplings G_j^i and cannot be predicted. Note that there is a coupling of the Higgs to the fermions which is proportional to the mass of the fermions. The higher-mass fermions couple most strongly to the Higgs. The mass of the Higgs is not predicted. However, the fact that the Higgs coupling to particles is proportional to their mass means that the Higgs will tend to couple with and to decay into the heaviest allowed particles; this serves to tag Higgs decays. At LEP searches have been made for $e^+e^- \to Z \to H + Z^*$; $Z^* \to \nu\bar{\nu}$. A lower limit of 65.2 GeV has been set on the Higgs mass.

11.6 THE STANDARD MODEL LAGRANGIAN

The various pieces of the Lagrangian have now been obtained. In this section the complete Standard Model Lagrangian will be written down for one generation. The mixing of generations will be ignored, but can easily be added by the interested reader. First, the Lagrangian will be written in I-spin, Y, notation.

Evaluating $(1/2)|i\mathcal{D}_\beta \phi i \mathcal{D}^\beta \phi| = (1/2)|i\partial_\beta - g(1/2)\tau \cdot W_\beta - g'(Y/2)B_\beta) \times (v+H)|^2$ can be simplified by writing $\mathcal{D}_\beta \phi$ in matrix notation:

$$i\mathcal{D}_\beta \phi = \begin{pmatrix} i\partial_\beta - (1/2)gW_\beta^3 + g'B_\beta & -(g/\sqrt{2})W_\beta^+ \\ -(g/\sqrt{2})W_\beta^- & i\partial_\beta + (1/2)(gW_\beta^3 - g'B_\beta) \end{pmatrix} \begin{pmatrix} 0 \\ v+H \end{pmatrix}$$
(11.17)

and the appropriate terms are then easily written down. Note that $v = 2M_W/g = \sqrt{M_H^2/(2\lambda)}$. X is a column vector of u, d, e, ν.

$$\begin{aligned}
\mathcal{L}_{SM} =\ & i\overline{X}_L \gamma^\mu \partial_\mu X_L + i\overline{X}_R \gamma^\mu \partial_\mu X_R && \text{Kinetic energy terms} \\
& -\frac{1}{2}(\partial_\beta W_\nu \cdot \partial^\beta W^\nu - \partial_\nu W_\beta \cdot \partial^\beta W^\nu) \\
& -\frac{1}{4}(\partial_\beta B_\nu - \partial_\nu B_\beta)(\partial^\beta B^\nu - \partial^\nu B^\beta) \\
& +\frac{1}{4}\partial_\mu H \partial^\mu H \\
& -m_e \overline{\psi}_e \psi_e - m_u \overline{\psi}_u \psi_u - m_d \overline{\psi}_d \psi_d && \text{Mass terms} \\
& +g[\overline{X}_L(I^{weak} \cdot W_\alpha)\gamma^\alpha X_L] && \overline{f}fW, Z, A \text{ couplings} \\
& +\frac{1}{2}\tan\theta_W (\overline{X}_L Y^{weak} \gamma^\alpha X_L \\
& \quad +\overline{X}_R Y^{weak} \gamma^\alpha X_R)]B_\alpha^0 \\
& +g\partial^\beta W_i^\nu \epsilon_{ijk}(W_j)_\beta(W_k)_\nu && WWZ, WWA \text{ couplings} \\
& -\frac{g^2}{4}\epsilon_{ijk}W_j^\beta W_k^\nu \epsilon_{ilm}(W_l)_\beta(W_m)_\nu && 4W, 3WZ, A \\
& && WWZZ, ZA, AA \text{ couplings} \\
& -\frac{m_e}{v}\overline{\psi}_e \psi_e H - \frac{m_u}{v}\overline{\psi}_u \psi_u H - \frac{m_d}{v}\overline{\psi}_d \psi_d H && \overline{f}f \text{ couplings}
\end{aligned}$$

$$+\frac{g^2}{4}(2vH + HH)(W^-)^\beta W_\beta^+ \quad HWW, HHWW \text{ couplings}$$

$$+\frac{g^2}{8\cos^2\theta_W}(2vH + HH)Z^\beta Z_\beta \quad HZZ\ HHZZ \text{ couplings}$$

$$-\frac{\lambda}{4}(4vHHH + HHHH) \quad 3H, 4H \text{couplings}.$$

(11.18)

The preceding Lagrangian does not include the strong interaction sector. The procedure for including these terms can be easily outlined. The strong interactions are a gauge theory based on $SU(3)$ as indicated in Chapter 9. A set of eight 3×3 matrices, λ_{ab}, which are a generalization of the three 2×2 Pauli matrices that enter $SU(2)$, and a set of eight structure constants, f_{abc}, which are a generalization of the ϵ_{ijk} entering $SU(2)$, are needed. The interested reader can find λ_{ab} and f_{abc} explicitly given in the Particle Data Group publication.[17]

Let G^a be the set of eight colored gluon fields and g_3 be the strong interaction coupling constant. The term $g_3 \sum_{a=1}^{8}(\lambda_{ab}/2)G_\beta^a$ should be added to the $i\mathcal{D}_\beta$ for the left-handed coupling and to the $i\partial_\beta$ for the right-hand coupling for the quarks, i.e., for the strongly interacting fermions. Each quark is regarded as a color triplet. This term provides the $\overline{f}fG$ coupling. The gluon field term $(-1/4)G_{\beta\nu} \cdot G_{\beta\nu}$ should also be added. Here $G_{\beta\nu}^a = \partial_\beta G_\nu^a - \partial_\nu G_\beta^a - g_3 f_{abc} G_\beta^b G_\nu^c$. This term provides the gluon field kinetic energy terms and gives $3G$ and $4G$ couplings. These terms will not be added in here, but can easily be included when needed.

Next the Lagrangian for the electroweak sector will be written with the physical charged states used. Consider terms coming from $(W^{\beta\nu} \cdot W_{\beta\nu})$. Several of these terms involve $\epsilon_{ijk}W_\beta^j W_\nu^k$. Write this subterm as a column vector in index i and use the relations obtained in Section 10.1 between the I-spin and charged components of the W's:

$$\begin{pmatrix} (W_2)_\beta(W_3)_\nu - (W_3)_\beta(W_2)_\nu \\ (W_3)_\beta(W_1)_\nu - (W_1)_\beta(W_3)_\nu \\ (W_1)_\beta(W_2)_\nu - (W_2)_\beta(W_1)_\nu \end{pmatrix} = \begin{pmatrix} (-i/\sqrt{2})[(W_\beta^+ - W_\beta^-)(W_3)_\nu - (W_3)_\beta(W_\nu^+ - W_\nu^-)] \\ (1/\sqrt{2})[(W_3)_\beta(W_\nu^+ + W_\nu^-) - (W_\beta^+ + W_\beta^-)(W_3)_\nu] \\ (-2i/2)[W_\beta^- W_\nu^+ - W_\beta^+ W_\nu^-] \end{pmatrix}. \quad (11.19)$$

Using this relation the relevant terms can be worked out.

11.6 The Standard Model Lagrangian

$$\mathcal{L}_{SM} = \overline{\psi}_e(i\not{\partial} - m_e)\psi_e + \overline{\psi}_\nu(i\not{\partial})\psi_\nu + \quad \text{KE, mass terms}$$
$$\overline{\psi}_d(i\not{\partial} - m_d)\psi_d + \overline{\psi}_u(i\not{\partial} - m_u)\psi_u$$
$$-(\partial^\beta(W^+)^\nu\partial_\beta W^-_\nu - \partial^\nu(W^+)^\beta\partial_\beta W^+_\nu) - \frac{1}{2}\partial^\beta(\cos\theta_W Z^\nu$$
$$+ \sin\theta_W A^\nu)\partial_\beta(\cos\theta_W Z_\nu + \sin\theta_W A_\nu)$$
$$-\partial^\nu(\cos\theta_W Z^\beta + \sin\theta_W A^\beta)\partial_\beta(\cos\theta_W Z_\nu + \sin\theta_W A_\nu)$$
$$-\frac{1}{4}\partial^\beta(-\sin\theta_W Z^\nu + \cos\theta_W A^\nu)$$
$$-\partial^\nu(-\sin\theta_W Z^\beta + \cos\theta_W A^\beta)$$
$$\partial_\beta(-\sin\theta_W Z_\nu + \cos\theta_W A_\nu)$$
$$-\partial_\nu(-\sin\theta_W Z_\beta + \cos\theta_W A_\beta) + \frac{1}{4}\partial^\beta H\partial_\beta H$$
$$-\frac{g}{\sqrt{2}}\left[(\overline{\psi}_u\gamma^\beta\frac{(1-\gamma^5)}{2}\psi_d + \overline{\psi}_{\nu_e}\gamma^\beta\frac{(1-\gamma^5)}{2}\psi_e)W^+_\beta \quad \overline{f}f'W \text{ couplings}\right.$$
$$\left. +(\overline{\psi}_d\gamma^\beta\frac{(1-\gamma^5)}{2}\psi_u + \overline{\psi}_e\gamma^\beta\frac{(1-\gamma^5)}{2}\psi_{\nu_e})W^-_\beta)\right]$$
$$+\frac{g}{\cos\theta_W}\frac{1}{2}\overline{\psi}_{\nu_e}\gamma^\beta\frac{(1-\gamma^5)}{2}\psi_{\nu_e} \quad \overline{f}fZ \text{ couplings}$$
$$+\overline{\psi}_d\gamma^\beta\frac{((2/3)\sin^2\theta_W - 1/2) + (1/2)\gamma^5}{2}\psi_d$$
$$+\overline{\psi}_u\gamma^\beta\frac{((-4/3)\sin^2\theta_W + 1/2) - (1/2)\gamma^5}{2}\psi_u$$
$$+\overline{\psi}_e\gamma^\beta\frac{(2\sin^2\theta_W - (1/2) + (1/2)\gamma^5)\psi_e}{2}]Z_\beta$$
$$+e\left[\overline{\psi}_e\gamma^\beta\psi_e + \frac{1}{3}\overline{\psi}_d\gamma^\beta\psi_d - \frac{2}{3}\overline{\psi}_u\gamma^\beta\psi_u\right]A_\beta \quad \overline{f}f\gamma \text{ couplings}$$
$$-ig(\partial^\beta(W^+)^\nu - \partial^\nu(W^+)^\beta)W^-_\nu(\cos\theta_W Z_\beta + \sin\theta_W A_\beta) \quad WWZ, WWA$$
$$-ig(\partial^\beta(W^-)^\nu - \partial^\nu(W^-)^\beta)W^+_\beta(\cos\theta_W Z_\nu + \sin\theta_W A_\nu) \quad \text{couplings}$$
$$-ig((W^-)^\beta(W^+)^\nu - (W^+)^\beta(W^-)^\nu)$$
$$\partial_\beta(\cos\theta_W Z_\nu + \sin\theta_W A_\nu)$$
$$-g^2(W^+)^\beta W^-_\beta(\cos\theta_W Z^\nu + \sin\theta_W A^\nu) \quad WWZZ, ZA,$$
$$(\cos\theta_W Z_\nu + \sin\theta_W A_\nu) \quad AA \text{ couplings}$$
$$+g^2(W^+)^\nu(W^-)^\beta(\cos\theta_W Z_\beta + \sin\theta_W A_\beta)$$
$$(\cos\theta_W Z_\nu + \sin\theta_W A_\nu)$$
$$+\frac{g^2}{2}(W^-)^\nu(W^+)^\beta(W^-_\nu W^+_\beta - W^-_\beta W^+_\nu) \quad 4W \text{ couplings}$$

$$-\frac{m_e}{v}\overline{\psi}_e\psi_e H - \frac{m_d}{v}\overline{\psi}_d\psi_d H - \frac{m_u}{v}\overline{\psi}_u\psi_u H \quad \overline{f}fH \text{ couplings}$$

$$+\frac{g^2}{4}(2vH + HH)(W^-)^\beta W_\beta^+ \quad WWH, WWHH$$
$$\text{couplings}$$

$$+\frac{g^2}{8\cos^2\theta_W}(2vH + HH)Z^\beta Z_\beta \quad ZZH,$$
$$ZZHH \text{ couplings}$$

$$-\frac{\lambda}{4}(4vHHH + HHHH) \quad HHH, HHHH$$
$$\text{couplings.}$$
$$(11.20)$$

11.7 Introduction to Study of Unitarity Restrictions

In the preceding sections, the Standard Model was developed with an emphasis on the symmetries of the couplings. It was indicated that the theory obtained did not have the divergence problems of earlier theories, and was a renormalizable theory, with good high-energy behavior.

In fact, one can, to a large extent, show the converse. With apparently reasonable assumptions, insisting that the bad high-energy behavior cancels forces the theory to have the form of a spontaneously broken gauge theory, although with the group structure not determined.

One of the major interests in particle physics is to try to go beyond the Standard Model. Because of this it is important to try to understand how the model comes about from as many points of view as possible. It is not yet known which one will be the most useful for extending the model.

The arguments limiting the couplings in order to try to remove bad high-energy behavior are complex. However, at least an outline of the arguments will be reproduced here, with indication of the assumptions made.

It might be noted that some divergences do not cause real problems. Whenever there are zero-mass particles, final states with parallel massless particles can lead to "mass divergences." These can be shown to be avoided by averaging the transition amplitudes over an appropriate ensemble of degenerate states (KLN theorem).[1] The related problems of high-energy behavior considered below are not of this kind; they put real restrictions on the form of the theory.

11.8 Preliminary Definitions and Assumptions

The discussion will mostly concern tree-level Feynman diagrams. These are diagrams without any loops. It is assumed that no basic interactions have coupling constants with dimensions of inverse mass or worse (such as would be the case if G, the Fermi weak interaction coupling constant, were a basic coupling constant). This is the case in all known renormalizable theories.

It is assumed that no particles of spin higher than 1 occur in the basic theory. There are serious divergence problems associated with particles of spin 3/2 or greater that are hard to handle at present. In this context it might be noted that gravity is a theory requiring a spin 2 graviton and with the usual point particle theory it does not seem possible to construct a renormalizable theory of gravity. The modern "string theory" attempts to solve this problem by making the basic particles one-dimensional open or closed strings rather than points and by increasing the dimensions of space–time beyond the usual four dimensions. This will not be treated in the present discussion.

In the next sections, some illustrative examples will be given, showing that the couplings of the Standard Model are extremely restricted if they are to possess some of the necessary cancellations. Following those sections, an outline of the general theorem will be given.

11.9 High-Energy Behavior

General forms of matrix elements must, of course, correspond to the correct dimensions. This does not indicate how fast elements increase with energy. For instance, if there are masses involved, $(E/m)^2$ has a quadratic behavior as E approaches infinity, but has units of E^0 = constant.

The high-energy behavior needed can be deduced with the help of the unitarity relation for the M matrix (Equation 5.29):

$$\Im\langle f \mid M \mid i\rangle = \frac{1}{2}\sum_n (2\pi)^{4-3n} \int \Pi_{i=1}^n \frac{d^3p_i}{2E_i} \delta^4(\sum_j p_j - P)$$
$$\times \langle p_1 \cdots p_n \mid M \mid f\rangle^* \langle p_1 \cdots p_n \mid M \mid i\rangle. \quad (11.21)$$

Consider a two-body initial and a two-body final state for the left-hand side of the relation. If, for n intermediate state particles, M varies as too high a power of energy, then the right-hand side of the relation will eventually become too large to be satisfied by the left-hand side. Examine the right-hand side of the unitarity relation for an n particle intermediate state. From the d^3p_i/E_i terms there is a factor of E^{2n}. The delta function has an effective factor of E^{-4}. For the special case of a two body intermediate state, it is thus seen that the maximum dependence is E^0. This will be the condition needed for the present analysis.

For the general case of n intermediate particles, suppose that the maximum allowable dependence of M on energy varies as E^{an+b}. Comparing right and left sides for the terms with n gives $a = -1$. Using that result, it is easily seen that $b = 2$. M can have a maximum factor of E^{2-n}. If N is the total number of particles, initial and final ($N = n + 2$), then the maximum allowable energy dependence of M to avoid bad high-energy behavior is E^{4-N}.

If a matrix element violates unitarity at tree level, then n-loop diagrams will have even worse behavior (unless there are cancellations) since they are iterations of tree-level diagrams. It is thus expected that tree-level unitarity is necessary for overall unitarity, i.e., necessary to have a finite perturbation theory. The present discussion will examine conditions such that tree-level diagrams for two particles in and two particles out increase with E at least as slowly as E^0.

11.10 Further Feynman Rules and Other Needed Formalities

In order to illustrate cancellations of specific sets of Feynman diagrams, it is necessary to specify the gauge. If all diagrams are included, the results must be gauge invariant, since the fundamental Lagrangian possesses this invariance, but the specific diagrams to be included and the form of the matrix elements for each diagram depend on the gauge.

For example, the propagator for a massive vector intermediate boson in the unitary gauge has the form

$$\frac{i(-g^{\mu\nu} + (q^\mu q^\nu)/M^2)}{q^2 - M^2}.$$

Recall that the unitary gauge is the gauge in which the extra component of the Higgs field is "gauged away." The most singular part of the above propagator is $q^\mu q^\nu / M^2$. But this term is gauge dependent.

However, in gauges other than the unitary gauge, there are additional graphs, since the extra component of the Higgs field is present. It does not correspond to a physical particle, since it can be gauged away, but if it is in the Lagrangian in some gauge, then, in that gauge, the Feynman graphs corresponding to it must be included. These particles are known as "ghost particles" and Feynman graphs such as shown in Figure 11.3 are present.

11.10 Further Feynman Rules

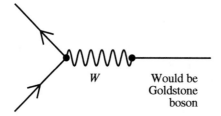

Figure 11.3 Example of ghost particle in a Feynman diagram.

For electromagnetism, suppose the propagator is written as

$$\frac{i(-g^{\mu\nu} + \lambda(k^\mu k^\nu)/k^2)}{k^2}.$$

In the Coulomb gauge, in which $\vec{A}$ is transverse, i.e., $\nabla \cdot A = 0$, the analogous term is not present, i.e., $\lambda = 0$. In another gauge, known as the Landau gauge, $\lambda = -1$.

For the present discussion the unitary gauge and Coulomb gauge will be used. The propagators for various spin particles can then be written as

$$\text{Spin 0} \quad \frac{i}{q^2 - m^2 + i\epsilon}, \tag{11.22}$$

$$\text{Spin 1/2} \quad \frac{i(\not{p} + m)}{p^2 - m^2 + i\epsilon}, \tag{11.23}$$

$$\text{Photon (Massless Spin 1)} \quad \frac{-ig^{\mu\nu}}{q^2 + i\epsilon}, \tag{11.24}$$

$$\text{Massive Spin 1} \quad \frac{+i(-g^{\mu\nu} + q^\mu q^\nu/m^2)}{q^2 - m^2 + i\epsilon}. \tag{11.25}$$

Here the $i\epsilon$ in the the denominator (where ϵ is an infinitesimal) is needed to maintain causality and to determine which sheet to use when integrals in the complex plane are taken. They will not be needed for the discussions in this text.

Some useful identities can be established for traces of gamma matrices. From the relation $\gamma^\mu \gamma^\nu + \gamma^\nu \gamma^\mu = 2Ig^{\mu\nu}$, it is seen that $\not{p}\not{q} + \not{q}\not{p} = 2p \cdot q$. If, furthermore, $p = q$, then $\not{p}\not{p} = p^2$. Other relations are given in Appendix E.

For an initial or final spin 1/2 particle, an appropriate u, v, $\bar{u}$, or $\bar{v}$ must be inserted as indicated in Chapter 7.

For a spin 1 particle, the polarization vector ϵ must be included for initial-state particles and ϵ^* for final-state particles. ϵ is a four-vector, which is defined to have unit size ($\epsilon \cdot \epsilon = 1$), and to have $q \cdot \epsilon = 0$. Here the four-dimensional dot product is to be understood, and q is the momentum of the particle. With this definition, there are three independent polarization vectors. For massive vector bosons, if the particle momentum is in the three direction, one possible set of three polarizations is (0,1,0,0), (0,0,1,0), ($|q|/m$,0,0,E/m). For this choice it is easily verified that the sum over the three polarization vectors gives

$$\sum (\epsilon^{*\mu} \epsilon^\nu) = -g^{\mu\nu} + q^\mu q^\nu / m^2. \qquad (11.26)$$

This relation is true using any set of three orthogonal polarization vectors. Note that the longitudinal polarization involves an extra power of energy. Because of this, the longitudinal bosons are most susceptible to high-energy behavior problems and in the next sections, the cross sections for longitudinal bosons will be considered.

The vertex functions basically can be read from the corresponding terms in the Lagrangian. However, a number of rules, extra factors, and phases are involved. Details of these vertex rules, a table of some needed vertex functions, and a summary of the overall Feynman rules is given in Appendix E.

Finally, after using the rules above and the appropriate vertex functions, an overall factor of $(-i)$ must be included to get the matrix element M. In the following sections some cancellations necessary to avoid bad high-energy behavior will be discussed. These sections will also provide examples of the methodology for obtaining matrix elements from a given Feynman diagram and Lagrangian.

11.11 Neutrino Electron Scattering

Consider first the tree-level diagram for neutrino electron elastic scattering in the absence of an intermediate boson. From the calculation of traces given in Section 7.5, it can be seen that $M \propto E^2$. From the preceding discussion of high-energy behavior, it is seen that the maximum allowable dependence on E for a two-body initial, two-body final state is E^0. This reaction violates this bound and will violate unitarity. A symptom of this problem is that the νN inelastic cross section calculated in Section 8.6 is proportional to E_{lab}^1 or to E_{cm}^2 in the absence of an intermediate boson. The addition of the boson introduces a Q^2 into the denominator reducing

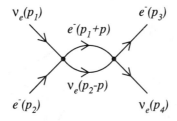

Figure 11.4 Second-order νe scattering Feynman diagram.

the energy dependence by the required factor of E^2. (Using the Dirac equation, the potentially dangerous part of the propagator, $q_\mu q_\nu/m_W^2$ is seen to be not proportional to E^2, but proportional to m_e^2/m_W^2 and hence negligible.)

One might ask whether going beyond tree-level could help the high-energy behavior in the absence of an intermediate boson. Consider the diagram shown in Figure 11.4.

The matrix element for this reaction can be written as

$$iM_{fi} = -8G_F^2 \int \frac{d^4p}{(2\pi)^4} \bar{u}(p_4)\gamma^\alpha \frac{(1-\gamma^5)}{2} \frac{i((\slashed{p}_1 - \slashed{p}) + m)}{(p_1 - p)^2 - m^2}$$
$$\times \gamma^\beta \frac{(1-\gamma^5)}{2} u(p_1) \bar{u}(p_3)\gamma_\alpha \frac{(1-\gamma^5)}{2} \frac{i((\slashed{p}_2 + \slashed{p}) + m)}{(p_2 + p)^2 - m^2} \gamma_\beta \frac{(1-\gamma^5)}{2} u(p_2).$$
(11.27)

Count the powers of the intermediate momentum p. One has $\int d^4p/p^2$, which is divergent, even for finite external momenta. If a cutoff Λ is introduced, then this is proportional to Λ^2. QED has infinities of the order of $\log\Lambda$ and can be renormalized. The bare mass and bare charge, etc., are set infinite and used to cancel the divergences. The observed mass, charge, etc., then come out finite. However, the present matrix element has a much worse divergence and the same tricks do not work. Going beyond tree level has only made the problems worse; divergences as well as bad high-energy behavior have entered. Thus, to avoid these problems a charged, vector intermediate boson is needed. The boson must be charged and vector because the beta decay current is a charged vector current. Without an intermediate boson, unitarity would be violated at about 250 GeV in the center of mass, i.e., essentially at $1/\sqrt{G}$. As seen in Chapter 7, with the inclusion of an intermediate boson, the cross section no longer increases indefinitely with energy and the matrix element is within the unitarity bound because of the propagator denominator.

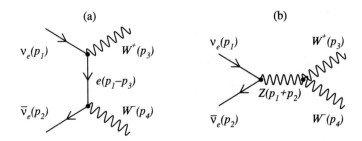

Figure 11.5 $\nu\bar{\nu} \to W^+W^-$. (a) No Z (b) Z exchange.

11.12 $\nu\bar{\nu} \to W^+W^-$

Consider the Feynman diagram shown in Figure 11.5a.[159] This reaction is not very practical to examine experimentally, but it does illustrate well a high-energy-behavior problem.

The matrix element can be obtained from the rules in the preceding sections. There is a $(-i)$ in front. There is a factor of $-ig/\sqrt{2}$ at each vertex from the vertex factors in Appendix E. The propagator is of the form $i(\slashed{p}_1 - \slashed{p}_3)/(p_1 - p_3)^2$. Polarization vectors must be introduced for the W's, and the usual gamma matrices and spinors are present giving

$$iM_{fi} = \frac{-g^2}{2}\epsilon^*_\mu(p_4)\epsilon^*_\nu(p_3)\bar{v}_2\gamma^\mu \frac{1-\gamma^5}{2}\frac{i(\slashed{p}_1 - \slashed{p}_3)}{(p_1-p_3)^2}\gamma^\nu\frac{1-\gamma^5}{2}u_1. \quad (11.28)$$

Next, the calculation of the appropriate averaged square of the matrix element will be obtained using the technique outlined in Section 7.5.

Before using the trace procedure, it is advisable to reduce the number of gamma matrices in the matrix element as much as possible. First anti-commute the two $(1-\gamma^5)/2$ terms to be side by side. Since they cross an even number of gamma matrices, they remain $(1-\gamma^5)/2$, not $(1+\gamma^5)/2$. The product of the two terms gives the term itself back. Next move the $\slashed{p}_1$ term to the right-hand side where the Dirac equation can be used giving $\slashed{p}_1 u_1 = 0$. Using the identity $\gamma^\mu\gamma^\nu + \gamma^\nu\gamma^\mu = 2Ig^{\mu\nu}$, the relation $\slashed{p}_1\gamma^\nu = -\gamma^\nu\slashed{p}_1 + 2p_1^\nu I$ is obtained. (This anti-commutation relation will be used repeatedly below in evaluating the trace.)

Note that the denominator term $(p_1-p_3)^2 = -2p_1\cdot p_3 + M_W^2$. The mass term can be ignored for the high-energy limit:

$$M_{fi} \approx \frac{g^2}{2}\epsilon^*_\mu(p_4)\epsilon^*_\nu(p_3)\bar{v}_2\gamma^\mu\frac{(2(p_1)^\nu - \slashed{p}_3\gamma^\nu)}{(2p_1\cdot p_3)}\frac{(1-\gamma^5)}{2}u_1. \quad (11.29)$$

When taking the trace the relation found earlier for the sum of products of polarizations is used, i.e., $\sum_{\text{pol}}\epsilon^*_\mu(p)\epsilon_\nu(p) = -g_{\mu\nu} + p_\mu p_\nu/M^2$. Furthermore, only the most singular term will be kept; the $g_{\mu\nu}$ term will be dropped.

$$|M_{fi}|^2 \approx \frac{g^4}{8M_W^4(2p_1 \cdot p_3)} \text{Tr}[\not{p}_4(2p_1 \cdot p_3)\not{p}_1\gamma^0(2p_1 \cdot p_3)\not{p}_4\gamma^0\not{p}_2(1+\gamma^5)]. \tag{11.30}$$

Using the preceding anti-commutation relation for the gamma matrices, one has $\gamma^0\not{p}_4 = -\not{p}_4\gamma^0 + 2E_4$ and $\not{p}_4\not{p}_1\not{p}_4 \approx 2p_1 \cdot p_4\not{p}_4$, where in the last relation a mass term was dropped. The matrix element squared then becomes

$$|M_{fi}|^2 \approx \frac{g^4}{8M_W^4} \text{Tr}[(2p_1 \cdot p_4)\not{p}_4\not{p}_2(1+\gamma^5) + 2E_4\not{p}_4\not{p}_1\gamma^0\not{p}_2(1+\gamma^5)]. \tag{11.31}$$

Next some of the identities for traces of gamma matrices listed in Appendix E are used. The trace of γ^5 with less than four gamma matrices is zero. The trace of an odd number of gamma matrices (with no γ^5) is zero. $\text{Tr}(\not{a}\not{b}) = 4a \cdot b$; $\text{Tr}(\not{a}\not{b}\not{c}\not{d}) = 4[(a \cdot b)(c \cdot d) - (a \cdot c)(b \cdot d) + (a \cdot d)(b \cdot c)]$; and $\text{Tr}(\gamma^5\gamma^\mu\gamma^\lambda\gamma^\sigma\gamma^\rho) = 4i\epsilon^{\mu\lambda\sigma\rho}$, where $\epsilon^{\mu\lambda\sigma\rho}$ is a completely anti-symmetric tensor with $\epsilon^{1230} = 1$.

Choose the center-of-mass system in which the energies of all four particles are the same and take the angle between particle one (ν_e) and particle three (W^+) to be θ.

The term involving the anti-symmetric tensor is zero. One index of the tensor is fixed at 0 because of γ^0. The other three correspond to the three-dimensional antisymmetric tensor and give the triple product of the vectors. However, in the center-of-mass system, the vectors are all in a plane and the triple product is zero. All terms but the first term in the trace evaluate to zero giving

$$|M_{fi}|^2 = \frac{g^4(p_1 \cdot p_4)(p_2 \cdot p_4)}{M_W^4}. \tag{11.32}$$

From the above equation, $M \propto E^2$, not constant, as is necessary for $N = 4$ to avoid high-energy behavior problems. Therefore, something more must be added to the theory. The coupling to the neutral vector boson, Z, will cancel the above bad behavior. Consider the diagram shown in Figure 11.5b.

Recall that the left and right coupling factors for fermions connected to the Z are given by $g_L = I_3^{weak} - q\sin^2\theta_W$; $g_R = -q\sin^2\theta_W$. Since the ν has $q = 0$, $I_3^{weak} = 1/2$, $g_L = 1/2$, $g_R = 0$. The ZWW coupling can be obtained from Table E.2 in Appendix E. Note that positive momenta are inward to the vertex. The matrix element is then

$$i(M_{fi})_Z = \frac{-g^2}{2}\epsilon^*_\beta(p_3)\epsilon^*_\nu(p_4)\left(\bar{v}_2\gamma^\alpha\frac{(1-\gamma^5)}{2}u_1\right)$$

$$\times\frac{i(g_{\alpha\rho}+q_\alpha q_\rho/M_Z^2)}{q^2 - M_Z^2}[(q-p_3)^\nu g^{\beta\rho} + (p_3-p_4)^\rho g^{\beta\nu} + (p_4-q)^\beta g^{\rho\nu}]. \quad (11.33)$$

Consider, first, the term involving $q_\alpha q_\rho$. Since $q = p_3 + p_4$, $q - p_3 = p_4$ and $q - p_4 = p_3$. Next, $\epsilon^*(p_3)\cdot p_3 = \epsilon^*(p_4)\cdot p_4 = 0$. Furthermore, $(p_3 - p_4)\cdot q = (p_3 - p_4)\cdot(p_3 + p_4) = 0$. Using these relations the $q_\alpha q_\rho$ term vanishes.

For the term involving $g_{\alpha\rho}$, consider only the parts of the polarizations that have the highest powers of energy, i.e., the longitudinal polarizations of the W's. Again two of the three terms are zero. For the third term in the high-energy limit $\epsilon(p_3) = p_3/M_W$; $\epsilon(p_4) = p_4/M_W$. Then,

$$(M_{fi})_Z = \frac{g^2}{4}\left[\bar{v}_2(\not{p}_3 - \not{p}_4)\frac{(1-\gamma^5)}{2}u_1\right]. \quad (11.34)$$

Next go back to Equation 11.28 for the matrix element of the diagram in Figure 11.5a. Again consider only the longitudinal W polarizations in the high-energy limit, ignore terms of order $(M_W/E)^2$, and note that in this limit $\not{\epsilon} \approx \not{p}/M_W$. Use the relation $\not{a}\not{b} = (a\cdot b) - i\sigma^{\mu\nu}a_\mu b_\nu$ that implies $\not{a}\not{a} = (a\cdot a)$. One then finds

$$M_{fi} = \frac{g^2}{2}\bar{v}_2\not{p}_4\frac{(1-\gamma^5)}{2}u_1. \quad (11.35)$$

Sum the contributions from the two diagrams.

$$(M_{fi})_{tot} = \frac{g^2}{4}\bar{v}_2(\not{p}_3 - \not{p}_4 + 2\not{p}_4)\frac{(1-\gamma^5)}{2}u_1. \quad (11.36)$$

Note that $\not{p}_3 + \not{p}_4 = \not{p}_1 + \not{p}_2$. Using the Dirac equation both of these terms are then zero and the bad high-energy behavior has cancelled. The

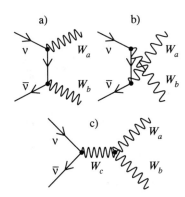

Figure 11.6 $\nu\bar{\nu} \to W^a W^b$. (a) and (b) Diagram and crossed diagram for no boson exchange. (c) Diagram with boson exchange.

existence of the Z with the Standard Model coupling is thus necessary to cancel the bad parts of these diagrams.

Furthermore, this cancellation is due to the Yang–Mills behavior of these couplings. To see this, go back to the Lagrangian in weak I-spin notation, Equation 11.15. Consider the diagrams shown in Figure 11.6. Modify the couplings of the previous calculations appropriately. In the non-boson exchange diagram, $\not{p}_4 \to [\not{p}_3 i\tau^a\tau^b + \not{p}_4 i\tau^b\tau^a]/4$. Similarly for the boson exchange diagram, the three-boson coupling changes $(\not{p}_3 - \not{p}_4) \to \epsilon_{abc}\not{p}_c\tau_c$. Use $\not{p}_4 = \not{p}_1 + \not{p}_2 - \not{p}_3$. Note that $\not{p}_1$ and $\not{p}_2$ give zero upon using the Dirac equation. Cancellation then requires the group structure constant relation $[\tau_a, \tau_b] = i\epsilon_{abc}\tau_c$.

Considerably more can be deduced from examining high-energy behavior in perturbation theory. The reaction $W^+ + W^- \to W^+ + W^-$ is discussed in Appendix F. There it is found to have very bad high-energy behavior if only vector boson exchange is included. If the Standard Model direct $4W$ coupling is introduced, the energy behavior is improved, but it still increases faster with energy than allowed by tree level unitarity. In addition to the $4W$ coupling, it is necessary to introduce the Higgs particle with Standard Model couplings to tame the high-energy behavior.

If the mass of the Higgs particle is too high, then the cancellation, which occurs only when $|q^2| \gg M_H^2$, fails and unitarity is violated before the cancellation mechanism can take effect. This can be viewed as putting an upper limit on the Higgs mass. This and similar calculations indicate the Higgs mass should be lower than about 1 TeV.

Also in Appendix F a general theorem on limitations of the Lagrangian because of these kinds of arguments is discussed. The Lagrangian is strongly limited to a form similar to the Standard Model. The conclusions from these arguments will be discussed after the following section.

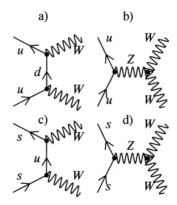

Figure 11.7 Diagrams illustrating the need for the GIM mechanism to cancel bad high-energy behavior.

11.13 The GIM Mechanism

Even part of the GIM mechanism can be viewed as driven by the need to avoid bad high-energy behavior. Consider the set of diagrams shown in Figure 11.7. It was noted previously that to cancel bad behavior in $\bar{\nu}\nu \to W^+W^-$ it was necessary to include diagrams with a lepton as well as diagrams with a Z intermediate state. A similar result, of course, is true for $\bar{u}u \to W^+W^-$ and for $\bar{s}s \to W^+W^-$. The cancellation will work as found previously for the diagram set 11.7 (a) and (b). However, it then must fail for the diagram set 11.7 (c) and (d). The charged strangeness-changing weak current is known to couple more weakly than the non-strangeness-changing current, lowering the contribution of 11.7 (c). However, the neutral current in 11.7 (d) is not strangeness changing and should have full strength. There must be another fermion that can serve as propagator in 11.7 (c). This is the charmed quark and to compensate, the coupling must be the full strength coupling $\times (d \sin\theta + c \cos\theta)$. To this extent, the GIM mechanism is forced by the need to eliminate bad high energy behavior.

11.14 Comments on These Results

From the examples discussed here and the general theorem given in Appendix F, the only theories which have well-behaved four-point (four external lines) Born amplitudes and satisfy the above assumptions can be put into a Yang–Mills form with an appropriate choice of field variables. It was necessary to introduce scalar particles; the relations found in Appendix F can be shown to imply that the couplings are intimately connected with the masses, which is a crucial feature of the Higgs particles.

11.14 Comments on These Results 279

Within the assumptions made here, tree unitarity excludes all interaction terms in the Lagrangian with dimension greater than four, unless they can be transformed away by a change of field variables.

To determine the dimension of various fields, note that spatial derivatives, $\partial/\partial x$, have energy dimension one. The Lagrangian must have dimension four since $\int \mathcal{L} d^4 x$ must be dimensionless. Since the kinetic energy term for fermions in the Lagrangian has the form $\overline{\Psi}\gamma^\mu \partial_\mu \Psi$, the fermion fields must have dimension 3/2. (The fermion field Ψ as defined in Chapter 7 seems dimensionless. However, once the implied integral of $\overline{\Psi}\Psi$ over $d^3 p$ is included, one obtains the present 3/2.) Similarly, one finds boson fields to have dimension one. Thus, for example, four boson fields or two boson fields plus one fermion field are allowed by tree unitarity, but additions to these are not possible.

There is a class of solutions to the equations of Appendix F which corresponds to the general case of the non-Abelian Higgs model of Kibble.[160] Are there solutions to these conditions other than those corresponding to a spontaneously broken gauge theory in the unitary gauge? Suppose the Lie algebra does not have Abelian invariant subgroups. (It is then called a semi-simple group.) An invariant subgroup is a subgroup consisting only of elements B_i, such that if A_j is any element of the full group, then $A_j B_i A_j^{-1}$ is an element of the subgroup. For this kind of algebra there are no other solutions.

However, if the Lie algebra is not semi-simple, then there can be other solutions. Some (and perhaps all) of these correspond to generalized Higgs models, but have extra gauge bosons corresponding to the Abelian invariant subgroups. For the physically interesting case with only the known leptons, the four vector mesons ($W^\pm, \gamma, Z^0$), and with the minimum number of scalar particles (one), there is a unique non-trivial solution, which corresponds to the Standard Model, even though the Lie algebra is not semi-simple.

It is certainly possible that a new theory will be found violating some of the assumptions of this derivation. For example, in string theory, the point interaction assumption is violated. Special heavy leptons, i.e., adding more particles, can help bad behavior,[161] but this scheme seems physically unreal. With the assumptions made here unitarity forces nature to make the choice of a spontaneously broken gauge theory.

One might speculate that if there were an ur-set of interactions between ur-objects which were very strong, nonetheless the present theory would be forced to exist as an effective Lagrangian. However, if the Standard Model is only an effective theory, then the requirements for good high-energy behavior could be relaxed. At higher energies, new terms, ignored here, may come into play to improve the very high energy behavior.

11.15 How Constant are the Coupling Constants?

Present theory has a number of constants such as Planck's constant, the velocity of light, the coupling constants, and so on. Some of these are now known not to be absolute constants but to vary in some way.

For example, the velocity of light is expected to vary depending on the value of the local gravitational field. The strong, weak, and electromagnetic coupling constants vary as a function of energy. Do some of these also vary as a function of time?

The existence of a naturally occurring nuclear reactor $\approx 2 \times 10^9$ years ago at the Okla Uranium mine in Gabon allows stringent limits to be set on this variation. From examining the ratio of ^{149}Sm/^{147}Sm it is estimated that the neutron capture probability of ^{149}Sm has changed by less than 10% in the past 2×10^9 years. This implies that the fractional changes per year in the strong, electromagnetic, and weak coupling constants have averaged less than 10^{-19}, 2×10^{-18}, 10^{-12}, respectively, over this time period.[162]

In this chapter the Higgs mechanism has been discussed. This mechanism makes a renormalizable theory possible. It has been shown how a combination of gauge symmetry and spontaneous symmetry breaking can allow for a theory with massive vector bosons. In this theory, a new particle, the Higgs particle is responsible for giving mass to the other fundamental particles. Applying this mechanism to the Weinberg–Salam Model gives the present Standard Model of particle physics. There are no known experimental contradictions to the Standard Model.

11.16 Exercises

11.1 Show explicitly that the Lagrangian given in Equation 11.6 is gauge invariant.

11.2 List the independent constants in the standard model Lagrangian. This gives an unpleasantly large number of independent constants in the theory.

12
Extensions of the Standard Model: Grand Unification

12.1 Introduction

The Standard Model outlined in the past two chapters is extremely successful. There are exceedingly few experimental problems with it and for these possible problems the experiments and/or the interpretations are in some question. In fact, the lack of experimental problems is itself a problem for particle physics. These considerations will be discussed in the last chapter.

However, the Standard Model is almost surely only a step toward an ultimate theory of nature, a "theory of everything." First of all there are too many arbitrary parameters. Including QCD and classical gravity, the minimal model has 25 free parameters.

To make matters worse some of these arbitrary parameters must have precise relations to others to very high precision to agree with observation. For example the magnitude of the charge on the electron and that on the proton are known to be numerically equal to about one part in 10^{19}.[163] It is difficult to make masses of physical particles as low as they are experimentally. The smallness of CP violation, the Higgs mass, and, in gravity, limits on the cosmological constant all require very fine tuning of the parameters.

Furthermore, it is not understood why there are four interactions, why one and only one violates parity, why there are a small finite number of generations of particles apparently identical except for mass, etc.

There must be new physics. In this chapter an attempt will be made to further unify the theory, to find a "Grand Unified Theory (GUT)."[135] As will be noted, it has experimental problems and, even if true, would not be a final theory itself. Nonetheless, it may help point a way toward future theories.

The Standard Model involves the group $SU(2) \otimes U(1)$. $SU(2)$ is the group of weak isospin and has coupling constant g. $U(1)$ is the group of Y^{weak} and has coupling constant g'. Furthermore, $g'/g = \tan\theta_W$. Can these groups be related? Is there a larger symmetry group?

For example, the lowest generation of particles has u, d, e, ν_e, and the other generations have analogous members. Should u, d, e, ν_e all be elements of a single symmetry group which then mixes leptons and quarks in a single representation? If this is possible, some features of the resultant group are necessary. First, it must be badly broken at normal energies, since the interactions of electrons and quarks are very different. Second, it opens up the possibility of having new particles containing both a lepton and a quark quantum number, "leptoquarks," and the possibility of having quark to lepton transitions. The latter may allow a proton, for example, to decay into leptons.

Such a larger group should include color as well as $SU(2) \otimes U(1)$. Thus, $G \supset SU(3) \otimes SU(2) \otimes U(1)$. It would be desirable to have a single symmetric coupling. However, in fact, at laboratory energies there are three separate coupling constants, α_s, g, g', all of which are functions of Q^2. Is it possible that at some high momentum they meet (see Figure 9.8)? One would have $g_i(Q) = g_G(Q)$ for $Q \equiv \sqrt{Q^2} > M_X$. Here new constants have been defined by $\alpha_s = g_3^2/4\pi$, $g = g_2$, and $g' = g_1/C$, where C is a Clebsch–Gordan coefficient different for each group G. For $G = SU(5)$, $C = \sqrt{5/3}$.

Previously, for α_s, one had (f = number of flavors of quarks)

$$\alpha_s(Q^2) = \alpha_s(\mu^2) \bigg/ \left[1 + \frac{\alpha_s(\mu^2)}{12\pi}(33 - 2f) \ln \frac{Q^2}{\mu^2}\right]. \quad (12.1)$$

This can be rewritten as

$$\frac{1}{g_3^2(\mu)} = \frac{1}{g_3^2(Q)} + 2b_3 \ln \frac{Q}{\mu}; \text{ where } b_3 = \frac{1}{(4\pi)^2}\left(\frac{2}{3}f - 11\right). \quad (12.2)$$

If $Q = M_X$, $g_3 = g_G$. Thus

$$\frac{1}{g_3^2(\mu)} = \frac{1}{g_G^2} + 2b_i \ln \frac{M_X}{\mu}, \text{ where } i = 3. \quad (12.3)$$

It turns out that the same equation works for g_i, $i = 1, 2, 3$ with different, but calculable, b_i. Let $n_g = f/2$ be the number of generations of fermions. (Within the Standard Model there are $n_g = 3$ generations unless some additional generations have heavy neutrinos.) Then the various b_i are[164]

$$b_1 = \frac{1}{(4\pi)^2}\frac{4}{3}n_g, \quad (12.4)$$

$$b_2 = \frac{1}{(4\pi)^2}\left(\frac{-22}{3}\right) + b_1, \quad (12.5)$$

$$b_3 = \frac{1}{(4\pi)^2}(-11) + b_1. \quad (12.6)$$

At present energies the fact that $g_3 > g_2 > g_1$ is understandable. The larger gauge groups have larger anti-shielding. Since the gauge particle of the lowest group, $U(1)$, does not carry the charge of the force, there is no anti-shielding for it.

There are three equations and three unknowns: M_X, g_G^2, and $\tan\theta_W (= g'/g)$. After some algebra one obtains (Exercise 12.1)

$$\ln \frac{M_X}{\mu} = \frac{6\pi}{11(1+3C^2)} \left(\frac{1}{\alpha} - \frac{1+C^2}{\alpha_s} \right). \tag{12.7}$$

In the next chapter it will be noted that for $\mu = 91$ GeV, $\alpha \approx 1/128$, and in Chapter 9 it was indicated that $\alpha_s = 0.115$ at this value of μ. Using these values, and taking $C^2 = 5/3$, one obtains $M_X = 9 \times 10^{14}$ GeV. This is a very large mass. The unification coupling constant is small, $g^2/(4\pi) = 0.023$.

As an aside, note this mass is still small compared to the energy at which gravitation is expected to be important. Consider two objects each of mass M so close together that the magnitude of the gravitational potential is approximately equal to the mass energy, Mc^2, i.e., let $GM^2/r = Mc^2$. Suppose these two particles are separated by the natural length unit $\hbar/Mc$, the Compton wavelength. Solving it is found that $Mc^2 = \sqrt{\hbar c^5/G} = 1.22 \times 10^{19}$ GeV. This mass is called the "Planck Mass" and is many orders of magnitude higher yet than M_X.

Much theoretical effort has gone into the study of $G =$ various $SU(N)$. An $SU(N)$ model has $N^2 - 1$ gauge particles, but most of them must be at very high masses. At normal energies, experimentally, there are 12. They are $\gamma, W^\pm, Z$, and eight gluons.

12.2 THE GROUP $SU(5)$

The simplest group G is $SU(5)$, the group of five-dimensional unitary matrices with determinant $= +1$. Group representations will be labeled either by their dimensionality (number of states) in $SU(5)$ or by their $SU(3)$(color) and their $SU(2)$(weak isospin) content. Consider only left-handed states and anti-states at first. (Note that under CP the anti-particle of e_L^- is e_R^+.) A useful identification of particles in representations is

$$\bar{5} = (1,2) \oplus (\bar{3},1) = (\nu_e, e^-) \oplus \bar{d}_L, \tag{12.8}$$

$$10 = (1,1) \oplus (\bar{3},1) \oplus (3,2) = e_L^+ \oplus \bar{u}_L \oplus (u,d)_L. \tag{12.9}$$

In matrix form

$$\bar{5} = \begin{pmatrix} \nu_e \\ e^- \\ \bar{d}_R \\ \bar{d}_B \\ \bar{d}_Y \end{pmatrix} ; 10 = \frac{1}{\sqrt{2}} \begin{pmatrix} 0 & e^+ & d_R & d_B & d_Y \\ -e^+ & 0 & u_R & u_B & u_Y \\ -d_R & -u_R & 0 & \bar{u}_Y & \bar{u}_B \\ -d_B & -u_B & -\bar{u}_Y & 0 & \bar{u}_R \\ -d_Y & -u_Y & -\bar{u}_B & -\bar{u}_R & 0 \end{pmatrix}. \tag{12.10}$$

This is somewhat awkward in that the basic states are in two separate representations with particles and anti-particles mixed. The electron antiparticle is in a separate representation from the electron. The couplings of the $\bar{5}$ and the 10 are taken equal to reproduce the Standard Model and to allow triangle diagram cancellations to occur.

There are $N^2 - 1 = 24$ bosons in $SU(5)$. These divide into

$$24 = [(8,1)] \oplus [(1,3) \oplus (1,1)] \oplus [(3,2) \oplus (\bar{3}, 2)]. \tag{12.11}$$

The first square bracket has the eight gluons, the second square bracket has the $W^\pm, Z, \gamma$, and the last has a set of 12 new bosons, $3X, 3Y$ and their anti-particles.

The X particles have charge $1/3$, and the Y particles have charge $-1/3$. They are weak doublets, they are colored, and they have both lepton and baryon quantum numbers. They can turn quarks into leptons. This is not surprising if quarks and leptons are in the same representation. One can have $(u, d)_L \to e_L^+ + (\bar{Y}, \bar{X})$. In color–weak isospin notation, this is $(3, 2) \to (1, 1) \otimes (3, 2)$. Neither the baryon number b nor the lepton number l is conserved. However, the combination $b - l$ is conserved.

The photon is a gauge boson of $SU(5)$. The charge operator is a generator of the group. One can show that for this type of group (simple group), the trace of a generator is 0. Consider what this means. For the $\bar{5}$, the trace of Q can be shown to be just the sum of the charges in this representation, $\text{Tr}(Q) = 3Q_{\bar{d}} + Q_\nu + Q_{e^-} = 0$. This implies that $Q_d = (1/3)Q_e$. The factor of 3 has come from the fact that there are three colors. For the 10, it can be shown that the trace condition implies $Q_u = -2Q_d$. Together this implies $Q_p = -Q_{e^-}$. With this symmetry group one can understand the fundamental reason for the equality of the magnitude of charges on the electron and proton, as well as understanding why the quark charges are $2/3, -1/3$. Charge quantization, at least within a single generation, is explained.

This model contains a number of Higgs bosons. There is an H at normal energies, plus a number at unification type energies: H_3, H_0, Ω^8 (a color octet), $H^\pm, H_{Y_i}^\pm$ (violates b, l). In fact, in the simplest versions of this model, there are no new fundamental particles above the standard W, Z, H until one reaches this very high unification energy regime. The range between normal and unification energies is called "the great desert."

Examine the weak mixing angle, θ_W in this model. Lowest-order mixing between the Z and the photon would occur through the diagram shown in Figure 12.1. However, since the Z and the photon are orthogonal, the loop must give zero when summed over the fermions in each multiplet.

12.2 The Group $SU(5)$

The coupling fermion–$Z \propto I_3 - Q\sin^2\theta_W$. The coupling fermion–$\gamma \propto Q$. Suppose one sums over the fermions in a multiplet. One wants

$$\Sigma Q(I_3 - Q\sin^2\theta_W) = 0 \Longrightarrow \sin^2\theta_W = \frac{\Sigma Q I_3}{\Sigma Q^2}. \qquad (12.12)$$

Consider the $\bar{5}$ representation. Note that $I_3 = 0$ for the various $\bar{d}_i$ since they are anti-particles of the right-handed d. Thus:

Table 12.1 I_3 and Q values for particles in the 5 representation.

Particle	Weak I_3	Q
ν_e	$1/2$	0
e^-	$-1/2$	-1
$\bar{d}_R$	0	$+1/3$
$\bar{d}_B$	0	$+1/3$
$\bar{d}_Y$	0	$+1/3$

Evaluating the above sum one gets

$$\sin^2\theta_W = \frac{\Sigma Q I_3}{\Sigma Q^2} = \frac{0 \times 1/2 + (-1) \times (-1/2) + 0}{(-1)^2 + 3 \times (1/3)^2} = \frac{1/2}{4/3} = 3/8. \qquad (12.13)$$

Thus, at the unification mass, it is expected that $\sin^2\theta_W = 3/8$. There are three coupling constants (strong, weak, electromagnetic) and two parameters, the unification energy and the unification coupling constant. The value of $\sin^2\theta_W$ at laboratory energies is then a prediction. The equations given in Section 12.1 for the running coupling constants can be used to work backward from g_G to lower energies (Exercise 12.2). At $Q = 10$ GeV, one obtains $\sin^2\theta_W = 0.21 \pm 0.004$. This is close to the measured value of $\sin^2\theta_W = 0.2323 \pm 0.0003$ found experimentally and is an important success of $SU(5)$.

It has been noted previously that hadrons can turn into leptons in this model. This leads to the question of whether hadronic matter is absolutely stable or whether the proton can spontaneously decay into leptons. "Are diamonds forever?"

286   12. Extensions of the Standard Model

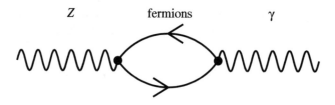

Figure 12.1 Feynman diagram for weak mixing angle.

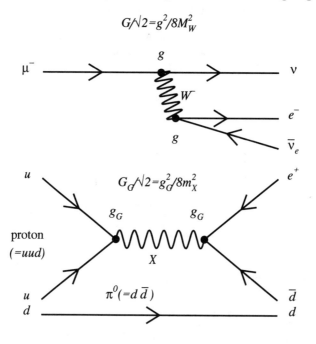

Figure 12.2 Muon decay and proton decay Feynman diagrams.

One can get a rough estimate of the proton decay lifetime by comparing the diagrams for muon decay and proton decay, in Figure 12.2. For the muon decay one has $4G/\sqrt{2} = g^2/(2M_W^2)$:

$$\Gamma(\mu \to e\nu\nu) = \cdots G^2 m_\mu^5 = \cdots \frac{m_\mu^5}{M_W^4}. \qquad (12.14)$$

For the proton decay shown in the figure one has $4G_G/\sqrt{2} = g_G^2/(2M_X^2)$. Since the evaluation of the matrix element is essentially the same as in the muon case, one gets

$$\Gamma(p \to \pi e) = \cdots G_G^2 m_p^5 = \cdots \frac{m_p^5}{m_X^4}. \tag{12.15}$$

The mass of the X particle is around the unification mass. Thus, the lifetime of the proton is

$$\tau_p \propto \frac{m_X^4}{m_p^5} \longrightarrow \tau_p \approx 10^{29} \text{ years.} \tag{12.16}$$

There is an uncertainty of several powers of ten in this estimate since $\tau_p \propto m_X^4$ and also $\tau_p \propto \Lambda^4$, where Λ is the strong scale parameter. The weak mixing angle, θ_W, also depends on m_X. A large lifetime corresponds to a small θ_W.

There have been major attempts to detect this decay. The most elaborate of these have been two underground experiments in which the walls, floor, and ceiling of a large body of water have been lined with photomultiplier tubes. These experiments looked at Cherenkov light from the decay particles and tried to reconstruct proton decay events. The Irvine–Michigan–Brookhaven (IMB) experiment[165] had 10,100 tons of water and 2048 8-inch photomultiplier tubes. The Japanese Kamioka experiment[166] had about a third as much water (3400 tons), but a greater density of photomultipliers. They had about 1000 photomultiplier tubes, each about 20 inches in diameter. They had 20% of the surface area of their water pool covered with light sensitive surface! No definite proton decay signals were seen. The current experimental lower limits on a number of the expected modes are greater than about 10^{32} years. This is sufficiently long that there is no worry about decaying away by that mechanism in one's own lifetime! It is also sufficiently long to be somewhat of a problem for $SU(5)$. In some variants of the model the main decays may be to K mesons which are harder to detect. Even here the lifetime limits are several $\times 10^{31}$ years.

Another question can be addressed by this model, which also concerns the existence of the Universe in its present state. Suppose that the Universe was created from some initial singularity and at the time of creation had no net charge or baryon number. The present Universe, however, seems to have more baryons than anti-baryons. If there are the same number of baryons as anti-baryons, then the ratio of baryons to photons can be estimated as follows. A 3K (microwave) background radiation is seen almost uniformly in the sky.[167] Assume that this is Doppler shifted radiation left over from the initial big bang. In the initial phases of the Universe, at times less than 10^{-4} s, $kT > 1$ GeV and there was a baryon–anti-baryon soup. The ratio of baryons to photons at these times can be estimated by statistical considerations, since they are continually changing into one another by collisions and annihilations.

As the Universe cooled, baryon–anti-baryon pairs stopped being produced. If it is assumed there are no baryon decays, but the baryons and anti-baryons are allowed to annihilate after this point, one would estimate that the present day ratio of the number of baryons to the number of photons should be about 10^{-18}. In fact, the experimental estimate is about 10^{-9}.

How has this come about? The $SU(5)$ model has a mechanism for this in the X particles. It turns out that the decay of these particles can have sizable CP violation. It is believed that CPT is a perfect symmetry of nature. It guarantees that the overall lifetime of a particle is the same as the overall lifetime of its anti-particle. It does *not* guarantee that the partial decay rates into specific modes are the same for particle and anti-particle. If CP is conserved, then, as noted in Section 4.12, these partial rates must be the same, but if it is violated they do not have to be the same. Thus, it can be imagined that in the very early Universe a baryon excess was created from the decays of the X particle. Since this excess cannot annihilate away with anti-baryons, there would be a much larger ratio of baryons to photons and 10^{-9} is possible. It is very hard to calculate this number as the details of the CP violating interaction are not known, although it is expected that the most important CP violation is in the interaction and decay of various Higgs particles.

Another possible source of baryon number violations, not associated with $SU(5)$, concerns "sphalerons." These are semi-classical non-perturbative solutions for W bosons within the Standard Model in which a very complicated vacuum is formed. Essentially, different parts of the vacuum have different baryon number and baryon number violation is possible. As is the case for $SU(5)$, specific calculations and predictions are lacking here.

12.3   FURTHER UNIFICATION

Much effort has gone into trying to find other even larger groups that will not have problems with proton decay and that will avoid the awkwardness of having the fundamental particles split into two fundamental representations with the d's split between both.

If one goes to $SO(10)$, all of the left-handed quarks and leptons fit into a single representation. [$SO(10)$ is the group of ten-dimensional orthogonal matrices with determinant $+1$. Orthogonal matrices are real, unitary matrices.] In addition, this representation includes one particle that has neither strong nor electromagnetic nor weak interactions. However, $SO(10)$ has the problems listed below for $SU(5)$ and its own problems as well.

$SU(5)$ hasn't solved many of the fundamental problems indicated at the beginning of this chapter. There is still fine tuning of the masses. Why is the mass scale of basic particles so many orders of magnitude less than m_X?

The ratio m_X/vacuum value (v) is approximately 10^{13}. This is sometimes known as the gauge hierarchy problem. Why are there generations? Why are the charges of different generations the same? Can gravity be added into the picture?

An elaborate attempt to go farther is called supersymmetry (SUSY). Here, an attempt is made to add not only elementary fermions, but also elementary bosons into the fundamental representations. In this theory, every boson (including Higgs particles) should have a fermion partner and vice versa. Supersymmetry can eliminate some of the fine-tuning problems and enable the intermediate bosons to have masses as low as 100 GeV. This occurs because in calculating masses of gauge particles, boson loops contribute with opposite sign to fermion loops. [In $SU(5)$ theory appropriate cancellations require adjustment of independent parameters to about 24 decimal places.] A local supersymmetry theory requires the existence of gravity, perhaps finally bringing a connection between gravity and the rest of the interactions. There are no existing candidates for partners of any of the present particles, but numerous searches for these particles have taken place and will take place at the new accelerators.

Since the symmetry is broken, the preceding cancellations in the loops occur only for virtual momenta such that the difference of m^2 of the partners is negligible. If this is taken seriously, it sets a maximum scale on the mass of supersymmetric partners if sufficient cancellation is to occur. This is known as low-energy supersymmetry. It implies the masses of the partners should be at the weak interaction scale or below. They should, therefore, appear in the next generation of hadron colliders, if not before.

In the minimal supersymmetric standard model (MSSM), SUSY requires at least two complex Higgs doublets (H_1, H_2). It thus has at least five physical neutral and charged Higgs bosons. At tree level at least one of these bosons has $M < M_Z$. Radiative corrections can raise this limit to about 130 GeV. This low value means that it may be found at LEP2 and certainly should be seen by the LHC if it exists. There are two free parameters in the theory that can be taken to be M_A, the mass of one of the Higgs bosons and $\tan\beta = \langle H_2\rangle/\langle H_1\rangle$, where $\langle H_2\rangle$ and $\langle H_1\rangle$ are the two vacuum expectation values of the complex doublets. Since $\langle H_2\rangle$ generates masses for t, c, and u while $\langle H_1\rangle$ generates masses for b, s, and d, it is suggested (but not required) that $\tan\beta$ be large.

Another argument for supersymmetry is that, experimentally, the coupling constant extrapolations do not meet at a single point for the Standard Model, but do under supersymmetry. Another, more dramatic way of expressing this is to evolve the coupling constants down from the unification energy to M_Z. The meeting of the extrapolations is then translated into a constraint between the value of α_s and $\sin^2\theta_W$, which is satisfied to a couple of percent for supersymmetry and badly violated for the normal Standard Model. However, the evolution of the coupling constants might

Table 12.2 Some SUSY partners.

Particle	SUSY partner	Name	spin
e	$\tilde{e}$	selectron	0
μ	$\tilde{\mu}$	smuon	0
ν	$\tilde{\nu}$	sneutrino	0
u	$\tilde{u}$	squark	0
d	$\tilde{d}$	squark	0
γ	$\tilde{\gamma}$	gaugino	1/2
g	$\tilde{g}$	gaugino	1/2
Z	$\tilde{z}$	gaugino (zino)	1/2
W	$\tilde{w}$	gaugino (wino)	1/2
H	$\tilde{h}$	higgsino	1/2

be sensitive to details of the higher-energy theory and not simply follow the evolution equations of the lower-energy theory.

If SUSY is correct, every particle has a SUSY partner. These are usually specified by adding an "s" in front of the particle name for fermions (electron → selectron) and adding "ino" at the end for bosons (W → wino). See table 12.2.

The lightest supersymmetric particle (lsp) is stable because there is no supersymmetric particle into which it can decay. It is thought that $\tilde{\nu}$ or $\tilde{\gamma}$ are the likeliest candidates for the lsp.

The most elaborate attempt for a theory of everything is superstring theory. Here it is assumed that the fundamental particles are not point particles but are one-dimensional strings that may be open or may be closed loops. This gets rid of some undesirable infinities in self-energy. If gravity and supersymmetry are added to this picture, one gets supergravity theory. Four spacetime dimensions present difficulties and it is often assumed there are many more, perhaps 12, but that all dimensions, except for the four known ones, have curled themselves up into very tiny closed paths. It looks like the theory can be unique. There are many people working on it. Unfortunately, the fundamental particles in this theory have no relation to the particles observed experimentally. If this is the correct theory, the symmetries apply only at super high energies around the Planck mass and are badly broken at normal and even at unification energies. It has not yet been possible to make any firm contact between these theories and any experimental data. There have been some phenomenological attempts to get toward predictions, but these attempts involve enough assumptions that they are not firm predictions.

In this chapter it was noted that there are serious flaws in the Standard Model. There are many independent fundamental constants that must be introduced into it and their values must be fine tuned with respect to each other to give an acceptable theory. Some attempts were discussed to further unify the forces of the Standard Model. The $SU(5)$ model was specifically described. Extending the Standard Model is an area of intense present activity. No model, so far, is satisfactory in that it makes contact with experiment and gives totally successful experimental predictions. Even so, some of the models may well hold real clues for future progress.

12.4 EXERCISES

12.1 Derive Equation 12.7 for M_X from the relations given in the preceding discussion.

12.2 Assuming that $\sin^2 \theta_w = 3/8$ at the unification energy, work back using the equations in Section 12.1, and obtain an estimate of $\sin^2 \theta_W$ for energies in the 1 GeV range.

13
Physics at the Z

In August 1989 the LEP accelerator at CERN first produced Z bosons. A year later each of the four experiments at LEP had recorded around 100,000 Z decays and analyzed them for a large number of physics results. The SLAC SLC accelerator had, in fact, produced Z's a few months before LEP. The MARK II group at SLC contributed strongly to the initial results, but later results have usually depended on the increased statistics available at LEP, of the order, now, of millions of Z's per experiment. SLC (with the SLD detector which replaced the MARK II dectector) is now again making real contributions with the aid of strongly polarized e^+ and e^- beams and intensity which, although still below that of LEP, is much improved.

Consider the process $e^+e^- \to Z^0 \to \bar{f}f$. (See Figure 13.1.) A general amplitude for this process can be written as

$$M = \sqrt{2} G M_Z^2 \rho^0 \left(\frac{1}{s - M_Z^2 + i\Gamma_Z s/M_Z} \right) \times X_{f_\mu} X_{e_\mu} + Q_e Q_f \frac{4\pi\alpha}{s} j_f^{em} j_e^{em}. \tag{13.1}$$

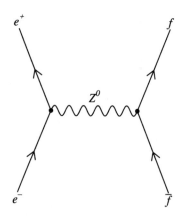

Figure 13.1 $e^+e^- \to Z^0 \to \bar{f}f$.

The denominator is essentially just the propagator term for an unstable particle. Small γ–Z interference terms have been ignored here. If only V and A couplings are assumed here, then for particle p

$$X_{p_\mu} = \frac{1}{2}\bar{u}_p(\gamma^\mu(g_V)_p - \gamma^\mu\gamma^5(g_A)_p)v_p. \tag{13.2}$$

What is obtained in the Standard Model (SM)? Consider the current that couples to the Z in the electroweak Lagrangian given in Chapter 10:

$$\mathcal{L} = \frac{g}{\sqrt{2}}((j^-)^\alpha W_\alpha^+ + (j^+)^\alpha W_\alpha^-) +$$

$$\frac{g}{\cos\theta_W}[(j^3)^\alpha - (j^{em})^\alpha \sin^2\theta_W]Z_\alpha + g\sin\theta_W(j^{em})^\alpha A_\alpha. \tag{13.3}$$

Use the definitions given there of $(j^3)^\alpha$, $(j^{em})^\alpha$, $(I_3)^{weak}$, Y^{weak}, and Q. (j^{em} = the electromagnetic current and j^3 = the third component of the weak I spin current.) Also, recall that

$$(u_L)_p = \left(\frac{1-\gamma^5}{2}\right)(u)_p; \quad (u_R)_p = \left(\frac{1+\gamma^5}{2}\right)(u)_p. \tag{13.4}$$

One can then obtain the relation

$$(j^3)^\alpha - (j^{em})^\alpha \sin^2\theta_W = \frac{1}{2}\bar{u}_p(\gamma^\alpha(g_V)_p - \gamma^\alpha\gamma^5(g_A)_p)v_p, \tag{13.5}$$

where for the SM:

$$(g_V)_p = (I_3)_p^{weak} - 2Q_p \sin^2\theta_W = \text{vector coupling const.} \tag{13.6}$$

$$(g_A)_p = (I_3)_p^{weak} = \text{axial vector coupling const.} \tag{13.7}$$

$$(X_{SM})_p = (j^3)_p^{weak} - j_p^{em} \sin^2\theta_W. \tag{13.8}$$

Next note that in the SM

$$M_Z^2 = \frac{M_W^2}{\cos^2\theta_W} = \frac{g^2\sqrt{2}}{8G\cos^2\theta_W}. \tag{13.9}$$

The Z part of the SM electroweak Lagrangian above becomes for particle p

$$\cdots + M_Z 2^{1/4}\sqrt{G}\bar{u}_p(\gamma^\alpha(g_V)_p - \gamma^\alpha\gamma^5(g_A)_p)v_p + \cdots. \tag{13.10}$$

Let u be the wavefunction for fermion f in the process $e^+e^- \to Z^0 \to \bar{f}f$. One can then calculate the matrix element for this process in the usual way to obtain:

$$M = \sqrt{2}GM_Z^2 \left(\frac{1}{s - M_Z^2 + i\Gamma_Z s/M_Z}\right) \times ((X_{SM})_f)_\alpha ((X_{SM})_e)_\alpha. \quad (13.11)$$

Compare this expression with Equation 13.1. It is seen that in the SM

$$\rho^0 = 1; \quad X_p = (X_{SM})_p. \quad (13.12)$$

ρ^0 is known as the Veltman parameter. More generally,

$$\rho^0 = \frac{M_W^2}{M_Z^2 \cos^2 \theta_W} \equiv \frac{G^{NC}}{G^{CC}}. \quad (13.13)$$

In the SM, $\sin^2 \theta_W \equiv 1 - M_W^2/M_Z^2$ and is not a free parameter.

The above expressions for the matrix elements are the first-order or Born approximation matrix elements. Interesting physics lies past the Born approximation. The corrections to the matrix element are of two kinds.

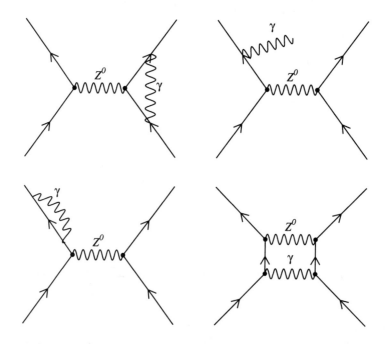

Figure 13.2 Typical photon corrections to $e^+ e^- \to Z^0 \to \bar{f}f$.

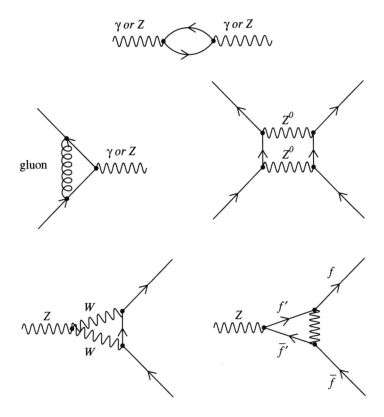

Figure 13.3 Typical non-photon corrections to $e^+e^- \to Z^0 \to \overline{f}f$.

The first kind of correction involves real and virtual photon corrections. Typical diagrams are shown in Figure 13.2. The photons can appear on any legs of the diagrams. The corrections from photons depend on the experimental cuts and must be taken into account by experimenters, but they contain no new basic physics. That is, the size of the cross section depends upon the experimental cuts made on visible photon energies and angles, etc.

The second kind of correction (Figure 13.3) involves non-photonic corrections. These are generally independent of the experimental cuts. They do bring in new physics such as the Higgs particle or the top. They affect the W mass.

Parametrize the corrections as follows (here, α is the fine structure constant):

$$M_W^2 \bigg/ \left(\frac{\pi \alpha}{\sqrt{2} G \sin^2 \theta_W} \right) \qquad (13.14)$$

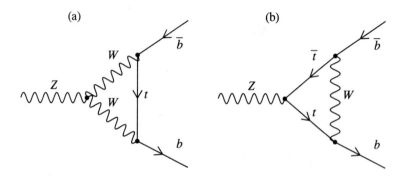

Figure 13.4 Vertex corrections involving top.

$= 1$ (Born)
$= 1 + \Delta r$ (1 loop)
$= 1/(1 - \Delta r)$ (bulk of higher order).

The corrections are all embodied in the parameter Δr. One has

$$\Delta r = \Delta \alpha + \Delta \rho \cot^2 \theta_W + \cdots . \qquad (13.15)$$

$\Delta \alpha$ corresponds to the running of α, which is induced by the photon corrections:

$$\alpha(M_Z^2) = \frac{\alpha(M_e^2)}{1 - \Delta \alpha} \approx \frac{1}{128}. \qquad (13.16)$$

$\Delta \rho$ has the "new physics" (g = generation):

$$\Delta \rho = \frac{\sqrt{2}G}{16\pi^2} \Sigma_g N_g^c |M_{gu}^2 - M_{gd}^2| + \cdots , \qquad (13.17)$$

where N_g^c is the color factor and $M_{gu}^2 - M_{gd}^2$ is the isodoublet mass splitting. Since the top has the largest mass, $\Delta \rho \propto M_t^2$:

$$\Delta \rho \equiv \frac{\sqrt{2}G}{16\pi^2} 3M_t^2 + \cdots . \qquad (13.18)$$

Also it turns out $\Delta \rho \propto \ln(M_H)$, not very sensitive to Higgs mass. For M_t varying from 40 to 200 GeV, Δr varies from 0.08 to 0.03. For M_H varying from 10 to 1000 GeV, Δr changes by only about 0.015.

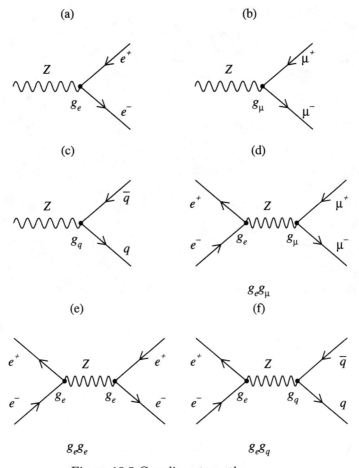

Figure 13.5 Coupling strengths.

Examine the effect of these corrections on the coupling constants. One can show that the non-photonic correction can be parametrized as follows:

$$(g_V)_f = I_{3_f} - 2Q_f \sin^2\theta_W \,(\text{Born}) \to I_{3_f} - K_f 2Q_f \sin^2\theta_W, \quad (13.19)$$

where $K_f = 1 + \Delta\rho \cot^2\theta_W + \cdots$. Redefine a modified $\sin^2\theta_W$:

$$\sin^2\bar\theta_W = K_f \sin^2\theta_W. \quad (13.20)$$

Look at $\rho = \rho^0 + \Delta\rho$. To first order in the correction

$$\sin^2\bar\theta_W = 1 - \frac{M_W^2}{\rho M_Z^2}. \quad (13.21)$$

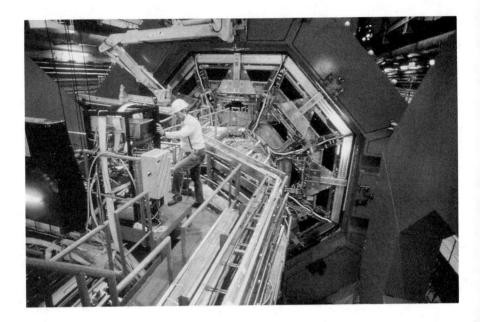

Figure 13.6 The L3 detector at the CERN LEP accelerator. *(Photograph courtesy of the CERN Photographic Service.)*

Measuring M_Z, M_W, $\cos^2\theta_W$ accurately gives information on the mass of the top quark and the Higgs particle. Since the top mass is known, these measurements give at least some information on the mass of the Higgs particle.

The "improved Born approximation" amplitude for $e\bar{e} \to Z \to f\bar{f}$ including these effects is (ignoring small γ-Z interference terms)

$$M = \sqrt{2}GM_Z^2\rho^0\left(\frac{1}{s - M_Z^2 + i\Gamma_Z s/M_Z}\right)$$

$$\times(j_f^3 - 2j_f^{em}sin^2\overline{\theta}_W)(j_e^3 - 2j_e^{em}sin^2\overline{\theta}_W)$$

$$+Q_eQ_f\frac{4\pi\alpha(M_z^2)}{s}j_f^{em}j_e^{em}. \tag{13.22}$$

All of the corrections have not yet been included. Imaginary parts of the self-energy, box corrections, and vertex corrections have been omitted. These are small, except for the vertex corrections when $f = b$-quark (due to the high mass of the t). See Figure 13.4. The effect of this correction is to produce a term in the decay width $Z \to b\bar{b}$, $\Gamma_{b\bar{b}}$ which is proportional to M_t^2. Thus, the ratio $\Gamma_{b\bar{b}}/\Gamma_{\text{hadronic}}$ has a small quadratic t dependence. Note that the Z propagator can also have a t loop introducing another quadratic dependence. This latter term cancels in the ratio, $\Gamma_{b\bar{b}}/\Gamma_{\text{hadronic}}$. However, the two effects have opposite signs and similar magnitudes in $\Gamma_{b\bar{b}}$. The absolute value of $\Gamma_{b\bar{b}}$ has only a small t dependence.

There are also QCD corrections to the hadronic σ, Γ for $f =$ any quark. The zeroth-approximation correction for these is to multiply M by

$$1 + \frac{\alpha_s(M_z^2)}{\pi}. \tag{13.23}$$

Before looking at experimental results, examine some other general features. Consider Figure 13.5. For Z decays to e^+e^- it is seen that the coupling of Z to e^+e^- is involved. For the moment call this g_e, suppressing all the complications above. For Z decay to $\mu^+\mu^-$, one has $M^{weak} \propto g_\mu$. Next consider $e^+e^- \to Z \to \mu^+\mu^-$. Here $M^{weak} \propto g_e g_\mu$. Similarly, for $e^+e^- \to Z \to e^+e^-$ one has $M \propto g_e^2$. For $e^+e^- \to Z \to \bar{q}q$, $M \propto g_e g_q$.

The full width of the Z includes the width into the "invisible" modes $\nu\bar{\nu}$, one for each type of ν. The visible total cross section depends on the visible modes only. The mismatch allows a measurement to be made of the number of types of ν and hence the number of fundamental families. If the SM is assumed, the width due to each of the quarks, leptons, etc., is known. With this assumption the number of neutrinos can be measured from the Z full width alone not depending on the absolute cross section. Of course, the method fails if there are neutrinos whose mass is higher than one-half of the Z mass.

Figure 13.6 shows the L3 detector, one of the large detectors at LEP, and Figure 13.7 (see the color insert) shows the SLD detector at SLAC.

The four LEP detectors and the SLD detector at SLAC each cover almost all of the solid angle, generally about 99%. The L3 detector consists of a barrel section and two endcap sections. The entire detector is immersed in a 5 kG axial magnetic field obtained using aluminum normal temperature coils and an 1100 ton iron return yoke. A particle leaving the interaction point first traverses the vacuum pipe at a radius of 5.5 cm. Just past this point it encounters a vertex detector containing a number of planes of silicon strips having a resolution in the 5–10 μm range. It then enters a time expansion chamber of radius 50 cm where approximately 60 points on the curving trajectory of a charged track are obtained with 100

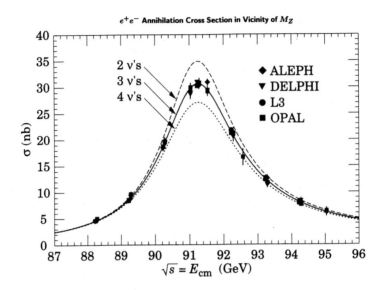

Figure 13.8 Fit to Z resonance for $e^+e^- \to$ hadrons within the Standard Model.[17] Data from the Mark II, ALEPH, DELPHI, L3, and OPAL collaborations.

μm resolution. The particle then enters an array of 11,000 bismuth germinate (BGO) crystals, each 22 radiation lengths long. This material is a scintillator with a very short radiation length (just over a centimeter). The full energy of electrons and the beginnings of hadronic showers are measured here with good angular resolution. The precision of electron and photon measurement is quite good. It is about 5% at 100 MeV, about 2% at 1 GeV and 1% or below above 10 GeV.

Non-electron charged tracks then pass through scintillation counters, which serve as one of the triggers and which serve to compare the track time of arrival to the time of beam arrival to discriminate against cosmic rays. [Even though the apparatus is located 50 m underground, cosmic rays are sufficiently frequent in this large [almost $(10\ \text{m})^3$] apparatus that they must be eliminated.] Following the scintillation counters, the track enters a uranium, proportional wire chamber sandwich calorimeter followed by a copper, proportional wire chamber calorimeter. Hadron showers are measured using these chambers and the BGO crystal information. The BGO and uranium, proportional wire chamber are 3.5 pion nuclear interaction lengths thick and the copper, proportional wire sandwich adds another in-

Table 13.1 Some LEP Results

Quantity	Value	SM value
M_z (GeV)	91.1885 ± 0.0022	
Γ_Z (GeV)	2.4963 ± 0.0030	$2.488^{+0.013}_{-0.011}$
Γ_e (MeV)	83.92 ± 0.17	83.8 ± 0.9
Γ_μ (MeV)	83.92 ± 0.23	83.8 ± 0.9
Γ_τ (MeV)	83.85 ± 0.29	83.8 ± 0.9
Γ_l (MeV)	83.93 ± 0.14	83.8 ± 0.9
Γ_h	1744.8 ± 3.0	$1738.9 \pm 19.$
Γ_{inv} (MeV)	499.9 ± 2.5	$500.3^{+2.4}_{-1.9}$
N_ν	2.991 ± 0.016	3
$\sin^2 \overline{\theta}_W$	0.23172 ± 0.00024	$0.2330^{+0.0015}_{-0.0019}$
M_t (GeV)	$178 \pm 8^{+17}_{-20}$	
g_A^2	0.25111 ± 0.00041	
g_V^2	0.00144 ± 0.000054	

teraction length. For hadronic decays of the Z, the resolution for the total measured energy is better than 10%.

At this point, almost all tracks except muons have been absorbed. Muons in the barrel part of the detector continue into a large (172 ton) muon detector. These are subdivided into an inner chamber with 16 layers of signal wires, a middle chamber with 24 layers and an outer chamber with 16 layers. The long-term alignment of this large structure is held to better than 30 μm. The multiple scattering induced sagitta error is also less than 30 μm for 50 GeV muons. The percentage momentum accuracy with all effects included is about 2.5% for 45 GeV muons. Muons in the end cap part of the detector are also measured, but because they are not traveling in a mainly perpendicular direction to the field, the accuracy is reduced. The magnet return flux field is used to assist in this measurement.

The other LEP and SLC detectors are similar. Each has emphasized different detection elements. The L3 detector has the highest precision electron, photon, and muon determination. All have silicon vertex chambers, but the ALEPH and OPAL detectors have much larger central tracking chambers outside the vertex chambers. The larger central chamber enables better momentum resolution on hadronic charged tracks and enables ionization measurements to be made to help in particle identification. The DELPHI detector uses ring imaging Cherenkov counters for particle identification. The SLD detector has a much smaller beam pipe allowing the initial silicon detector to be nearer the vertex. This provides an advantage

302   13. Physics at the Z

in measuring decay distances of short lived particles such as B mesons, charmed particles, and τ particles.

Table 13.1 gives the best LEP results as of August 1995.[49]

Figure 13.8 shows a fit to the hadronic line shape. The invisible decay width (Γ_{inv}) translates into 2.991 ± 0.016 families with light neutrinos. This is the single most significant result of the giant accelerator LEP.

From the mass and width of the Z, the mass of the W as obtained at $\bar{p}p$ colliders, and information on $\sin^2\theta_W$ obtained from neutrino nucleon experiments it is found[49] that

$$M_t = 178 \pm 8^{+17}_{-20} \text{ GeV}. \qquad (13.24)$$

As indicated in Chapter 6, the CDF and D0 collaborations at the Fermilab $\bar{p}p$ collider have directly measured[47][48] the mass of the top quark. The mass, from these direct observations, is $M_t = 176 \pm 8(\text{stat.}) \pm 10(\text{syst.})$ GeV from the CDF group and $M_t = 199^{+19}_{-21}(\text{stat.}) \pm 22(\text{syst.})$ GeV from the D0 group. This is in excellent agreement with the indirect e^+e^- results quoted above.

The LEP cross section values are all normalized by looking at the e^+e^- scattering at small angles in which the purely electromagnetic ("Bhabha") terms should dominate. In principle, this scattering is easy to calculate. However, the experimental accuracy of the results is now at the 1% range and better, and the accuracy of the calculations of higher-order corrections is already an important source in the quoted error on results. Better calculations are in progress.

Precision values of g_V and g_A are also shown in Table 13.1. These are obtained by looking at the decay channels $Z \to e^+e^-$, $Z \to \mu^+\mu^-$, and $Z \to \tau^+\tau^-$.[168][150] See Figure 13.9.

For the channel $Z \to e^+e^-$, both s- and t-channel amplitudes must be taken into account as shown in Figure 13.9a. This complicates the analysis. As with the low-angle Bhabha scattering, the radiative corrections to this process have an uncertainty at present comparable to the experimental uncertainties.

The decay width for each of these channels depends only on $g_V^2 + g_A^2$. As can be seen from the numbers in Table 13.1, lepton universality is well tested. To about 1% the widths into the various lepton channels are the same and all agree well with the width predicted within the standard model.[168][150]

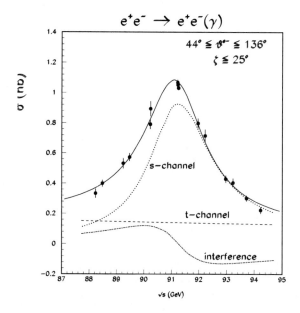

Figure 13.9 (a) $e^+e^- \to e^+e^-(\gamma)$ cross section.

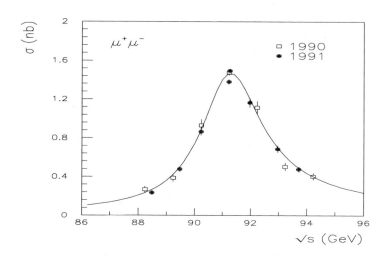

Figure 13.9 (b) $e^+e^- \to \mu^+\mu^-(\gamma)$ cross section.

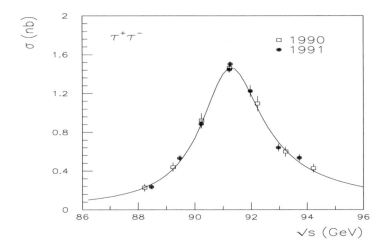

Figure 13.9 (c) $e^+e^- \to \tau^+\tau^-(\gamma)$ cross section.

Figure 13.9 Fit to (a) $e^+e^- \to e^+e^-(\gamma)$, (b) $e^+e^- \to \mu^+\mu^-(\gamma)$, and (c) $e^+e^- \to \tau^+\tau^-$ across the Z peak.[168]

To find g_V and g_A separately the asymmetries in the decay process must be looked at. For $e^+e^- \to l^+l^-$, let θ be the angle between the incoming e^- and the outgoing l^-. Note that for the s-channel diagrams, this process can have either a Z or a photon intermediate state. Then (ignoring the electron t-channel diagrams) the cross section can be shown to be (Exercise 13.2):

$$\frac{d\sigma}{d\Omega} = \frac{\alpha^2}{4s}[\{1 + 2g_V^e g_V^l \Re(\chi) + [(g_V^e)^2 + (g_A^e)^2][(g_V^l)^2 + (g_A^l)^2]|\chi|^2\}$$
$$\times (1 + \cos^2\theta) + \{2g_A^e g_A^l \Re(\chi) + 4g_V^e g_V^l g_A^e g_A^l |\chi|^2\} \times 2\cos\theta], \quad (13.25)$$

with

$$\chi(s) = \frac{GM_Z^2}{2\sqrt{2}\pi\alpha} \frac{s}{(s - M_Z^2) + is\Gamma_Z/M_Z}. \quad (13.26)$$

The first curly bracket corresponds to the terms symmetric in $\cos\theta$ and the second to the asymmetric terms. First-order radiative and electromagnetic corrections have been included here. The last term in the second curly brackets is the V−A interference term and involves $g_V^2 g_A^2$. (α^2 cancels out

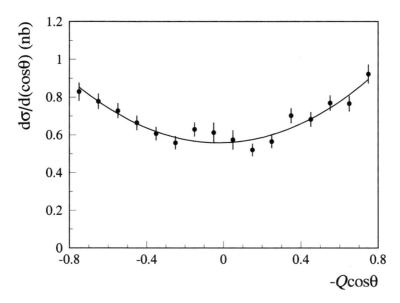

Figure 13.10 Differential cross section for $e^+e^- \to \mu^+\mu^-(\gamma)$ at the Z peak.[168]

of this term.) The other asymmetry term is the interference between the diagram with a photon and that with a Z intermediate state. This interference gives an asymmetry only for the axial part of the weak interaction. If one is far off the peak, this latter term will dominate the asymmetry. By measuring the overall cross section and measuring the asymmetry as a function of energy one can disentangle the individual contributions of g_V and g_A. The muon asymmetry has been the most convenient one to measure. See Figures 13.10 and 13.11 from the L3 experiment.[150] [168] The signs of g_V and g_A are inferred from other experiments. The very accurate agreement with the Standard Model predictions for g_V and g_A seen in Figure 13.12 constitutes a major confirmation of the Standard Model.

The decays of the Z into b quarks provide a different test of the standard model. Before the top was found, people speculated on various topless theories. In some of these, the coupling of the Z to the b is not the standard one. At LEP this coupling has now been studied. The decays of b-mesons, (B's), into final states involving muons or electrons have been studied first. These leptons originate from $b \to cW \to cl\nu$. In these decays, the b-quark transfers a large fraction (x) of its energy to the b-meson ($x = 0.686 \pm 0.006 \pm 0.016$), and the b-meson is a heavy particle. These two facts mean that decay leptons from primary b's tend to be fast and have high transverse momentum relative to the jet axis. b-quarks are then often identified by

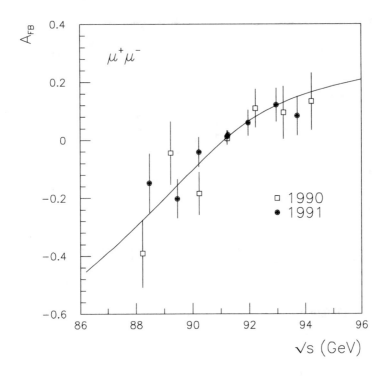

Figure 13.11 Forward backward asymmetry (A_{FB}) for $e^+e^- \to \mu^+\mu^-(\gamma)$ across the Z peak.[168]

a lepton with high transverse momentum relative to a hadronic jet. The LEP results are[49]

$$\Gamma_{b\bar{b}} = 387 \pm 3 \text{ MeV}. \tag{13.27}$$

This result is about 3.7 standard deviations from the theoretical value of $\Gamma_{b\bar{b}} = 376$ MeV in the Standard Model. This is one of the very few results in disagreement with that model. While this deviation may be due to experimental biases, it has lead to considerable speculations.

The Z has been a very useful source of secondary particles. b-quark decays are quite interesting in their own right. With new silicon microvertex detectors, the actual secondary decay vertex can sometimes be separated from the primary vertex. At energies lower than the Z, the ARGUS[169] and CLEO[170] groups have reported that in decay the neutral B and $\bar{B}$ mesons mix in a manner analogous to the $K - \bar{K}$ system. Studies of this mixing are of great interest. Furthermore, there is then the possibility of looking

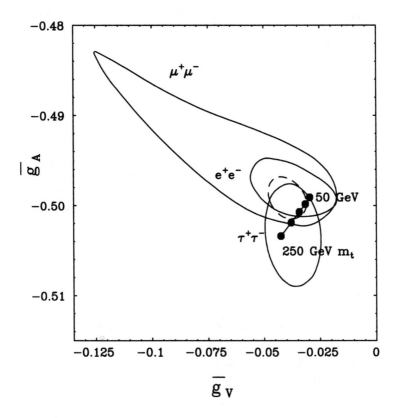

Figure 13.12 The 68% confidence level contours for $\bar{g}_V$ versus $\bar{g}_A$ along with Standard Model predictions for different values of M_t.[168] . The dashed line corresponds to the contour assuming lepton universality.

at $B - \bar{B}$ oscillations and having a second system in which to examine CP violation. Theoretical estimates are that this will take many millions of B decays, more than present accelerators will get. Either extensive improvements to present machines or new machines are required. Because of the extreme interest in a new CP violating system, it is important to get as much information on the mixing as possible now to help guide the designs now underway for B-factories. There are two kinds of neutral B mesons, those in which the other quark is a d-quark (B_d) and those in which the other quark is an s-quark (B_s). ARGUS and CLEO have obtained information on the mixing of B_d. At LEP information on the mixing of B_s is also available because the produced B's have an appreciable number of B'_ss. The mixing is identified by having two high momentum leptons (muons or electrons) of the same sign, one in each of the two jets in the event. An example of $B - \bar{B}$ mixing is seen in Figure 13.13. The event shown has two high energy, high transverse momentum leptons with the same sign electric charge. Normal decays of a B and a $\bar{B}$ lead to leptons of opposite charges. Parametrize the mixing by χ_B, the probability that the b-quark decays into the wrong sign lepton because of oscillation:

$$\chi_B = \frac{\Gamma(b \to \bar{b} \to l^+ \cdots)}{\Gamma(b \to l^\pm \cdots)}. \tag{13.28}$$

The LEP experiments[172] obtain the value $\chi_B = 0.1145 \pm 0.0061$. From this result and from the measurement of the b-hadron lifetimes limits can be put on CKM matrix elements. ARGUS/CLEO results directly measure $|V_{ub}|/|V_{cb}|$. Putting these measurements together one obtains the results shown in Figure 13.14. These two matrix elements are now reasonably well determined.

The ALEPH group[173] has examined the fraction of $B_d^0 \bar{B}_d^0$ decays leading to leptons with the same sign of charge, as a function of decay time. From this one can obtain the mass difference of the B_d^0 mass eigenstates! They obtain this difference to be

$$\Delta m_d = (3.3^{+0.5}_{-0.4} \pm 0.7) \times 10^{-4} \text{ eV}/c^2. \tag{13.29}$$

They are even able to obtain a limit on the mass difference of the B_s^0 mass eigenstates to be

$$\Delta m_s > 12 \times 10^{-4} \text{ eV}/c^2 \tag{13.30}$$

The charge asymmetry in $e^+e^- \to Z \to b\bar{b}$ decay has been measured. The observed value reported by LEP-SLD[172] is

$$A_{b\bar{b}}^{\text{obs}} = \frac{N_{\text{forward}} - N_{\text{backward}}}{N_{\text{forward}} + N_{\text{backward}}} = 0.0724 \pm 0.0020.$$

This must be modified for the effects of mixing (Exercise 13.1) according to

$$A_{b\bar{b}} = A_{b\bar{b}}^{\text{obs}}/(1 - 2\chi_B) = 0.0939 \pm 0.0030.$$

This value can be used to determine $\sin^2 \bar{\theta}_W$. In the improved Born approximation the forward-backward asymmetry on the peak is given by $A_{b\bar{b}}^{\text{Born}} = \frac{3}{4} A_e A_b$, where $A_i = 2g_{Vi}g_{Ai}/(g_{Vi}^2 + g_{Ai}^2)$. Some QED and QCD corrections must be included and the shift between the Z mass and the effective center-of-mass energy of the experiment must be taken into account. The result of an overall LEP fit[49] is $\sin^2 \bar{\theta}_W = 0.23205 \pm 0.00051$ to be compared with the SM prediction for a top quark mass of 178 GeV and a Higgs mass of 300 GeV of $\sin^2 \bar{\theta}_W = 0.23172$.

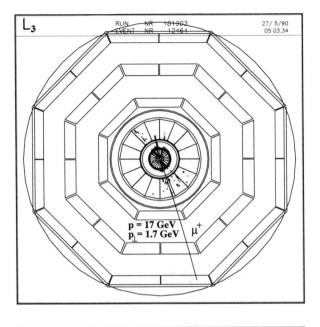
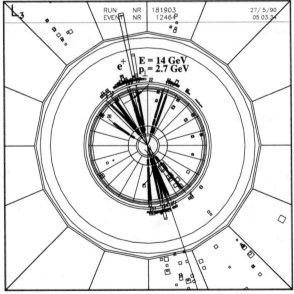

Figure 13.13 Example of $B\overline{B}$ mixing; an event $e^+e^- \to$ hadrons at the Z peak with like sign high $p_\perp$ dileptons.[171]

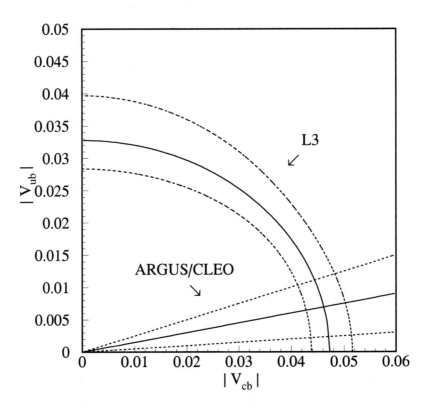

Figure 13.14 Contour plot of $|V_{ub}|$ versus $|V_{cb}|$.[168] The dashed lines are one standard deviation errors including theoretical uncertainties.

The decay $Z \to q\bar{q}\gamma$ has also proven interesting.[174][175] By comparing this decay to the decay without a γ one can look at the coupling of the Z to the u-type-quarks and d-type-quarks separately. The OPAL group has verified that these separate couplings are consistent with standard model values. As more data become available, this channel will allow precision tests of these couplings and of QCD "form-factor" effects.

The study of the $Z \to l\bar{l}\gamma$ ($l = e, \mu, \tau$) allows one to look for excited lepton states and to do a QED check (lepton form factor). No excited leptons have been seen yet. The study of the lepton form factors will be especially interesting for the τ, since one might expect deviations, if present, to show up first in this, by far the heaviest of the leptons.

The study of $ee \to n\gamma's$ is interesting. Two gamma events cannot come from the Z. This is basically due to conservation of angular momentum. See the discussion concerning the spin of the neutral pion in Section 4.7.

13. Physics at the Z 311

These events come from the purely electromagnetic annihilation interaction. Thus, measurement of this process allows a QED test. It is customary to parametrize possible deviations in terms of a cutoff, Λ, in the propagator, a term of the form $1/[q^2(1 \pm q^2/\Lambda^2)]$. This corresponds to

$$\frac{d\sigma}{d\Omega} = \frac{\alpha^2}{s} \times \frac{1+\cos^2\theta}{1-\cos^2\theta} \times \left[1 \pm \frac{s^2}{2\Lambda_\pm^4}(1-\cos^2\theta)\right]. \tag{13.31}$$

At 95% C.L., L3 reports[176] $\Lambda_+ > 103$ GeV and $\Lambda_- > 118$ GeV. The 3γ and higher $n\gamma$ events serve as a search for exotic things. For example, in some models, compositeness of particles could show up here as events with 3γ's and missing energy. None are yet seen at the level of a few in 10^{-4}.

The four LEP experiments have searched very hard for the Higgs boson without success. The mode $Z \to H\nu\nu$ has been especially important. Searches in which the Higgs would be accompanied by a pair of leptons $l\bar{l}$, (l = e,μ,τ), or a $b\bar{b}$ pair have also been used. Putting the various LEP results together in 1995 limited the Higgs mass to be greater than 65.2 GeV at the 95% confidence level.[177]

Note that $\rho^0 = 1$ and the masslessness of the photon follow naturally from a single Higgs. They are either accidental or follow from new symmetries in theories with several Higgs.

Nonetheless, a number of other exotic particles have been searched for including charged Higgs, supersymmetric Higgs, and other supersymmetric particles, excited or heavy charged leptons or neutrinos and other particles. Some limits, mostly from LEP[178] (the $Z \to e\mu$ limit is from the limit on $\mu \to eee$ by the SINDRUM group), are given in Table 13.2. In addition to those in the table, limits on leptoquarks have been placed up to close to 45 GeV.

In 1996 LEP is expected to increase its beam energy and pass the WW threshold. Even below the threshold, effects of possible WW resonances might show up at the few times 10^{-6} level. M. Veltman notes that WW resonances, if seen, would probably involve mass states in the 3 TeV region or so and would start to shed light on existence and properties of a new set of high-mass states.

In this chapter a part of the wealth of information coming from the examination of $e^+e^- \to Z$ has been described. The number of generations has been found to be three, very accurate tests of electroweak theory and accurate measurements of electroweak parameters have been performed. The mass of the top particle has been bracketed. Limits on Higgs particles, both standard and non-standard, have been obtained. Interesting B-physics measurements have been made. A great many (so far unsuccessful) searches for extensions of the Standard Model have been performed.

Table 13.2 Limits on particles and rare interactions.[178]

a) Limits on Pair Produced Particles from LEP

Limits on Pair Produced Particles from LEP (GeV)	
"Conventional" Particles	
Particle	Best Published Limit
t	45.8
b'	46.0
$L^\pm$ (Stable)	44.6
$L^\pm$ (Unstable)	44.3
L^0 (Dirac)	46.4
L^0 (Majorana)	45.1
Excited Particles	
e^*, μ^*	46.1
τ^*	46.0
ν^*	47
q^*	45
Supersymmetric Particles	
$\chi^\pm$	45.2
e, μ, τ	45
l (Stable)	40
d	43
u	42
t	37
Charged Higgs Particle	
$H^\pm$	41.7
$H^{\pm\pm}$	45.6

b) Limits on Higgs and flavor violating decay branching ratios.

Standard Model Higgs	
H^0	62.5
Limits on Flavor Violating Decay Branching Ratios	
$Z \to e\mu$	6.6×10^{-8}
$Z \to e\tau$	1.4×10^{-5}
$Z \to \mu\tau$	2.2×10^{-5}
$Z \to e\gamma$	12×10^{-5}
$Z \to \mu\gamma$	8×10^{-5}
$Z \to e\pi^0$	15×10^{-5}
$Z \to \mu\pi^0$	9.5×10^{-5}

13.1 Exercises

13.1 Derive the relation used in discussing the effect of mixing on the $b\bar{b}$ charge asymmetry: $A_{b\bar{b}} = A^{obs}_{b\bar{b}}/(1 - 2\chi_B)$. This correction arises since, generally, only one b leptonic decay per event is seen. If the $b \to \bar{b}$, the wrong sign lepton is seen; a forward event appears to be a backward event and vice versa.

13.2 The equations 13.27–13.28 can, of course, be derived from the Lagrangian by calculating matrix elements, traces, etc. However, almost all of the terms can be derived from simple considerations. Do this, using as a basis the vertex couplings and the electromagnetic scattering calculated in Equation 8.12. The vertex couplings are $g_{V,A}$ and $g/(2\cos\theta_W)$.

14
High-Energy Processes at Low Q^2

For hadron–hadron interactions most of the cross section for high-energy collisions consists of events that do not have high Q^2. Asymptotic freedom does not apply; the strong interactions are strong and perturbation theory cannot be used. In this realm it is difficult to make contact with fundamental theory. There are a few results from general principles, but for most of the data only phenomenological pictures, i.e., ad hoc theories based on intuition for the process in hand and not rooted in a fundamental theory, are used.

At the heart of most of the general results is the concept of analyticity. The scattering amplitude is imagined to be an analytic function, except for well-defined singularities, as a function of one or another variable taken as complex. Given this assumption relations can be obtained between reactions in differing domains, e.g., between s and t-channel processes. In principle, if an analytic function is known over some continuous region and if the singularities are known, the function is determined over all space. There are some questions that occur with this method.

The first is that it is usually not proven that the singularity structure is completely established. It will be seen below that well-defined singularities occur with particles, resonances, and thresholds. However, the behavior of the function as the complex variables approach infinity is still an assumption, although a plausible one, and in some applications other singularity structures are assumed, not proven, to be absent.

The second question is how well analytic continuation determines a function in another region. A small experimental discrepancy in one region might grow exponentially in another and reduce or eliminate predictability.

Another concept used is "crossing symmetry." For individual Feynman diagrams in perturbation theory, every diagram for $A + B \to C + D$ has a crossed diagram for $\overline{C} + B \to \overline{A} + D$ as shown in Figure 14.1. The amplitudes for each diagram are the same functions, but they are in different kinematic regions. Crossing corresponds to interchanging s and t, i.e., s in the first reaction is t in the second. It can be shown that crossing symmetry holds in the full theory also.

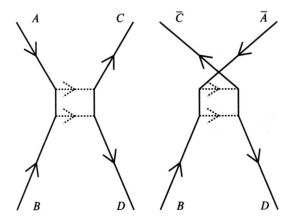

Figure 14.1 Regular and crossed Feynman diagrams.

Since general methods at least try to make contact with first principles, they have been popular. This chapter is intended as a very brief introduction to some high-energy methodology. More general methods will be described first, followed by a few of the phenomenological pictures and relations often used.

14.1 POMERANCHUK THEOREM; FROISSART BOUND; REGGE ANALYSIS

First the Pomeranchuk Theorem will be examined. This theorem states that under fairly general assumptions the total cross sections for scattering of a particle or its anti-particle on the same target must approach equality at sufficiently high energies. I. Ia. Pomeranchuk[179] proved that $\sigma_{\text{tot}}^{AB}(s) = \sigma_{\text{tot}}^{\overline{A}B}(s)$ in the limit $s \to \infty$ if the cross sections approached a constant at high energies and approached it sufficiently rapidly. A few years later the proof was generalized by Weinberg.[180] Suppose that $\sigma_{\text{tot}}^{AB}(s) - \sigma_{\text{tot}}^{\overline{A}B}(s)$ does not change sign an infinite number of times. Consider $C = [\sqrt{\sigma_{\text{tot}}^{AB}(s)} - \sqrt{\sigma_{\text{tot}}^{\overline{A}B}(s)}]/E_A$. If C is bounded, then Weinberg was able to show that $(\sigma_{\text{tot}}^{AB}(s) - \sigma_{\text{tot}}^{\overline{A}B}(s)) \ln s \to 0$ as $s \to \infty$. If C is not bounded, but $C = O(\log E)$, then $\sigma_{\text{tot}}^{AB}(s)/\sigma_{\text{tot}}^{\overline{A}B}(s) \to 1$ as $s \to \infty$.

It is fortunate that this generalization was made, since, as will be seen, the data are not consistent with cross sections approaching a constant at high energy.

316   14. High-Energy Processes at Low Q^2

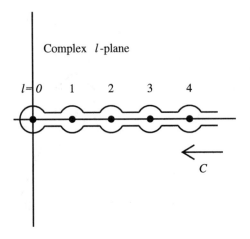

Figure 14.2 Contour C for integration in the complex l-plane.

Even if they are not constant, cross sections are bounded at high energy. Under very general conditions it has been shown[181] that for high energy

$$\sigma_{\text{total}} < \text{constant} \times (\log s)^2. \tag{14.1}$$

This is known as the "Froissart bound."

Another general result, after being mixed with a large dose of phenomenology, has become known as Regge theory. As originally presented[182] this theory applied rigorously to the non-relativistic Schrödinger equation for potentials that can be represented as a superposition of Yukawa potentials. Some results that can be obtained under these assumptions will be outlined here. Again take $Q^2 \equiv -q^2$. Consider the wave optics equation discused in Chapter 5, now broadened to include inelastic scattering, i.e.,

$$kF(\theta) \longrightarrow f(Q^2, \cos\theta) = \sum_{l=0}^{\infty}(2l+1)f_l(Q^2)P_l(\cos\theta). \tag{14.2}$$

Suppose, as a mathematical convenience, one imagines the integer parameter l to be a continuous complex variable. Then the above sum can be written as an integral over the contour C shown in Figure 14.2 in the complex angular momentum plane (Sommerfeld–Watson transform):

$$f(Q^2, \cos\theta) = \frac{i}{2}\int_C (2l+1)f(l,Q^2)\frac{P_l(-\cos\theta)}{\sin(\pi l)}dl. \tag{14.3}$$

This equivalence is easily seen since $1/\sin(\pi l)$ has poles at integer l with residue $(-1)^l/\pi$. The contour shown in Figure 14.2 can be viewed as

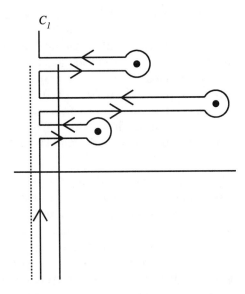

Figure 14.3 Deformed contour C_1 in the complex l-plane.

dividing space into two parts. If the contour is thought of as covering the larger region, it can be deformed into that shown in Figure 14.3 where the poles shown are poles in $f(l,Q^2)$. It is necessary to stay away from the line $\Re(l) = -1/2$, which is a branch-cut line. The poles in $f(l,Q^2)$ are, in general, functions of Q^2. Thus, near the poles, $f(l,Q^2)$ can be written in the form $f(l,Q^2) \approx \beta_n(Q^2)/[l - \alpha_n(Q^2)]$. The scattering amplitude can be written as a sum over the poles plus an integral over the cut. Let C_2 be the vertical part of C_1 near the branch-cut line:

$$f(Q^2, \cos\theta) = \frac{i}{2} \int_{C_2} \frac{(2l+1)}{\sin(\pi l)} f(l,Q^2) P_l(-\cos\theta)$$

$$- \pi \sum_n \frac{(2\alpha_n(Q^2)+1)\beta_n(Q^2)}{\sin \pi \alpha_n(Q^2)} P_{\alpha_n(Q^2)}(-\cos\theta). \quad (14.4)$$

The first term above is known as the "background integral" and the second consists of the "Regge poles." Thresholds, which produce cuts, are a minor complications that will be ignored here. In use the background term is often neglected. $\alpha_n(Q^2)$ are the trajectory functions determining the position of the poles. Regge was able to show that for $Q^2 \to -\infty$, the poles are all on the real axis and at $l \leq -1$. As Q^2 increases they stay on the real axis in the complex l plane until $Q^2 = 0$, after which they travel in the upper half plane, eventually returning to the line $l = -1$ for sufficiently large Q^2. If an α_n crosses $l = 0$ while still on the real axis, i.e., for $Q^2 < 0$, then from Equation 14.3 above, the amplitude has a pole, corresponding to a bound S-wave state. Suppose for $Q^2 > 0$ a trajectory passes near an integral value of l with Im $\alpha_n(Q^2) > 0$. If $f(Q^2, \cos\theta)$ is dominated by that pole, then by decomposing f into the various f_l, it can be shown that the lth partial wave scattering amplitude has a Breit–Wigner resonance form. In this representation $\cos\theta$ can also be made complex and this is useful as will be noted below.

This is quite intriguing, but established only for non-relativistic potential theory. It was soon applied, without proof, to the general relativistic case.[183] A troubling problem for analytic continuation could then be averted. If there was a resonance in the s-channel with $J \geq 1$, then the s-channel amplitude was proportional to $P_J[1 + 2t/(s - s_0)]$. Here s_0 is the threshold value of s for the reaction. If the analytic continuation of the same amplitude is imagined to be the scattering amplitude in the t-channel, then in the t-channel, where $\sqrt{t}$ is then the center-of-mass energy, the scattering amplitude increases as t^J, since P_J is a polynomial of order J. This increase eventually violates the unitarity constraint. If, instead of P_J, one has $P_{\alpha(s)}$, then in the t-channel, $s < 0$ and it is possible that $\alpha(s)$, for $s < 0$, can be < 1, thus avoiding the dilemma.

For relativistic theory there are exchange potentials, potentials involving exchanges of coordinates of a pair of particles. This means that odd and even l are treated differently, the pole contribution being proportional to $(P_{\alpha_n(s)}(-\cos\theta) \pm P_{\alpha_n(s)}(\cos\theta))/\sin(\pi\alpha_n(s))$. Thus the trajectories have a "signature." Depending on that signature, physical states associated with that trajectory will either have only odd or only even l.

Suppose a continuation is next made in $\cos\theta$. Let $\cos\theta \to \infty$. Large $\cos\theta$ implies large values for $2t/(s - s_0)$. Consider α defined in the crossed channel with t-channel poles so that $\alpha = \alpha(t)$, i.e., interchange s and t. Then for $s \to \infty$ and $t < 0$, the scattering amplitude is

$$f \propto \sum_n \left(\frac{s}{s_0}\right)^{\alpha_n(t)} \beta_n(t). \tag{14.5}$$

In this equation, s_0 is just a parameter. One can then speak of an exchange of Regge poles as determining the scattering. From the behavior of forward elastic scattering as a function of s, it is found that the leading pole at $t = 0$ must have $\alpha_P(0) \approx 1$. The path of this pole is known as the Pomeranchuk trajectory, since, if one supposes it has the quantum numbers of the vacuum, the Pomeranchuk theorem can be easily shown.

Next consider a plot of $\Re[\alpha_P(t)]$ versus t. Imagine the slope of α_P to be small, $\alpha_P(t) \approx 1 + \epsilon t$. From Equation 14.4 it is seen that

$$\frac{d\sigma}{dt} \approx \frac{d\sigma}{dt}\bigg|_{t=0} e^{(\alpha_P(t)-1)\ln(s/s_0)}. \tag{14.6}$$

Noting that the differential cross section falls as $-t$ increases, ϵ must be positive. Because of the log term, the width of the elastic scattering cross section peak should shrink as s increases. Unfortunately, experiments at present accelerator energies are not in good agreement with this prediction. Shrinkage is observed for $p-p$ scattering, but "anti-shrinkage" is observed for $\bar{p}-p$, and no clear effect is seen for $\pi-p$ scattering. Within this model, this must be due to the fact that asymptotically large energies have not yet been approached.

In the early 1960s, there was a major attempt to identify poles with resonances, a Pomeranchuk pole (the "Pomeron"), a ρ pole, a π pole, etc. Searches were made for "Regge recurrences," resonances with spins two higher than the original resonance, i.e., a spin 3 ρ, etc. A few meson recurrences were found, and candidates for recurrences of the proton and $\Delta(1232)$ poles were identified. The trajectories tended to lie on straight lines. In potential theory this corresponds to potentials linear in r, i.e., forces independent of r. This kind of force is also postulated in the more modern string theory.

The trajectories were then used in elastic and exchange scattering to try to understand the Q^2 dependence of the scattering. Thus, in principle, Regge trajectories could be used for understanding both scattering and resonances. However, the physical picture was somewhat abstract and the actual applications somewhat limited. When the quark theory became successful, much of the attention shifted to that theory.

Recent data from the Fermilab collider and from HERA indicate a substantial cross section for events which seem as if they were produced according to the reactions shown in Figure 14.4. These events have a gap in the distribution of the rapidities (a term defined in the next section) of the final state particles. This has renewed interest in the Regge formalism. It is not yet clear how broad this renewed interest will be, but it seems that Regge terminology, at least, will remain in the field for some time.

14.2 PHENOMENOLOGY OF HIGH-ENERGY COLLISIONS

Many particles are produced in the final state for high-energy inelastic scattering. Two variables often used to describe these particles are Feynman x and y, which are different from the Bjorken x and y introduced in Chapter

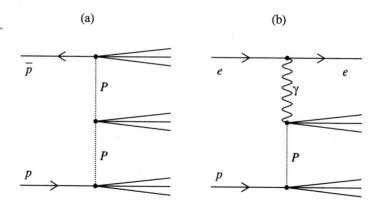

Figure 14.4 Triple vertex diagrams involving Pomeron (P) exchange for (a) $\bar{p}p$ interactions and (b) ep interactions.

8. Let z be the longitudinal direction and $m_\perp \equiv \sqrt{m^2 + p_x^2 + p_y^2}$ be the "transverse mass." Then for the individual final-state particles

$$x(\text{Feynman}) = \frac{p_z}{(p_z)_{max}} \approx \frac{E + p_z}{(E + p_z)_{max}}, \qquad (14.7)$$

$$y(\text{Feynman}) = \frac{1}{2}\log\left(\frac{E + p_z}{E - p_z}\right) = \log\left(\frac{E + p_z}{m_\perp}\right) = \tanh^{-1}\left(\frac{p_z}{E}\right). \qquad (14.8)$$

y is also known as the "rapidity." For a Lorentz transformation along z to a frame with velocity β, $y \to y + \tanh^{-1}\beta$. This means that the shape of the rapidity distribution is invariant under Lorentz transformations.

In the usual case that the mass of the produced particle is small compared to its energy, then the expression for y simplifies further. Let θ be the angle of the particle with respect to the longitudinal direction:

$$y = \frac{1}{2}\log\left(\frac{E + p_z}{E - p_z}\right) = \frac{1}{2}\log\left(\frac{\left(p(\sqrt{1 + (m/p)^2}\right) + p_z}{p(\sqrt{1 + (m/p)^2}) - p_z}\right)$$

$$\approx \frac{1}{2}\log\left(\frac{1 + (m^2/2p^2) + \cos\theta}{1 + (m^2/2p^2) - \cos\theta}\right).$$

Use $1 + \cos\theta = 2\cos^2(\theta/2)$ and $1 - \cos\theta = 2\sin^2(\theta/2)$:

$$y \approx -\ln\tan(\frac{\theta}{2}) \equiv \eta. \qquad (14.9)$$

η is known as the "pseudo-rapidity" and is more often used at high energies

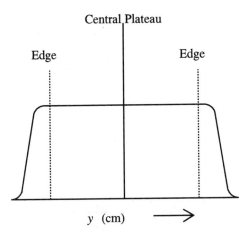

Figure 14.5 Number of produced particles versus rapidity (y) in the center of mass.

than y. Note that from the definition, $\sinh\eta = \cot\theta$, $\cosh\eta = 1/\sin\theta$, and $\tanh\eta = \cos\theta$.

In the center of mass, $x \approx 2(p_z)_{cm}/\sqrt{s} \approx (2m_\perp \sinh y_{cm})/\sqrt{s}$. $(y_{cm})_{max} = -(y_{cm})_{min} = 1/2 \ln(s/(m^2 + p_\perp^2))$. If $e^{-2y_{cm}} \ll 1$, then $x \approx m_\perp e^{y_{cm}}/\sqrt{s}$, and $(y_{cm})_{max} = \ln(\sqrt{s}/m)$.

The invariant cross section for the production of a secondary particle with momentum, energy (E, p) is $E d^3\sigma/dp^3 = d^3\sigma/dy dp_\perp^2 = F(x, y, s)$. To first order, experimentally, F factorizes, i.e., $F(x, y, s) \approx A(x) B(y, s)$. A depends on the type of secondary particle. Generally $\langle p_\perp \rangle$ increases somewhat with the mass of the secondary and is about 0.35 GeV for pions.

If one plots the number of particles produced versus y, one obtains a result similar to that shown in Figure 14.5 for high energies. There is a plateau in the central region and a fall off near the edges. The size of the edge regions stays constant as the energy increases, but, from the formulas for y, the length of the plateau increases as $\ln s$. Note that the integral over x and y of the invariant cross section above is proportional to $\langle n \rangle \sigma_{total}$. It will be noted later that asymptotically the average multiplicity, $\langle n \rangle$, varies as $\ln s$. Thus, for inclusive reactions, i.e., the totality of the absorptive cross section, the height of the plateau is expected to vary asymptotically with s in the same manner as the total cross section, i.e., as $\ln s$ to some power.

This approximate shape is very convenient for simple Monte Carlo estimates.

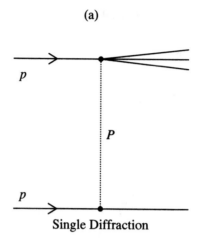

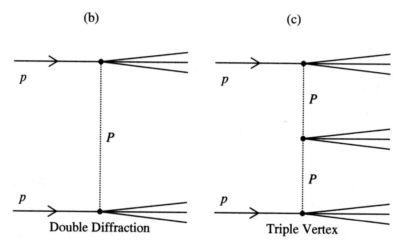

Figure 14.6 Diffraction Scattering Diagrams. (a) Single Diffraction Scattering; (b) Double diffraction scattering; (c) Triple vertex scattering.

A few percent of the time the collision seems to be a "diffractive" one. For these events the square of the momentum transfer, $-t$, is small ("peripheral collision"), and no quantum numbers are exchanged between the target and incident particles. In single diffractive events, only the particle at one vertex is excited into a resonance, while in double diffractive events, both particles are excited. These processes may be viewed as Pomeron exchange. See Figure 14.6.

Some phenomenological success was obtained by imagining that peripheral events in general, even those involving quantum number exchange, could be treated by a one-particle-exchange Feynman graph, with arbitrary form factors introduced at the vertices to absorb out the low partial waves. This was done because the one-particle exchange usually corresponds to a long-range force. The low partial waves can penetrate the centrifugal barrier and cause non-peripheral strong interactions that must be suppressed experimentally, and for which many other diagrams would contribute theoretically.

14.3 TOTAL CROSS SECTIONS AND ELASTIC SCATTERING

For pp and $\bar{p}p$ scattering, data have now accumulated up to $\sqrt{s} = 1800$ GeV. The total cross section at first decreases, but, between 10 and 50 GeV incident particle energy on a stationary target, it turns around and starts rising again. See Figure 14.7. The rise is compatible with $\ln s$ and possibly $\ln^2 s$ behavior.[184] In the reference quoted, the result was obtained that it was consistent only with $\log s$ behavior, but, since that time, new data at the highest energies obtain a larger cross section than previous results and the question is again perhaps open. There is no need for odd crossing amplitudes ("odderons") to fit the data.[184] However, the existence of a dip in the differential elastic scattering cross section around $Q^2 = 1$ GeV^2 for pp but not $\bar{p}p$ may be mild evidence for such an amplitude.

Consider a QCD inspired analysis.[185] One considers gluon ladder diagrams such as the one shown in Figure 14.8. One obtains an effective Pomeron-like trajectory, but with an intercept at $t = 0$ of somewhat greater than one, i.e., $J = \alpha_P(0) = 1 + 12(\ln 2)\alpha_s/\pi$. It corresponds to an exchange of at least two gluons. This gives a rising cross section, but one that violates unitarity since it rises as a power of s.

Continuing with phenomenology, one "eikonalizes" this.[186] That is, one considers an impact parameter representation transform of the scattering amplitude. At each value of the impact parameter b, one imagines that the amplitude is exponentially absorbed as one particle passes through the other, with the absorption constant determined by the cross section being calculated. This can be done in a consistent way. It is equivalent to summing up classes of diagrams which have several ladders and interconnected ladders of gluons. It leads to a cross section similar to that of a black disk with the radius of the black disk growing with energy. If a black disk is imagined to completely absorb all partial waves ($\eta_l = 0$), up to some l and then have no effect on the others, and if one sets $\sigma_{elastic} = \pi R^2$, one can easily see from the derivations in Chapter 5 that $\sigma_{elastic} = \sigma_{absorption}$ and therefore $\sigma_{total} = 2\pi R^2$:

$$\sigma_{tot} = 2\pi R^2 = 2\pi \left(\frac{J-1}{\mu}\right)^2 \ln^2\left(\frac{s}{s_0}\right), \qquad (14.10)$$

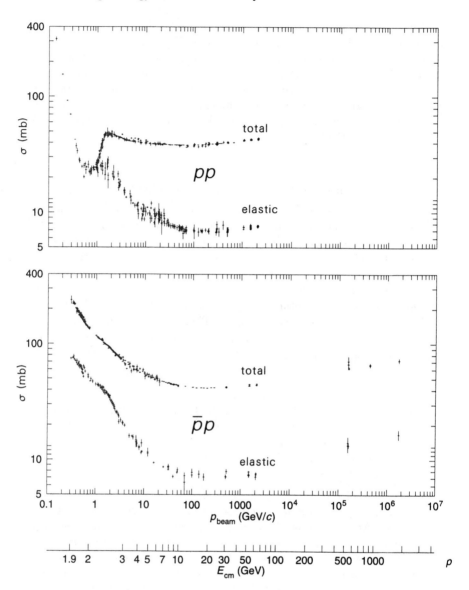

Figure 14.7 Total and elastic cross sections for (a) pp and (b) $\bar{p}p$ scattering versus beam momentum.[17]

$$R = \frac{J-1}{\mu} \ln\left(\frac{s}{s_0}\right) \qquad (14.11)$$

Here J, μ, and s_0 are constants. s_0 can be related to a $(p_\perp)_{min} \approx 1/R_{hadron}$ cutoff. This occurs since one is considering colored gluon exchange. When

14.3 Total Cross Sections and Elastic Scattering

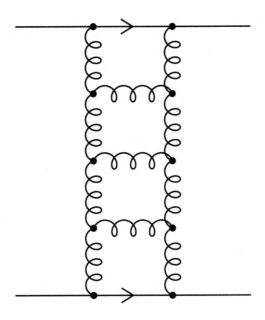

Figure 14.8 Scattering involving the exchange of a gluon ladder.

the wavelength, $\lambda \approx (p_\perp)_{min}$ gets to be about the size of the whole proton, then the gluon starts to see the colorless proton, not its constituents, and the interaction is switched off. Thus, in this model, the cross section grows as $\ln^2 s$. The black disk analogy is not perfect. Experimentally, $\sigma_{elastic}$ is considerably less than $\sigma_{absorption}$ at present accelerator energies. In a grey disk model, R increases with energy, but the blackness decreases, in agreement with experiment.

We are not yet in the asymptotic region, so even if the cross section were, at present, increasing only as $\ln s$, it would not be a serious blow to the preceding considerations. At very high energies, jet phenomena and "mini jets" should dominate, produced mainly by soft partons, i.e., mainly by gluons. At present energies jet production is rising but still not completely dominant. Jets with $p_\perp > 3$ GeV account for perhaps 30% of the events in 800 GeV center of momentum p–p collisions. At even the highest accelerator energies reached so far, the asymptotic gluon terms contribute less than one-half the total cross section.

Because of the unitarity relation given in Chapter 5 linking the imaginary part of the forward elastic scattering amplitude with the total cross section, the imaginary part of the elastic scattering amplitude can be viewed as the shadow of the inelastic scattering. If, at high enough energies, this were the only elastic scattering, the amplitude would be totally imaginary. By measuring elastic scattering at angles small enough that Coulomb scattering, whose phase is known, is important, one can determine the phase of the elastic scattering for small Q^2. For $\bar{p}p$ scattering, it perhaps retains a significant real part even at $\sqrt{s} = 1800$ GeV, i.e., $\rho = \Re(F)/\Im(F) = 0.134 \pm 0.069$.[184] At $\sqrt{s} = 550$ GeV, $\rho = 0.135 \pm 0.015$.[187] This latter group, extrapolating to $\sqrt{s} = 40$ TeV, obtains $\sigma_{total} = 130 \pm 13$ mb and $\rho = 0.127 \pm 0.019$.[188]

Suppose, for small t, one parametrizes the elastic scattering as

$$\frac{d\sigma_{el}}{dt} \propto e^{B|t|+Ct^2}. \tag{14.12}$$

Up to present energies C is positive. However, it is turning over and just about zero at the Fermilab collider. The preceding QCD inspired model predicts a negative value for C, and the change of sign may imply that the threshold of the asymptotic region is now being approached. A sharp-edged black disk has $C = -R^4/192$ and $B = R^2/4$.

Diffractive dissociation can also be described in the model listed previously, leading to

$$s\frac{d\sigma}{dtdM^2} \propto e^{bt}\left(\frac{s}{M^2}\right)^{2(J-1)}. \tag{14.13}$$

The CDF experiment finds $J - 1 = 0.115 \pm 0.013$. This matches a value of about 0.1 obtained from the phenomenological value of forward scattering, but is somewhat smaller than the value from the preceding QCD inspired model.[189]

One can make qualitative connections between the highest accelerator energies and still higher energy (but poor statistics) cosmic ray observations. For instance, if an event has two jets plus some projectile and target fragments, three separated groups of particles in a plane of fixed azimuth should appear. Events of this type are indeed seen in cosmic ray experiments. If three jets are produced, four groups will be seen, and since, in QCD, the additional jet is preferentially produced in the plane of the first two, this leads to four groups of particles essentially in a fixed azimuth plane. Events of this sort are also seen.

Finally, in cosmic rays a few events are seen in which there are clusters of large numbers of either almost all charged, or almost all neutral particles. If, for very-high-energy collisions in nuclear matter, one could obtain temperatures where the vacuum returns to its unbroken chiral symmetry state locally, then when it cools it will undergo a phase transition to its broken phase. The vacuum inside the fireball may not have the same orientation as its external value and it will have to relax, mainly by emitting soft pions. If there is a large mismatch between the local and the overall vacuum, these might be predominantly neutral (or predominantly charged) pions and thus large fluctuations can be produced. This is, of course, extremely speculative.

14.4 Multiplicity Distributions

The average multiplicity for hadronic final states from a large variety of incident particles and targets at high energy is given by $\langle n \rangle = A + B \ln s$, i.e., the multiplicity varies as $\ln s$. For moderate energies (fixed target accelerator energies), if one plots $\langle n \rangle \sigma_n / \sigma_{tot}$ versus $n/\langle n \rangle$, one obtains an approximately universal curve. This is known as Koba–Nielsen–Oleson (KNO) scaling.[190] Furthermore, the dispersion, $D = \sqrt{\langle n^2 \rangle - \langle n \rangle^2}$ is approximately given by $D = A(\langle n \rangle - B)$, where $A \approx 0.6$ and $B \approx 1$. Because D is not equal to $\langle n \rangle$, a binomial or a normal distribution does not fit the data. The data can be fit with a distribution known as a negative binomial distribution. At higher energies, deviations from KNO behavior are observed. A high multiplicity tail associated with gluons appears. These events also cause a growth in $\langle p_\perp \rangle$ and introduce a correlation between multiplicity and $p_\perp$, high multiplicity events tending to be high $p_\perp$ events, whereas kinematics alone would have indicated the opposite correlation.

14.5 Exercises

14.1 Find the approximate maximum center of mass rapidity for the Fermilab collider and for a pp collider with 40 TeV in the center of mass.

14.2 It has been hypothesized that much of the dependence on particle mass of the transverse momenta distributions of produced particles is taken into account if the distributions are assumed to vary as $d\sigma/dp_\perp \propto e^{-am_\perp}$. a is taken to be independent of particle type. Assuming this form, if $\langle p_\perp \rangle$ is 0.35 GeV/c for pions, what would it be expected to be for kaons? For anti-protons? Numerical integration is a reasonably easy way to approach this problem.

14.3 Diffractive scattering was measured in several Fermilab fixed target experiments using 400 GeV protons. By what factor would the diffractive cross section change at the Fermilab collider and at a collider with 40 TeV in the center of mass, based on Equation 14.12?

14.4 Suppose it is assumed that in the central part of the rapidity region (about 4 units away from a boundary) the multiplicity per unit rapidity is constant, ≈ 6, and that it falls rapidly near the boundaries of the rapidity region.[191] Calculate the multiplicity expected at a Fermilab fixed target interaction utilizing 1 TeV protons, at the Fermilab collider, and at a collider with 40 GeV in the center of mass. Assume the multiplicity to be zero for rapidity within 2 units of the boundary and multiplicity constant otherwise. Note that, for the fixed target experiment, there is only a small central region. Until the hadron colliders started operation, it was hard to see more than end effects!

15
Heavy Quark Effective Field Theory

In this chapter, decays of particles involving heavy quarks (c, b) will be discussed. There are several reasons for studying these decays. The Standard Model predicts small, but measurable, violations of CP invariance in B decays. This is a major reason for the building of "B-factories", e^+e^- storage rings with high luminosity for the study of B decays. Rare decays of B and D mesons and baryons are sensitive to departures from the Standard Model. Decays of heavy mesons and baryons can provide precise determinations of the parameters of the CKM matrix.

In Chapter 6 it was found that for light quarks, the mass could be neglected to first order and a number of results obtained as if the quarks were massless. Their masses were too small to matter. They were small compared to the appropriate strong interaction scale. In this chapter it will be seen that similar results can be found for heavy quarks. Their masses are too large to matter!

At low energies, the effective radius of strongly interacting particles is about $r = 1/\Lambda_s \approx 1$ fermi. This sets a scale of $\Lambda_s \approx 200$ MeV. The heavier quarks, c, b, and t are well above this value. To first approximation the mass can be taken as infinite and then corrections for finite mass can be added. The t is so heavy it usually decays before forming a hadron. Heavy quark theory is not very useful for the heaviest quark of all.

Consider a hadronic system containing one heavy quark of mass m_Q and light quarks. If the quark is very heavy, then in the hadron containing that quark, the quark and the hadron will have essentially the same velocity. In the rest system of the hadron, the quark will be at rest. The typical amount by which the quark will be off mass-shell will be Λ_s. Since this is small compared to the mass of a heavy quark, that quark will be almost on mass-shell.

The heavy quark is surrounded by a very complicated cloud of light quarks and gluons, which is sometimes referred to as a "brown muck." [192] The properties of this brown muck cannot be easily calculated, but the muck is essentially independent of the type of heavy quark present. The soft gluons that couple to the brown muck only resolve distances larger than $1/\Lambda_s$, which is much larger than the distance scale set by the heavy quark mass. These soft partons depend only on having a color triplet source; they are independent of the flavor and even the spin of the heavy quark. This latter statement is true since the spin couples through a "color magnetism" term, similar to the magnetism term in electromagnetism. As in electromagnetism it is a relativistic effect and is proportional to $1/m_Q$; therefore, the spin also decouples.

If the brown muck is independent of the type of heavy quark, then there are relations between the properties of hadrons containing the same light quark(s) but differing heavy quarks. The heavy quark is almost on mass shell. There are corrections since the quark mass is not infinite; these corrections are of the order of Λ_s/m_Q. An outline of only the simplest results will be given here. The monograms of Neubert[193] and of Grinstein[194] are excellent reviews of the area. The present discussion follows the treatment given by Neubert.

Note that the above discussion applies to a familiar system in a different branch of physics, namely the atom. The nucleus is the analogue of the heavy quark and the light degrees of freedom are the electrons and electromagnetic field. In atomic physics, different isotopes have the same chemical properties. The atomic wavefunction, to first order, is independent of the mass of the nucleus. Furthermore, if the nuclear spin is s_N, the atoms have a $2s_N + 1$ degeneracy. This degeneracy is broken by the hyperfine splitting, whose place is taken, in the present discussion, by the color magnetic coupling. The hyperfine splitting in atoms is small because the nuclear mass is far greater than the binding energies of the electrons. The atomic system thus provides a familiar model to help build intuition for the present discussion.

It is now necessary to approximate the QCD Lagrangian for heavy quarks, $\mathcal{L}_Q = \bar{Q}(i\slashed{D} - m_Q)Q$, into a form in which the limit $m_Q \to \infty$ can be taken. Here, the strong interactions only are included and $\mathcal{D}_\mu = \partial_\mu - i(g_3/2)\lambda_a G^a_\mu$, where a is the color index going from 1 to 8. It is understood that the usual Lagrangian $\mathcal{L}_{\text{light}}$ for gluons and light quarks is added to $\mathcal{L}_Q$. Weak and electromagnetic interactions will be added as external currents. Note that the symmetries that will appear here will not be symmetries of the QCD Lagrangian; they are symmetries of the S matrix in a certain region of kinematics and of an effective theory, to be written down shortly, which approximates QCD in that same region.

Since the heavy quark moves with approximately the velocity of the hadron, the momentum, p_Q, of the heavy quark is

$$(p_Q)_\mu = m_Q v_\mu + k_\mu , \qquad (15.1)$$

where v is the four-velocity of the quark, defined as $(\gamma, \gamma \vec{\beta})$, satisfying $(v \cdot v) = 1$. The "residual momentum," k_μ, is small, of the order of Λ_s.

Define

$$P_\pm = \frac{1 \pm \slashed{v}}{2}. \tag{15.2}$$

This quantity is a positive (negative) energy projection operator. It can easily be shown that $P_\pm^2 = P_\pm$ and $P_\pm P_\mp = 0$ (Exercise 15.1). Define the effective heavy quark field by $h_v(x)$, where

$$h_v(x) = e^{im_Q v \cdot x} P_+ Q(x). \tag{15.3}$$

"v" is written as a subscript in h_v to indicate that it is a parameter, not a variable. The Lagrangian is to be written for a specific v, that of the heavy hadron. Because of the projection operator, P_+, $h_v(x)$ is effectively a two-component field. The phase factor has removed most of the space–time dependence from h_v except for the "residual momentum" term. Since it also can be shown (Exercise 15.2) that $P_+ \gamma^\mu P_+ = P_+ v_\mu P_+$ (Exercise 15.2), the Lagrangian above can be rewritten and simplified. The Lagrangian can now be approximated as

$$\mathcal{L}_{eff} = \bar{h}_v i v \cdot \mathcal{D} h_v = \bar{h}_v \left(i v^\mu \partial_\mu + \frac{g_3}{2} \lambda_a v^\mu G_\mu^a \right) h_v. \tag{15.4}$$

This Lagrangian is only an approximation to the QCD Lagrangian. The heavy quark field, Q also contains a component H_v, which satisfies $\slashed{v} H_v = -H_v$. In the rest frame, h_v corresponds to the upper two components and H_v to the lower two components of Q. The H_v terms lead to corrections of order $1/m_Q$ to the effective Lagrangian above. The Lagrangian is, however, valid even if the heavy quark has a relativistic velocity. This is important in systems containing more than one heavy quark, in which it may not be possible to have both of them moving with non-relativistic velocities. Since P_+ was used in the definition, this equation describes heavy quarks and not anti-quarks. The field h_v destroys a heavy quark of four-velocity v; it does not create the corresponding anti-quark. Pair creation of heavy quarks does not occur in effective theory.

Note that the effective Lagrangian has no Dirac matrices. This means that the interactions of the heavy quark with gluons do not rotate the two remaining components of the spinor h_v. Because of this invariance, there is an $SU(2)$ symmetry group for the Lagrangian. This is really an internal symmetry, not "spin" in the usual sense, although it is strongly related to the spin. Also, as promised, the mass of the heavy quark does not appear in this Lagrangian. If there is more than one heavy quark, the Lagrangian can be written

$$\mathcal{L}_{eff} = \sum_{k=1}^{N_h} \bar{h}_{v_k}^k i(v_k)_\mu \mathcal{D}_\mu h_{v_k}^k. \tag{15.5}$$

Recall that the brown muck is the same independent of the heavy quark. This is seen here since, if the velocities of the heavy quarks are the same, the Lagrangian is invariant under rotations in flavor space. Note that this symmetry is good to the extent that $m_c \gg \Lambda_s$ and $m_b \gg \Lambda_s$. It is valid even if $m_b - m_c \gg \Lambda_s$. This is in contrast to isospin symmetry, which requires that $m_d - m_u \ll \Lambda_s$. Including this symmetry and the above internal symmetry, the Lagrangian has a symmetry group of $SU(2) N_h$.

The preceding Lagrangian is not invariant under Lorentz transformations. The color fields seen by the gluons change with the velocity of the quark. This is expected since the Lagrangian was expanded about one particular velocity. The Lorentz covariance is recovered if this velocity is boosted along with the Lorentz boost given to the rest of the equation. This emphasizes the fact that heavy quark symmetry is not a symmetry of the QCD Lagrangian.

15.1 Mass Relations

In the Lagrangian of the last section, it was noted that the spin of the heavy quark was unaffected by the interactions. Hence, the spin of the heavy quark and the angular momentum of the light degrees of freedom are decoupled. For a hadron containing a single heavy quark, there should be two degenerate states with total spin $j \pm 1/2$, where j is the angular momentum of the light degrees of freedom. Heavy quark hadrons should come in pairs.

For ground-state mesons containing a heavy quark, the light quark degrees of freedom should have $j = 1/2$ and one expects degenerate pseudoscalar ($J = 0$) and vector ($J = 1$) states. For charm and bottom mesons there are indeed pairs of states with reasonably small splitting:

$$m_{B^*} - m_B \approx 0.046 \text{ GeV},$$
$$m_{D^*} - m_D \approx 0.142 \text{ GeV},$$
$$m_{D_s^*} - m_{D_s} \approx 0.142 \text{ GeV}. \tag{15.6}$$

One can do even better. If the corrections are assumed to be $\propto 1/m_Q$, then since $m_Q \times (m^* - m) \approx ((m^*)^2 - m^2)/2$, one expects $(m^*)^2 - m^2$ to be independent of m_Q. In fact,

$$m_{B^*}^2 - m_B^2 \approx 0.49 \text{ GeV}^2,$$

$$m_{D^*}^2 - m_D^2 \approx 0.55 \text{ GeV}^2,$$
$$m_{D_s^*}^2 - m_{D_s}^2 \approx 0.58 \text{ GeV}^2. \tag{15.7}$$

The flavor of the light degrees of freedom is different for the last pair than for the others. Thus, experimentally, the brown muck is in first order independent of its flavor.

These arguments can also be applied to excited meson states if the j of the light degrees of freedom can be identified. The $D_1(2420)$ with $J^P = 1^+$ and the $D_2(2460)$ with $J^P = 2^+$ have only a 35 MeV mass difference. If they are identified with the doublet corresponding to $j = 3/2$, then one would predict a bottom doublet with

$$m_{B_2^*}^2 - m_{B_1}^2 \approx m_{D_2^*}^2 - m_{D_1}^2 \approx 0.17 \text{ GeV}^2. \tag{15.8}$$

The splitting seems less for the excited states than for the ground states. This result seems reasonable in the context of a non-relativistic quark model. In such a model the light anti-quark is in a P-wave state for the excited meson. The wavefunction for a P-wave state vanishes at the origin, where the heavy quark is located. This suppresses the corrections.

Another prediction of this flavor symmetry is that the mass differences for pairs of states with different light degrees of freedom should be the same when the heavy quark is changed, since the mass difference is just due to the change in the brown muck. One observes $m_{D_S} - m_D \approx 100$ MeV, and $m_{B_S} - m_B \approx 96$ MeV. One predicts

$$m_{B_1} - m_B \approx m_{D_1} - m_D = 0.557 \text{ GeV},$$
$$m_{B_2^*} - m_B \approx m_{D_2^*} - m_D = 0.593 \text{ GeV}. \tag{15.9}$$

15.2 Form Factors for Weak Decay

Heavy quark effective field theory provides important relations between the form factors for decay of heavy mesons. Consider meson states of a heavy meson as a function of the meson velocity, v. It is convenient to define these states with a mass independent normalization, $\langle M(v)| = m_M^{-1/2} \langle M_{\text{conventional}}|$. These states are then determined entirely by the light degrees of freedom with shape and normalization independent of the heavy quark mass.

Consider elastic scattering of a pseudoscalar meson $P(v) \to P(v')$ produced by a vector current (not in $\mathcal{L}_{eff}$), which couples to the heavy quark in the meson. This external current replaces the color source represented by the heavy quark moving at velocity v, with a source moving at velocity v' starting at $t = t_0$. If the velocity is not changed, then, as far as the light degrees of freedom are concerned, nothing happens. If the velocity changes, the brown muck must rearrange itself to accommodate to a source moving with a different velocity. This leads to a form-factor suppression, which must depend exclusively on $(v \cdot v')$, the only scalar available. The form factor, to first order, will not depend on the mass of the heavy quark since in this rearrangement the light partons carry momenta typically $\Lambda_s v$ and $\Lambda_s v'$ corresponding to typical momentum transfers of $q^2 \approx -\Lambda_s^2(v \cdot v' + 1)$, which is independent of m_Q. Thus, the hadronic matrix element describing the scattering process can be written

$$\langle P(v')| \bar{h}_{v'} \gamma^\mu h_v |P(v)\rangle = \xi(v \cdot v')(v + v')^\mu + \chi(v \cdot v')(v - v')^\mu. \quad (15.10)$$

χ can be seen to be zero by contracting the matrix element with $(v - v')_\mu$ and using $\slashed{v} h_v = h_v$ and $\bar{h}_{v'} \slashed{v}' = \bar{h}_{v'}$. The function $\xi(v \cdot v')$ is known as the Isgur–Wise function[192] and is real. This last fact is seen by noting that the matrix element is invariant under complex conjugation followed by an interchange of v and v'.

If the vector current changes the heavy quark, Q, into another heavy quark Q', the same argument applies; the effective Lagrangian does not distinguish Q and Q'. Since this is a symmetry transformation in $SU(2)\,N_h$, one obtains the *same* form factor $\xi(v \cdot v')$ as obtained previously; the Isgur–Wise form factor is a universal form factor

$$\langle P'(v')| \bar{h}_{v'} \gamma^\mu h_v |P(v)\rangle = \xi(v \cdot v')(v + v')^\mu. \quad (15.11)$$

Consider next the limit of equal initial and final velocities, $v = v'$. This limit is known as the zero recoil limit, since, in this limit, the final-state meson P' is at rest with respect to the initial-state meson P. It is expected that for this situation, $\xi = 1$, since the brown muck is insensitive to the change in flavor of the heavy quark. This is indeed the case.

In this limit the vector current $J^\mu = \bar{h}'_v \gamma^\mu h_v = \bar{h}'_v v^\mu h_v$. For the simple Dirac equation, the Lagrangian is $\mathcal{L} = \bar{Q}(i\slashed{\partial} - m)Q$ and the equation of motion is $(i\slashed{\partial} - m)Q = 0$. Similarly, the Lagrangian for the effective theory is $\mathcal{L}_{eff} = \bar{h}_v i(v \cdot \mathcal{D}) h_v$ and the equation of motion is $i(v \cdot \mathcal{D}) h_v = 0$. By applying this equation of motion it can be shown that $\partial_\mu J^\mu = 0$, and hence the current is conserved (See Exercise 15.3). There is then a conserved charge

$$N_{Q'Q} = \int d^3x J^0(x) = \int d^3x (h'_v)^\dagger(x) h_v(x). \quad (15.12)$$

Here there are both diagonal and off diagonal charge elements. The diagonal ones count the number of heavy quarks, $(N_{QQ} |P(v)\rangle = |P(v)\rangle)$, and the off-diagonal ones replace one heavy quark with another, $N_{Q'Q}|P(v)\rangle = |P'(v)\rangle$. Thus, with the normalization of states given previously,

$$\langle P'(v)| N_{Q'Q} |P(v)\rangle = \langle P(v) | P(v)\rangle = 2v^0(2\pi)^3\delta^3(0). \tag{15.13}$$

By comparison with Equation 15.11, it is seen that at the point of equal velocities, the expected result is obtained:

$$\xi(1) = 1. \tag{15.14}$$

The relations for ξ are expected to break down for high momentum transfer, when the recoil kinetic energy of the light particles [which is $\approx -\Lambda_s((v \cdot v') + 1)$] is of the order of magnitude of the heavy quark mass. For the important decay $\bar{B} \to D l \bar{\nu}$, the recoil energy $-\Lambda_s((v \cdot v') + 1)$ has a maximum value of $\Lambda_s(m_B - m_D)^2/(2m_B m_D) \approx 0.6\Lambda_s \ll m_Q$. The relations for ξ should be valid for this decay.

Consider then decays of B and D mesons. A more standard parametrization of the form factors[195] will be used. Using the conventional normalization of states ($\langle M| = \langle M_{\text{conventional}}|$) and $q = p - p'$, one has

$$\langle B(p')| \bar{b}\gamma^\mu b |B(p)\rangle = F_{el}(q^2)(p + p')^\mu,$$

$$\langle D(p')| \bar{c}\gamma^\mu b |B(p)\rangle = F_1(q^2)\left[(p + p')^\mu - \frac{m_B^2 - m_D^2}{q^2}q^\mu\right]$$

$$+ F_0(q^2)\frac{m_B^2 - m_D^2}{q^2}q^\mu. \tag{15.15}$$

To avoid a spurious pole at $q^2 = 0$, one requires $F_0(0) = F_1(0)$. In the limit that $m \to \infty$, with $(v \cdot v')$ staying finite, these form factors are related to ξ. By comparing this equation with Equations 15.10 and 15.11, one sees for the elastic form factor that

$$\xi(v \cdot v') = \lim_{m_b \to \infty} F_{el}(q^2), \tag{15.16}$$

$$(v \cdot v') = 1 - \frac{q^2}{2m_B^2}. \tag{15.17}$$

For the inelastic form factor, using m for m_B or m_D, one has

$$\xi(v \cdot v') = \lim_{m \to \infty} RF_1(q^2) = \lim_{m \to \infty} R\left[1 - \frac{q^2}{(m_B + m_D)^2}\right]^{-1} F_0(q^2), \tag{15.18}$$

where

$$R = \frac{2\sqrt{m_B m_D}}{m_B + m_D} \approx 0.88,$$

15.2 Form Factors for Weak Decay

$$(v \cdot v') = \frac{m_B^2 + m_D^2 - q^2}{2 m_B m_D}. \tag{15.19}$$

Note that the zero-recoil-limit relations correspond to relating the form factors at different values of q^2. ξ relates elastic form factors at spacelike momentum transfers ($q^2 < 0$) to weak form factors at timelike momentum transfers ($q^2 > 0$). The elastic form factor is normalized at $q^2 = 0$ and the weak form factors are normalized at $q_{max}^2 = +(m_B - m_D)^2$. In both instances this corresponds to the zero recoil limit.

Relations can also be obtained between matrix elements for vector mesons and for pseudoscalar mesons. Since for the appropriate pseudoscalar–vector meson pairs, the only difference within heavy quark effective theory is in the spin of the heavy quark the two are closely related. Suppose that in the quark rest frame one selects, for spinors, the basis $u_\alpha^1 = \delta_{1\alpha}$ and $u_\alpha^2 = \delta_{2\alpha}$ as spin up and spin down, respectively, and for anti-spinors selects the basis $v_\alpha^1 = -\delta_{4\alpha}$ and $v_\alpha^2 = \delta_{3\alpha}$ and spin up and spin down. The combination $u_Q^1 \bar{v}_q^2 + u_Q^2 \bar{v}_q^1$ then is the spin 0 combination and $u_Q^1 \bar{v}_q^1$, $u_Q^1 \bar{v}_q^2 - u_Q^2 \bar{v}_q^1$, $u_Q^2 \bar{v}_q^2$ correspond to the spin 1 combinations with third component 1, 0, and -1, respectively.

Using this, one can show (Exercise 15.4) that if a vector meson has longitudinal polarization ϵ_3, it is related to the pseudoscalar meson by means of a hermitian spin operator acting on the spin of the heavy quark. The operator will be denoted S_Q^3 and has a matrix representation $(1/2)\gamma^5 \slashed{\epsilon}_3$. In the rest frame of the final-state meson, the matrix representation becomes $S^3 = (1/2)\gamma^5 \gamma^0 \gamma^3$. One has

$$|V(v, \epsilon_3)\rangle = 2 S_Q^3 |P(v)\rangle. \tag{15.20}$$

Let Γ be any combination of Dirac matrices. One has, in general,

$$\langle V'(v', \epsilon_3)| \bar{h}_{v'}' \Gamma h_v |P(v)\rangle = \langle P'(v')| 2[S_{Q'}^3, \bar{h}_{v'}' \Gamma h_v] |P(v)\rangle$$
$$= \langle P'(v')| \bar{h}_{v'}' (2 S^3 \Gamma) h_v |P(v)\rangle. \tag{15.21}$$

In the rest frame of the final-state meson, one can obtain the commutation relations for the components of the weak current $(V - A)_\mu = \bar{h}_{v'}' \gamma^\mu [(1 - \gamma^5)/2] h_v$:

$$2[S_{Q'}^3, (V - A)^0] = -(V - A)^3,$$
$$2[S_{Q'}^3, (V - A)^3] = -(V - A)^0,$$
$$2[S_{Q'}^3, (V - A)^1] = -i(V - A)^2,$$

$$2[S^3_{Q'}, (V-A)^2] = i(V-A)^1. \tag{15.22}$$

From Equations 15.21 and 15.22 the matrix element of the weak current between a pseudoscalar and a vector meson is related to the matrix element between two pseudoscalar mesons. In covariant form one can show

$$\langle V'(v',\epsilon)| \bar{h}'_{v'}\gamma^\mu(1-\gamma^5)h_v |P(v)\rangle = i\epsilon^{\mu\nu\alpha\beta}\epsilon^*_\nu v'_\alpha v_\beta \xi(v\cdot v')$$
$$-[\epsilon^{*\mu}((v\cdot v')+1) - v'_\mu(\epsilon^*\cdot v)]\xi(v\cdot v'). \tag{15.23}$$

Here $\epsilon^{\mu\nu\alpha\beta}$ is the completely anti-symmetric tensor with $\epsilon^{1230} = 1$.

Consider $B \to D^*l\nu$, which is a decay of the preceding type. A conventional form factor basis for this decay[195] (with conventional, not heavy quark normalization of wavefunctions) is given by

$$\langle D^*(p',\epsilon)| \bar{c}\gamma^\mu(1-\gamma^5)b |B(p)\rangle = \frac{2i\epsilon^{\mu\nu\alpha\beta}}{m_B+m_{D^*}}\epsilon^*_\nu p'_\alpha p_\beta V(q^2)$$
$$- \left[(m_B+m_{D^*})\epsilon^{*\mu}A_1(q^2) - \frac{(\epsilon^*\cdot q)}{m_B+m_{D^*}}(p+p')^\mu A_2(q^2) \right.$$
$$\left. -2m_{D^*}\frac{(\epsilon^*\cdot q)}{q^2}q^\mu A_3(q^2)\right] - 2m_{D^*}\frac{(\epsilon^*\cdot q)}{q^2}q^\mu A_0(q^2). \tag{15.24}$$

The preceding form factors are not independent. A_3 is an abbreviation for a linear combination of A_1 and A_2:

$$A_3(q^2) \equiv \frac{m_B+m_{D^*}}{2m_{D^*}}A_1(q^2) - \frac{m_B-m_{D^*}}{2m_{D^*}}A_2(q^2). \tag{15.25}$$

In addition the spurious pole at $q^2 = 0$ needs to be removed, which requires that $A_0(0) = A_3(0)$. Comparing Equations 15.23 and 15.24:

$$\xi(v\cdot v') = \lim_{m\to\infty} R^*V(q^2) = \lim_{m\to\infty} R^*A_0(q^2) = \lim_{m\to\infty} R^*A_2(q^2)$$
$$= \lim_{m\to\infty} R^*[1 - \frac{q^2}{(m_B+m_{D^*})^2}]^{-1}A_1(q^2). \tag{15.26}$$

Here $R^* \approx 0.89$ and $v\cdot v'$ are related by Equation 15.19 with m_D replaced by m_{D^*}.

Any heavy meson matrix element of the form $\langle M'|\bar{Q}\Gamma Q|M\rangle$ can be expressed in terms of the one universal Isgur–Wise function.

15.3 HEAVY BARYON SEMILEPTONIC DECAYS

Consider the simplest states, in which the angular momentum of the light degrees of freedom is zero. With conventional normalization there are three form factors,[194] F_i for the vector current and three more G_i for the axial vector current for the decays of Λ_b or Λ_c states:

$$\langle \Lambda_c(v',s')| \bar{c}\gamma^\mu b |\Lambda_b(v,s)\rangle = \bar{u}^{s'}(v')[\gamma^\mu F_1 + v^\mu F_2 + (v')^\mu F_3]u^s(v), \quad (15.27)$$

$$\langle \Lambda_c(v',s')| \bar{c}\gamma^\mu\gamma^5 b |\Lambda_b(v,s)\rangle = \bar{u}^{s'}(v')[\gamma^\mu G_1 + v^\mu G_2 + v'^\mu G_3]\gamma^5 u^s(v). \quad (15.28)$$

All six of these are given in terms of the universal form factor ξ. Let Γ be an arbitrary combination of gamma matrices. One finds that, in the effective theory, the matrix element is given by

$$\langle \Lambda_c(v',s')| \bar{c}_{v'}\Gamma b_v |\Lambda_b(v,s)\rangle = \xi(v \cdot v')\bar{u}^{s'}(v')\Gamma u^s(v). \quad (15.29)$$

From comparing terms in the last three equations

$$F_1 = G_1; \quad F_2 = F_3 = G_2 = G_3 = 0. \quad (15.30)$$

Furthermore, to zeroth order $G_1(1) = 1$.

15.4 PSEUDOSCALAR AND VECTOR MESON DECAY CONSTANTS

In the decay of a pseudoscalar meson M containing a heavy quark, the matrix element of the current between the meson and the vacuum occurs and is denoted the effective pseudoscalar decay constant f_M. This is exactly analogous to pion decay treated in Chapter 7. For the heavy quark effective theory, one then has

$$\langle 0| A_\mu |M(v)\rangle = \bar{f}_M v_\mu. \quad (15.31)$$

The bar on the decay constant indicates that this is the constant in the effective theory, in which the wavefunction has a different normalization than in the conventional theory. In the effective theory, to first order, $\bar{f}_M$ is independent of the heavy quark mass. Converting back to the conventional normalization requires multiplying the above equation by $\sqrt{M_M}$ to restore the normalization of states and writing $v_\mu = p_\mu/M_M$. One then obtains

$$f_M = \bar{f}_M/\sqrt{M_M}. \tag{15.32}$$

Applying this to B and D mesons:

$$\frac{f_B}{f_D} = \sqrt{\frac{M_D}{M_B}}. \tag{15.33}$$

For vector mesons M^*, the decay constant for the effective theory is defined by

$$\langle 0| \bar{V}_\mu(0) |M^*(v,\epsilon)\rangle = \bar{f}_{M^*}\epsilon_\mu. \tag{15.34}$$

Consider a pseudoscalar–vector meson pair. The pseudoscalar constant is proportional to

$$\text{Tr}[\gamma^\mu \gamma^5 M(v)] = \text{Tr}\left[\gamma^\mu \gamma^5 \left(\frac{1+\slashed{v}}{2}\right)\gamma^5\right] = -2v^\mu, \tag{15.35}$$

and the vector meson constant is proportional to

$$\text{Tr}[\gamma^\mu M^*(v,\epsilon)] = \text{Tr}\left[\gamma^\mu \left(\frac{1+\slashed{v}}{2}\right)\slashed{\epsilon}\right] = 2\epsilon^\mu. \tag{15.36}$$

Both have the same constant of proportionality and hence

$$\bar{f}_{M^*} = \bar{f}_M. \tag{15.37}$$

Restoring the conventional normalization yields

$$f_{M^*} = f_M M_M. \tag{15.38}$$

These results have not yet been able to be tested experimentally because of the small expected branching ratios. Calculations within lattice gauge theory indicate there may be large $1/M_Q$ corrections to these relations.

15.5 Comments on Corrections to Zeroth Order

Most of the results above have been derived from the general picture, without the use of the details of the Lagrangian. In actual practice, $1/M_Q$ corrections and others are needed. These involve more detailed discussions, beyond the scope of this text. Part of the $1/M_Q^2$ correction comes from including the small parts of the wavefunction (the P_- parts) into the Lagrangian. Including quark loop diagrams introduces logarithmic corrections, essentially the running of the strong coupling constants. In the loop diagrams, the momentum of the quarks is uncontrolled, not necessarily small and advanced means are used to calculate the corrections induced by these diagrams. The corrections due to this usually involve $\alpha_s(M_Q)/\alpha_s(\mu)$ to some power, where μ is a lower mass which depends on the decay. Finally, there are corrections of the order of α_s. The corrections of order Λ_S/m_c to baryon decays can be made accurately, but the equivalent corrections for meson decays are not well determined, leading to new form factors and uncertainties of the order of 10–20%. The interested reader can examine the details of the corrections in references [193] and [194].

15.6 Determining CKM Matrix Elements

It is fortunate that for meson decays the unknown corrections indicated above of order Λ/m_c vanish at the symmetry point $(v \cdot v') = 1$. This allows the mixing angle $|V_{cb}|$ to be determined with high precision by measurements at the end of the spectrum for $B \to D e \nu$ and $B \to D^* e \nu$.

Determining $|V_{ub}|$ is much more complicated. The appropriate decays are hard to observe experimentally because $|V_{ub}|/|V_{cb}| \ll 1$ and because charm decays quickly to light hadrons. The first effect makes the decays rare and the second effect makes them difficult to distinguish from background. As noted in Chapter 7, the process has now been definitely observed and in the future it is hoped that it will be possible to examine decays such as $B \to \omega e \bar\nu$. The problem with such decays is that the calculation of either exclusive or inclusive rates has been unreliable.

However, using heavy quark effective field theory, there may be a method of obtaining reasonably precise results. The idea is to measure the decays of another meson that has the same brown muck as the one of interest. Then the form factors should be the same. For instance, the matrix elements of $B \to \rho$ and $D \to \rho$ should be related in the effective theory. One might even be able to use the $SU(3)$ flavor symmetry of the light quarks to relate $D \to \rho$ to the Cabibbo allowed decay $D \to K^*$. Unfortunately the spectrum for B decay extends far past the end point for D decay and extrapolations might be difficult.

The same trick as above may help in examining rare decays. For example the matrix element for the rare decay $B \to K e^+ e^-$ can be related to that for a process more accessible experimentally, $D \to K e \nu$. Again this involves continuing some quantities into the physical region of the decay of interest.

15.7 Exercises

15.1 Show that $P_\pm^2 = P_\pm$ and $P_\pm P_\mp = 0$, where $P_\pm = (1\pm\slashed{v})/2$. Show also that in the rest frame, only the upper two components are selected from the spinor u by P_+.

15.2 Show that $P_+\gamma^\mu P_+ = P_+ v_\mu P_+$.

15.3 Using the equations of motion that follow from $\mathcal{L}_{eff}$, show that in the limit of equal velocities, the vector current, $J^\mu = \bar{h}'_v \gamma^\mu h_v = \bar{h}'_v v^\mu h_v$, is conserved.

15.4 Show that, in the rest frame, $2S^3$ operating on the heavy quark in a pseudoscalar meson gives the vector meson partner with $m = 0$.

16
Monopoles

16.1 Monopoles in Electromagnetic Theory

An asymmetry in Maxwell's equations has long been noted. Maxwell's equations in Gaussian units are

$$\nabla \cdot \vec{D} = 4\pi\rho, \tag{16.1}$$

$$\nabla \cdot \vec{B} = 0, \tag{16.2}$$

$$\nabla \times \vec{E} = -\frac{1}{c}\frac{\partial \vec{B}}{\partial t}, \tag{16.3}$$

$$\nabla \times \vec{H} = -\frac{1}{c}\left(\frac{\partial \vec{D}}{\partial t} + 4\pi \vec{J}\right). \tag{16.4}$$

There is a source term for the electric field, $4\pi\rho$, and none for the magnetic field. There is no magnetic analogue of the $4\pi \vec{J}$ that appears in the fourth Maxwell equation. If one tries to make Maxwell's equations symmetric in the electric and magnetic fields, then magnetic charge, i.e., isolated magnetic north or south poles, "monopoles" have to be introduced. The only argument against doing this is that it is not clear nature has chosen this path; monopoles have not yet been observed.

In 1931, a very beautiful quantization condition on monopoles[196][197] was found by Dirac. Consider only electromagnetism at first. Suppose a monopole exists with magnetic charge g. Then $\vec{B} = g\hat{r}/r^2$. [The Coulomb law in these units is $\vec{E} = q\hat{r}/(4\pi r^2)$.] Suppose here that the particle is a pure monopole, not a particle with both an electric and a magnetic charge ("dyon"). The quantization is harder for a dyon.

16. Monopoles

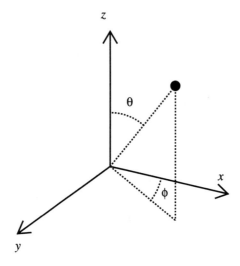

Figure 16.1 Definition of coordinates.

Consider the vector potential $\vec{A}$, where $\vec{B} = \nabla \times \vec{A}$. For a monopole $\vec{A}$ is always singular. For example, a possible $\vec{A}$ for the point monopole is

$$\vec{A} = g \frac{1 - \cos\theta}{r \sin\theta} \hat{\phi}. \tag{16.5}$$

See Figure 16.1. This potential is singular along the $-z$ axis. It can be looked at as the potential of a long thin solenoid running from the origin to $-\infty$ in z and carrying a flux $f = 4\pi g$. This is called a "Dirac string."

Digress for a moment. Suppose that one has a narrow, long real solenoid. The magnetic field outside is essentially zero. However, outside the solenoid, $\vec{A}$ is given by essentially the same expression as previously given and forms circles around the solenoid. From measurements outside the solenoid only, can one detect whether the current in the coil is on or off, i.e., whether there is a magnetic field in the solenoid or not? The question hinges on whether the vector potential $\vec{A}$ is just a mathematical convenience or a physical quantity that can be directly measured.

Classically, since there is no magnetic or electric field outside the solenoid, there are no forces on a test charge and no way to examine whether the solenoid is on or off.

Quantum mechanically, the situation is quite different. In the studies of gauge invariance it was previously noted that $\vec{A}$ enters into the Lagrangian in combination with the momentum as $p - eA/c$. Y. Aharonov and D.

Bohm[198] have shown that one can detect a real solenoid from the vector potential $\vec{A}$. To do this one uses quantum mechanical interference effects. Arrange an experiment such that an electron can go either around the right side or the left side of the solenoid and the two amplitudes can interfere. (See Figure 16.2.) The phase for each path will involve the integral of $\vec{A}$ over that path. The phase difference will involve the integral of $\vec{A}$ over the closed path surrounding the solenoid. This phase difference is in general measurable for charged particles, q, unless $e^{iqf} = e^{4\pi iqg} = 1$. This phase difference has been measured in several clever experiments and certainly really exists.

Resume the discussion on monopoles. For this case, the solenoid is not real. The Dirac string can be placed in an arbitrary position and it should not be detectable. There is then the amazing conclusion that if a monopole exists, the phase difference must vanish or be an integral multiple of 2π. This can happen only if all charge is quantized! The existence of a monopole requires electric and magnetic charge quantization. $e^{4\pi iqg} = 1$ implies $qg = 0, \pm 1/2, \pm 1, \cdots$ ($\hbar = c = 1$). The minimum monopole charge is $g_{\min} = g_{\text{Dirac}} = 1/(2q_{\min}) = \hbar c/2e$

Thus, monopoles are allowed in an extension of the Maxwell equation, and, if they exist, provide an explanation of charge quantization.

16.2 MONOPOLES IN EXTENDED GAUGE THEORIES

Consider the modern extended gauge theories. The quark charge can be 1/3 of the electron charge. Does the monopole then have three times the charge indicated above? It turns out the answer is not that simple. For the Standard Model or for GUT, the gauge group is more complicated than for QED, and the phase difference depends on other gauge fields as well as $\vec{A}$. Suppose that the unbroken gauge symmetry is $G = SU(3)_{color} \otimes U(1)_{em}$. When the old $SU(3)$ theory with the u, d, s quarks was discussed, $\vec{I}$ and Y were defined. Similarly, for the color $SU(3)$ analogous quantities, $\vec{I}_{color}$ and Y_{color} can be defined. One can show[197] that the magnetic monopole field can be written as $\vec{B} = G\hat{r}/r^2$, where G is a matrix in the group G corresponding to the full set of particle species. This matrix can be diagonalized to give $G = G_{em} + G_{color} = a(Q_{elec}/2e_{em}) + (1/2e_{color})[b(I_3)_{color} + cY_{color}]$, where a, b, and c are constants determining the effective size of the charges. e_{em} is the magnitude of the electric charge on an electron. The vector potential is much the same as before, $\vec{A} = G[1 - \cos\theta]/(r\sin\theta)]\hat{\phi}$.

If one supposes that the test "particle" for the Aharonov–Bohm experiment (which is now less than 10^{-13}cm in size) has: $e_{em}Q_{em}, e_{color}I_3^c$,

344   16. Monopoles

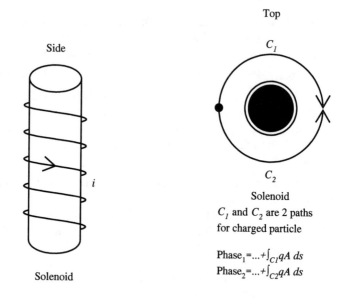

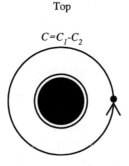

Figure 16.2 Aharonov–Bohm experiment.

$e_{color}Y^c$, then the condition to make the solenoid unobservable is

$$e^{2\pi i[aQ_{em}+bI_3^c+cY^c]} = 1. \tag{16.6}$$

By examining the quantum numbers of the electromagnetically charged particles in the Standard Model, one can show that a, b, and c are integers, and that in some gauges $b = 0$ and $a = c$ (mod 3). In some cases the monopole will carry color! Whether they do or not depends on which extension of the Standard Model is used.

16.3 T'HOOFT POLYAKOV MONOPOLES

Stable monopolelike solutions do not readily occur in the Standard Model. However, monopole solutions do exist in many of the extensions of the Standard Model. One needs a simple group for the gauge, i.e., one with a single gauge coupling constant. In 1974 t'Hooft[199] and Polyakov[200] independently noted that there exists a solution of extended Standard Model Lagrangians which looks like a monopole at distances larger than a few fermi, i.e., past these distances the magnetic field looks like a hedgehog, i.e., is radial. Specifically, these solutions exist if the electromagnetic group $U(1)$ is a subgroup of a larger compact group. These correspond to solutions forming a topological knot, where the Higgs field ϕ cannot be continuously deformed into the vacuum. In the simplest model the mass of this monopole comes out to be $M \approx (4\pi/e^2)M_W \approx M_W/\alpha \simeq 10^{13}$ eV. In a GUT theory one gets $M \approx M_X/\alpha_{GUT} \approx 10^{15}/0.1 \approx 10^{16}$ GeV. More generally, two kinds of monopoles appear in these theories. One kind has M in the 10^{16}–10^{19} GeV range and has a magnetic charge $1/(2e)$, and the second kind has a magnetic charge an integral multiple of the first and a mass in the 10^{10}–10^{16} GeV range or even lower. This second kind cannot participate in the catalysis of proton decay described below and is not present at all in $SU(5)$.

16.4 OTHER TOPOLOGICAL DEFECTS; COSMIC STRINGS

The monopole solutions are one of a class of solutions corresponding to topological defects in the vacuum, in this case a point defect.

Another type of defect is a line defect. Shortly after the big bang initiated the Universe, different parts of the vacuum had different spontaneously broken values, since the breaking between regions not in causal contact was independent. The values then relaxed to a common value. Consider the phase of the complex ϕ field. In some regions, when going around a loop, the phase might not return to the same phase, but to the same phase plus some multiple of 2π. If this happens, there must be a place within the loop where ϕ is zero, i.e., the vacuum is at its unbroken value. This cannot relax away and a string of unbroken vacuum forms. There is a small region around this string in which the vacuum is different from its normal broken value. The size of this region is set by the competition between the extra vacuum energy in the unbroken solution and the gradient energy coming from the $(\nabla\phi)^2$ term in the Lagrangian. The string cannot end, so it would probably form into a closed loop, although infinite strings are not ruled out. Similar classes of defects occur in real condensed matter situations and have been seen in liquid He.

These strings could have provided the initial perturbations in the Universe needed to seed galaxy formation. Estimates suggest that there are only of the order of one of these strings within our event horizon today. They are hard to detect. They cannot provide enough mass to significantly affect the amount of matter in the Universe. They would be essentially superconducting; a flux of current would be trapped in them, but no clear tests for them have been found from that effect.

One possibility is the use of the gravitational lensing effect. Suppose, within our field of view from the earth, a non-luminous body comes between a star and our telescope. The photons from the star are gravitationally bent by the non-luminous object and the star image is distorted. Two images of the star might be seen or a change of intensity noted, etc. This is known as gravitational lensing. If there is a string within our horizon, there would be a gravitational lensing line across the sky and it might be possible to see it.

16.5   Catalysis of Proton Decay by Monopoles

This is an effect first noted by Rubakov[201] and discussed by Rubakov and by Callan.[202] For energies much less than the monopole mass one might expect that wavefunctions wouldn't penetrate to the $SU(5)$ core of the monopole wavefunction and $SU(5)$ physics would not then be relevant. However, the core is singular. For even an abelian monopole with electromagnetic interactions only, one can show that it can transform an electron with left-handed helicity into one with right-handed helicity with finite cross section. It turns out that in non-perturbative field theory one can show that particle helicity is not necessarily conserved if $\vec{E}$ and $\vec{B}$ fields are simultaneously present and if $\vec{E} \cdot \vec{B} \neq 0$. This is true in S-wave scattering. The monopole has induced an apparent conservation law violation at a rate independent of monopole size.

For $SU(5)$ similar considerations lead to the expectation that there may be a violation of baryon number conservation, that the monopole can catalyze proton decay! If a monopole passes through a proton decay experiment, one or more proton decays may result. Some estimates of the cross section have been as high as $10^{-26}/\beta_{monopole}$ cm^2, but these high estimates have been somewhat controversial. However, if the cross section is reasonably high, one might have a single experiment showing both the existence of monopoles and baryon number non-conservation.

16.6 EXPERIMENTAL AND OBSERVATIONAL LIMITS ON MONOPOLE FLUX

If monopoles are as heavy as 10^{17}eV, they would not yet have been produced in the laboratory. However, they would have been produced in the initial stages of the Universe and one might be able to detect them now. Note, though, that in one popular model, the inflationary model of the Universe, it is possible that the number of monopoles in the Universe is zero or one.

Monopoles, if they exist in interstellar space, would be accelerated by the galactic magnetic fields and would therefore remove energy from the fields. Since it is known observationally that galactic magnetic fields exist and have not been completely eaten away by monopoles, a limit can be put on the flux of monopoles. This limit is $F < 10^{-15} \text{cm}^{-2} \text{s}^{-1} \text{sr}^{-1}$ and is known as the Parker bound[203][204] (see Exercise 16.1) and a recent extension is called the extended Parker bound.[205] The monopoles would be expected to now have a velocity, β, in the range from 10^{-4} to 10^{-3}.

Somewhat weaker limits are given just by the total mass of the monopoles in the universe.[206] The limits are somewhat different depending on whether the monopoles are dispersed uniformly or are clumped.

Limits can also be obtained[203][207] from considering magnetic plasma oscillations of the galactic magnetic fields. This limit is somewhat controversial.

One can get better limits from the fields in white dwarfs (WD)[208] and in neutron stars[209] (NS) *if* one assumes catalysis of proton decay by monopoles. Of these limits, the WD one is more secure than the NS limit since the WD equations of state are believed better known than the NS ones. Even more speculative is the still better limit from monopoles captured by the main sequence progenitor of the neutron star (NS w/ MS accretion).[210] A review of these limits can be found in the book of Kolb and Turner.[206] Some of these limits are shown in Figure 16.3.

There have been a number of earthbound searches for monopoles as well. A number of references for earthbound and other searches for monopoles are listed in reference 17. Searches have been made at most of the standard accelerators for low-mass monopoles. The cosmic radiation has been searched for higher-mass monopoles. So far none have been seen. A few upper limits on monopole flux not depending on proton catalysis are given in Figure 16.4.

In this chapter we noted that magnetic monopoles are allowed to exist by using an extension of Maxwell's equations. If they exist, they provide an explanation of the quantization of charge. In extended gauge theories they can carry other quantum numbers, perhaps color. Monopoles may, by a peculiar mechanism first noted by Rubakov, be able to catalyze proton decay. Experimentally, monopoles have not yet been seen.

16. Monopoles

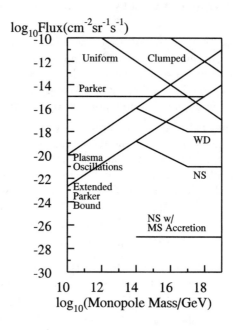

Figure 16.3 Theoretical limits on monopole flux vs monopole mass.

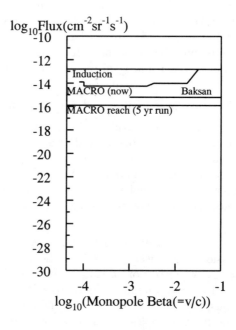

Figure 16.4 Experimental limits on monopole flux vs β.

16.7 Exercises

16.1 The simplest version of the Parker bound on monopoles can be derived with the following assumptions. If monopoles were present in sufficient numbers, they would drain the energy contained in the galactic magnetic fields, which, in fact, average about 3 microgauss. These fields replenish themselves through a galactic dynamo mechanism as the galaxy rotates. The time scale for this is about 200 million years. Assume the monopoles are light and travel at about the velocity of light. Assume they travel parallel to the B fields about half of the time and perpendicular to the B fields about half of the time. If the flux F of monopoles is isotropic, it is related to the monopole density ρ by $F = \rho v/(4\pi)$. From this information show that the monopole flux should be less than about $0.9 \times 10^{-16} \text{cm}^{-2} \text{sr}^{-1} \text{s}^{-1}$.[211]

17
Present Status of Particle Physics and Outlook

There is no strong evidence against the Standard Model. A recurring nightmare is that the t and a Higgs will be found–and nothing else. If the great desert is real, the energies needed to get some direction from nature on what lies beyond the Standard Model may not be reachable. However, for many generations, physicists have had some variant of this worry and nature has always been kind enough to continue yielding its secrets. It is hoped that this guidance will continue.

17.1 PRESENT RESULTS THAT MIGHT HINT AT FURTHER PHENOMENA

1. The solar neutrino problem. For many years Ray Davis and associates[114] have been measuring the flux of neutrinos from the sun. They find fewer than the standard solar models predict. The Kamioka proton decay group has examined solar neutrinos with their detector and confirms the Davis et al. results. A more sensitive version of the Kamiokande detector (super Kamiokande) is now being installed. However, all of the preceding detectors have the disadvantage that they can look only at the highest-energy neutrinos from the sun. These neutrinos are only a few percent of the total flux and are not from the main energy-producing reactions. Further experiments, sensitive to some of the lower-energy neutrinos, utilizing gallium detectors, help clarify the problem. The results from the gallium detectors confirm the generally smaller flux seen by the Davis and Kamioka groups. The presumed shortage of neutrinos has been the subject of much discussion. Neutrino oscillations could affect the observed rate as could the presence of various exotic particles in the sun.

2. Dark matter. Many astronomical observations lead to the conclusion that the amount of matter in the Universe is at least about 10 times the amount of visible matter[212][213] and about 100 times the visible matter if the universe is a closed universe. It is not clear what this dark (i.e., non-luminous) matter is. It is possible that it is in large planet sized chunks, and, indeed, some of these have recently been observed.

Most other standard forms of matter are ruled out. It is possible that if neutrinos have mass, the dark matter could be neutrinos. It could be monopoles. Note, that from Figure 16.3, because of the extended Parker bound, very heavy (greater than about 10^{17} GeV) monopoles could not provide sufficient density to close the universe. The dark matter could also be many other exotic types of particles. Experimental results from LEP have eliminated many previously possible candidates. A popular view at present is that a mixture of two kinds of dark matter is present. These consist of hot dark matter, matter consisting of particles traveling at relativistic velocities, and cold dark matter, heavier particles or other objects moving at non-relativistic velocities.

3. Quasars and AGN's. Quasars are bright objects in the sky with an apparent high red shift. If they are as far away as their red-shift indicates, they have enormous energy. At present quasars are thought to be active galactic nuclei (AGN's), galaxies with a very hot center, probably having a massive black hole there. TeV gamma rays have been detected from nearby AGN's.[214] Blazars are another class of objects also believed to be AGN's. Quasars are considered as AGN's seen edge on and blazars as AGN's seen head on. Whether near or far, these are very interesting objects in the sky.

17.2 SOME THEORETICAL QUESTIONS

The present Standard Model, although very beautiful, leaves a number of unanswered questions and problems as we have indicated. There are a large number of arbitrary parameters that have to be fine tuned to get masses of observed particles to be as low as they are observed to be. There are a large number of elementary particles.

Are the strong, weak, and electromagnetic charges the same for each generation? They seem to be the same experimentally. Why is this? Why do all fermions feel the weak force? Like gravity, it seems to be universal. There are no fermions with weak charge zero, although a particle such as the $SO(10)$ particle, with no strong, electromagnetic, or weak interactions, could only be detected astronomically.

Are the quarks and leptons the final elementary particles or are they composite, made of "preons" or "rishons" or whatever?

Will colored particles appear or is color completely hidden? Can there be W or Z bosons with color?

Will heavier W and Z bosons coupling to right-handed fermions appear, i.e., is the ultimate theory a right–left symmetric theory, which is (spontaneously?) broken? In models, the right-handed bosons would have to have $M \geq 10$ TeV to agree with the measured limits on the ratio of ϵ'/ϵ in K^0 decay. Nonetheless, there are experimental ways of examining this hypothesis. CP violation in B decays is expected to be larger than CP violation in K^0 decays in the Standard Model and smaller for many left–right symmetric models. Electric dipole moments, to be discussed shortly, also may be relevant for this hypothesis.

Where does gravity and the cosmological constant fit into the picture? (Supergravity attempts to answer this.)

We have discussed some possible extensions to the Standard Model, such as larger gauge groups, technicolor, and supersymmetry. As of now, there are no firm indications that nature has chosen one of these paths, but experimentalists are searching hard for evidence for some extension to the Standard Model.

17.3 PRESENT EXPERIMENTAL CHALLENGES

1. Looking past the Standard Model

    (a) New particles. Many different kinds of new particles are predicted in various theories or suggested by various analogies. Many of them have been discussed in this text. Some examples of new particles are:

    (i) Preons. (If quarks are composite, these are quark constituents.)

    (ii) Higgs particles.

    (iii) Axions. (These are like light Higgs particles with masses in the eV range.)

    (iv) More Z's, W's. Are there heavy bosons coupling to right-handed fermions?

    (v) Supersymmetric particles. As noted in Section 12.3, some versions of the latest coupling constant extrapolations[215] don't seem to meet at a common point around 10^{15} GeV. Does this mean there is new physics between present energies and these high energies? A supersymmetric model has all three coupling constants meeting at about 10^{17} GeV.

    (vi) More generations (with heavy neutrinos).

    (vii) Heavy leptons in old generations. M. Perl and others have searched for these for a number of years at e^+e^- machines. There are few theoretical hints that these particles should exist, (Who ordered that?) but it is nonetheless interesting to search for them.

    (viii) Monopoles.

17.3 Present Experimental Challenges

(ix) Majorana neutrinos. As discussed in Sections 7.1 and 7.15, these are a special kind of neutrino in which particle and antiparticle are in the same four-component object. If present, it is possible neutrinoless double β-decay would occur. As noted in Section 7.15, searches for these decays have been vigorously pursued, so far with no success.

(b) New states of matter

  (i) Quark–gluon plasma. At very high density of matter, a new phase of matter, a kind of quark–gluon plasma might exist. People are searching for this phenomenon in the collisions of heavy atom ions at high energies. Dedicated runs have taken place at CERN and at Brookhaven with accelerated heavy ion beams. Brookhaven is building a large heavy ion colliding beam facility (RHIC).

  (ii) Hyperon nuclei. It is possible that quite heavy nuclei exist with many strange quarks in them.[216]

  (iii) Quark nuggets. It has been suggested[217][218] that almost macroscopic bodies consisting of quarks, not in the form of nucleons, might exist with enormous numbers of strange quarks in them. These can be looked for only in cosmic ray experiments and astronomical observations.

  (iv) Primordial black holes. Primordial black holes are black holes left over from shortly after the big bang. Black holes of less than 10^{15} g would have radiated away via Hawking radiation before now. Anything bigger than 10^6 solar masses would have been detected by seeing time-dependent distortions in star images (gravitational lensing effect) as the black holes passed across the field of view of stars. Black holes of about Jupiter mass are good dark matter candidates and could be the planet sized chunks seen in recent searches for gravitational lensing (MACHO search) events.

  (v) Cosmic strings. These strings were discussed in Section 16.4. They and other topological defects in the vacuum would have been formed soon after the big bang in many extensions of the Standard Model. They may have seeded the initial perturbations in the Universe needed for galaxy formation. Except for a possible gravitational lensing line in the sky, they are hard to detect now. They are too sparse (of the order of one within our horizon) to be a significant source of dark matter.

(c) Other forces, phenomena.

  (i) CP violation. This is still a very mysterious phenomenon. It is not understood why it is so weak. Is it due to some new "superweak" force?

(ii) Fifth, sixth...forces. E. Fischbach and coworkers[219] have pointed out that, until a few years ago, tests of the law of gravitation were incomplete and a new force with a range of a few to a few hundred meters was not yet strongly ruled out. Their remarks have stimulated a number of experiments. There is, as yet, no convincing evidence for a fifth kind of force.

(iii) HERA (ep collisions) opens a new era in very high Q^2 weak interaction observations. Is anything there?

2. Fixed target physics. This and the next points are interwoven with the preceding themes.

(a) Neutrino oscillation experiments and other lepton mixing tests. Neutrino oscillation tests are very high risk experiments, but quite interesting. Small neutrino masses appear naturally in many generalizations of the Standard Model and are suggested as an explanation of the solar neutrino puzzle. Neutrinos with masses in the few to 30 eV range could account for the dark matter problem. The observed ratio of ν_e/ν_μ at ground level from showers of particles produced by cosmic rays interacting in the earth's atmosphere disagrees with calculations by about a factor of 2. Neutrino oscillations are a possible explanation of this effect. A related question to neutrino oscillation is lepton number conservation, i.e., whether electron lepton number is exactly conserved. For example, can $\mu \to eee$? The branching ratio for this decay is measured to be below 10^{-12}. $\mu \to e\gamma$ is another decay mode of considerable interest. The present limit for that branching ratio is 4.9×10^{-11}.

(b) Further CP violation studies with kaons. It is very important to see if the "superweak" hypothesis can be ruled out or not.

(c) Rare K decays. These are also very high risk experiments. One is looking for small low-energy effects which are the tails of much higher energy things. Even positive results are not easy to interpret; nonetheless, a first indication of new processes might be seen in these decays.

(d) Polarization measurements in hadron scattering experiments. It is not clear if these are fundamental or "noise," but interesting patterns seem to appear. The results cannot yet be interpreted.

(e) Hard scattering. So far breaks in slopes of cross sections versus Q^2 have been interesting although they haven't been crucial for theoretical understanding.

(f) Mass spectrometer, chemical, etc., studies of material near p–p interaction regions at Fermilab or near the CERN target for nucleus-nucleus studies are worth performing. Is something new being produced? These are very high risk experiments.

17.3 Present Experimental Challenges   355

(g) Heavy nuclei study for quark–gluon plasma. It is hard to find definitive evidence ("a smoking gun") for a quark–gluon plasma or to use the plasma once it is found. Will it be fundamental or only an interesting sidelight?

3. Non-Accelerator Physics

   (a) Electric dipole moments of particles. These moments, (which violate T conservation), could be very useful clues to new physics and to the source of CP violation. The moments come from penguin diagrams such as that shown in Figure 17.1. In the Standard Model, the fact that both vertices marked "L" in the figure, are left-handed turns out to inhibit the process. In many Standard Model extensions, right-handed vertices occur, which enhance the size of the dipole moment and bring it near the present limits of observation. For many years there have been searches for neutron dipole moments.[220] Nuclei dipole moment searches are now approaching the neutron moment searches in sensitivity to new phenomena[221] and may well surpass them soon.

   (b) Search for dark matter. This is a very high risk, but very high reward effort.

   (c) Monopole studies. Many of the present theories have the possibility that magnetic monopoles might exist. There are several very interesting but very high risk experiments searching for monopoles in cosmic radiation. The predicted mass range is very high so that they would not be seen at accelerators. Nonetheless, experimenters wisely ignore that restriction and also search for monopoles at every accelerator around.

   (d) Solar neutrino studies. It is very important to understand what is going on with the solar neutrinos. This might be only a problem in solar calculations, but of the various strange results, it is one of the most promising. Even if it is a problem in solar calculations, clarifying this would be very important for astronomy. The new round of gallium experiments[117][116] and a future planned hydrogen and deuterium experiment[222] should, within the next few years, provide considerably more understanding of the problem.

   (e) General relativity tests and gravitational radiation tests. A beautiful experiment examined the variation of the period of a binary pulsar over a time span of about fifteen years. The pulsar consisted of two neutron stars in orbit about each other. The orbital speed is about 0.1% c for the pulsar measured. The experiment found[223] a variation of period in precise agreement with general relativity constituting an excellent new confirmation of the theory and providing the first observational evidence for the existence of

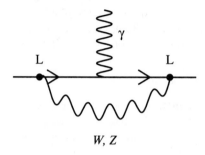

Figure 17.1 A type of diagram for a neutron dipole moment.

gravitational radiation. J. Taylor and R. Hulse were awarded the 1993 Nobel Prize for this work. Ambitious experiments are now under preparation to try to directly detect gravitational radiation.

(f) Very-high-energy cosmic ray photons from point sources. Some sources have now been seen. It is believed that for AGN's, a large fraction of all the energy that is emitted is in the form of photons of 50 MeV or more. These photons are inside of jets from the quasar and are emitted at a rate of perhaps 10^{45} ergs/sec. These photon observations explore general relativity in the strong field limit, i.e., near black holes of perhaps the order of 10^8 solar masses. It is possible that all of the extragalactic cosmic rays initiate in these objects. However, if sufficiently high energy cosmic rays exist, they cannot come from this source. At about $10^{19.7}$ eV, protons have enough energy to photoproduce the $\Delta(1232)$ resonance[224] by reacting with the cosmic $3K$ microwave background radiation. Depending on the intragalactic magnetic fields, cosmic rays above this limit originating from distant sources will not arrive at the earth. The cutoff should be at about 10^8 years of transit time from the source. The fly's eye experiment has seen one event at 3×10^{20} eV which is above this limit. What models can explain the enormous rates of high-energy particles from AGN's? What models can explain the presence of enormous energy cosmic rays?

(g) Gamma ray burst events. Bursts of gamma rays are seen from many point sources uniformly distributed over the sky.[225] These sources have a variability range between 0.2 ms and 1000 seconds. The energy of the gamma rays extends into the GeV range. It is not yet clear what objects produce these bursts or even whether these objects are close or distant.

4. Collider Physics

(a) CP violation in B decay. Several $\times 10^6$ B decays are needed.

(b) Precision measurements measurements of Standard Model parameters continue to be very useful.

(c) Search for Higgs particles. This is an ongoing search. It will be extended at LEP and the Fermilab collider. If not found in those searches, it will be the main effort at the LHC accelerator to either find the Higgs or find some evidence for deviations from the Standard Model.

(d) Top studies. The present Fermilab collider runs should give us further information on the top.

(e) Scan of energy for high-energy e^+e^- collisions to examine σ vs s. A search will also be made of the sphericity (jet-likeness) of the events. The search is for anything that might be around.

(f) Continue to examine rare Z decay modes. Are there some non-standard modes?

(g) Are there any hints of higher Z's–or any of the panoply of particles listed above such as supersymmetry partners, etc.?

(h) In LEP II one examines $e^+e^- \to W^+W^-$. This process tests a number of important elements of the standard model. It is very sensitive to the weak–electromagnetic mixing and allows a probe of anomalous WW interactions or resonances. Gauge invariance suggests these effects should be small at LEP II, but they must be searched for anyway. WW processes are also being searched for now via $\bar{p}p \to WW$.

(i) Further form factor studies, including studies of τ and quark form factors.

17.4 FINAL COMMENT

It is exceedingly likely that the crucial experiment will turn out to be none of the above experiments, but something simply unexpected. The most important thing to remember in experimental physics is to avoid tunnel vision. Do not concentrate so hard on the result you are trying to get that you ignore the accidents or the little things that do not quite fit in. The number of great discoveries that have come from accidents and unexpected apparent dirt effects is legion. The Davisson–Germer effect, CP violation, and the τ discovery are three that jump immediately to mind. Whatever you are doing always keep your mind open to the possibility of something unexpected. Good hunting!

APPENDIX A
Review of Lagrangians and Perturbation Theory; the Heisenberg and Interaction Pictures

A.1 Review of Classical Lagrangian and Hamiltonian Formalism

In classical (Newtonian) physics the Hamiltonian (H) is defined as

$$H = \Sigma p\dot{q} - L = fn(p_1, \ldots, p_n, q_1, \ldots, q_n, t), \tag{A.1}$$

where the independent variables are the p's, q's, and t. The p's and q's are called generalized momenta and generalized coordinates respectively. L is the Lagrangian where

$$L = \Sigma(T - V) = fn(q_1, \ldots, q_n, \dot{q}_1, \ldots, \dot{q}_n, t). \tag{A.2}$$

Here T and V are the kinetic and potential energies of the various particles and $\dot{q}_i$ is the time derivative of q_i. For the Lagrangian the independent variables are considered to be the $q's$, the $\dot{q}'s$ and t. One finds

$$p_i = \frac{\partial L}{\partial \dot{q}_i}; \quad \dot{p}_i = -\frac{\partial H}{\partial q_i}; \quad \dot{q}_i = \frac{\partial H}{\partial p_i}. \tag{A.3}$$

Lagrange's equations are

$$\frac{d}{dt}\left(\frac{\partial L}{\partial \dot{q}_i}\right) - \frac{\partial L}{\partial q_i} = 0. \tag{A.4}$$

For continuous coordinates $q_1, q_2, \cdots \to \phi(x)$ and a Lagrangian density, $\mathcal{L}$, is defined:

$$L = \int \mathcal{L} d^3x.$$

$\mathcal{L} = fn(\phi, \partial\phi/\partial x_\mu, x_\mu)$. The Euler–Lagrange equation in covariant notation with summation over μ understood is

$$\frac{\partial}{\partial x_\mu}\left(\frac{\partial \mathcal{L}}{\partial(\partial\phi/\partial x_\mu)}\right) - \frac{\partial \mathcal{L}}{\partial \phi} = 0. \tag{A.5}$$

The Lagrangian density for free Dirac particles is

$$\mathcal{L} = i\overline{\Psi}(\gamma^\mu \partial_\mu - m)\Psi. \tag{A.6}$$

A.2 Perturbation Theory; the Heisenberg and Interaction Pictures

In the Schrödinger picture in quantum mechanics the wavefunctions change with time, but the operators are usually fixed (unless there is an explicit time dependence, e.g., an external potential, which is explicitly time varying). In the Heisenberg picture, which is derived below, the wavefunctions are fixed and the operators change with time. The interaction representation is a compromise; the wavefunctions are fixed for the main part of the Hamiltonian and vary only due to the perturbation part. Start with the Schrödinger equation:

$$i\frac{\partial \Psi(t)}{\partial t} = H\Psi(t). \tag{A.7}$$

Now let $U(t, t_\circ)$ be an operator such that

$$\Psi(t) = U(t, t_\circ)\Psi(t_\circ). \tag{A.8}$$

Thus, U takes Ψ from $t_\circ$ to t. Divide this time interval into n small, equal parts, δt. From the Schrödinger equation:

$$\Psi(t + \delta t) = \Psi(t) + \delta\Psi,$$

where $\delta\Psi = (H\delta t/i)\Psi$. Thus,

$$\delta U = 1 + (H\delta t/i); \quad U = \lim_{(n\to\infty)}(\delta U)^n = \lim_{(n\to\infty)}(1 - iH\delta t)^n$$

$$\equiv e^{[-iH(t-t_\circ)]}.$$

This can be considered the *definition* of the exponential for operators. Thus,

$$U(t, t_\circ) = e^{[-i(t-t_\circ)H]}; \quad U^{-1} = U^\star = e^{[+i(t-t_\circ)H]}; \quad \Psi^\dagger(t) = \Psi^\star(t_\circ)U^{-1}(t, t_\circ). \tag{A.9}$$

Next let q be any observable, and let $Q_\circ$ be the Schrödinger and Q be the Heisenberg operators for q. The physical results must be the same in either picture. In the Schrödinger and Heisenberg pictures, respectively, one then has

$$\langle q \rangle = \int \Psi^\dagger(t) Q_\circ \Psi(t) d\tau = \int \Psi^*(t_\circ) Q \Psi(t_\circ) d\tau. \quad (A.10)$$

This shows one must have

$$\Psi^*(t_\circ) Q \Psi(t_\circ) = \Psi^*(t_\circ) U^{-1} Q_\circ U \Psi(t_\circ).$$

Thus, $Q = U^{-1} Q_\circ U$ gives the time dependence of Q if $Q_\circ$ is constant.

Let us try to write the equivalent of the Schrödinger equation in the Heisenberg picture:

$$i\frac{dQ}{dt} = i\frac{dU^{-1}}{dt} Q_\circ U + iU^{-1} Q_\circ \frac{dU}{dt} + iU^{-1} \frac{\partial Q_\circ}{\partial t} U,$$

$$i\frac{dQ}{dt} = -HU^{-1} Q_\circ U + U^{-1} Q_\circ U H + iU^{-1} \frac{\partial Q_\circ}{\partial t} U. \quad (A.11)$$

Define:

$$\frac{\partial Q}{\partial t} \equiv U^{-1} \frac{\partial Q_\circ}{\partial t} U. \quad (A.12)$$

This is the Heisenberg representation of the operator $\partial Q_\circ/\partial t$. Then,

$$i\frac{dQ}{dt} = [Q, H] + i\frac{\partial Q}{\partial t}, \quad (A.13)$$

where $[Q, H] = QH - HQ$ is called the *commutator* of Q and H. This is the Heisenberg equation of motion for any operator Q.

In the interaction picture, the wavefunctions are fixed for the main part of the Hamiltonian. Suppose an interaction can be considered as a small perturbation. Let $H = H_\circ + \epsilon H_{int}$ (ϵ is used to make visible the expansion parameter. In the literature it is often just incorporated into H_{int}):

$$i\frac{\partial \Psi(t)}{\partial t} = (H_\circ + \epsilon H_{int}) \Psi(t_\circ).$$

Analogously to the Heisenberg picture one can write

$$U_{H_\circ} = e^{-i(t-t_\circ) H_\circ},$$

$$\Psi(t) = U_{H_\circ}(t, t_\circ) \Phi(t, t_\circ).$$

Note that this reduces to the Heisenberg picture for the unperturbed but

A.2 Perturbation Theory; the Heisenberg

not for the full Hamiltonian:

$$\Phi(t) = e^{i(t-t_\circ)H_\circ}\Psi(t), \tag{A.14}$$

$$i\frac{\partial \Phi(t)}{\partial t} = i\frac{\partial}{\partial t}e^{i(t-t_\circ)H_\circ}\Psi(t) = -H_\circ\Phi + e^{i(t-t_\circ)H_\circ}H\Psi$$
$$= -H_\circ\Phi + H_\circ e^{i(t-t_\circ)H_\circ}\Psi + e^{i(t-t_\circ)H_\circ}\epsilon H_{int}e^{-i(t-t_\circ)H_\circ}e^{i(t-t_\circ)H_\circ}\Psi.$$

Let

$$H_{int}(t) = e^{i(t-t_\circ)H_\circ}H_{int}e^{-i(t-t_\circ)H_\circ}. \tag{A.15}$$

Then

$$i\frac{\partial \Phi(t)}{\partial t} = \epsilon H_{int}(t)\Phi(t). \tag{A.16}$$

This is known as the interaction representation[226] (Tomonaga and Schwinger). Note that if $H_{int} = 0$, Φ is constant.

The formal solution of the Φ equation is

$$\Phi(t) = U_I(t, t_\circ)\Phi(t_\circ). \tag{A.17}$$

Note that U_I and $U_{H_\circ}$ are different operators:

$$i\frac{\partial \Phi(t)}{\partial t} = i\frac{\partial U_I(t, t_\circ)}{\partial t}\Phi(t_\circ) = \epsilon H_{int}(t)U_I(t, t_\circ)\Phi(t_\circ); \quad U_I(t_\circ, t_\circ) = 1. \tag{A.18}$$

This is equivalent to

$$U_I(t, t_\circ) = 1 - i\int_{t_\circ}^{t} dt[\epsilon H_{int}(t)U_I(t, t_\circ)]. \tag{A.19}$$

Suppose the system is in state Φ_i at time $t_\circ$. Then the probability of finding it in Φ_f at time t is given by the intersection of Φ_f and the state which Φ_i has become by time t, i.e.,

$$|(\Phi_f, U_I(t,t_o)\Phi_i)|^2 \equiv |\langle f| U_I(t,t_o) |i\rangle|^2 \equiv |U_I(t,t_o)|_{fi}^2. \qquad (A.20)$$

The quantity $\langle f| U_I - 1 |i\rangle$ is called the transition amplitude for i goes to f. (This corresponds to removing the amplitude for staying in the same state.) If t_o approaches $-\infty$, that means that the beam is switched on and continuous and that the scattering is continuous. The wavefunctions are assumed normalized to one particle per unit volume.

Let w = the transition probability/unit time/unit volume and $t_o = -T$; $t = +T$:

$$w = \frac{1}{(t-t_o)\Delta V}|U_I(t,t_o)|_{fi}^2 \longrightarrow \lim_{(T\to\infty)} \frac{1}{2T\Delta V}|U_I(T,-T)|_{fi}^2. \qquad (A.21)$$

To find U_I, apply iteration, i.e., keep replacing U_I by the integral in the preceding equation:

$$U_I(t,-\infty) = 1 - i \int_{-\infty}^{t} dt'\epsilon H_{int}(t')$$

$$+ (-i)^2 \int_{-\infty}^{t} dt' \int_{-\infty}^{t'} dt'' \epsilon^2 H_{int}(t') H_{int}(t'') + \cdots. \qquad (A.22)$$

This is a perturbation series in powers of the (small) parameter ϵ. A useful further quantity is the S-matrix. In the interaction representation

$$S = U_I(\infty,-\infty) = \lim_{(t\to\infty)} U_I(t,-\infty). \qquad (A.23)$$

APPENDIX B
Proof of the Noether Theorem

The Noether Theorem states that every continuous symmetry of a Lagrangian implies that there is a conserved "current." To prove this theorem consider variations of an action S integrated between two invariant fixed points in space–time:

$$S = \int_{\tau_1}^{\tau_2} \mathcal{L} d^4 x. \tag{B.1}$$

If the action is invariant under the specific continuous variations chosen, then it will be shown that there is a conserved current. A variation in the Lagrangian $\mathcal{L}$ at a spacetime point P may involve a variation in the functional form and a variation in the spacetime coordinates of P since a change in the inertial frame will be allowed:

$$\delta \mathcal{L} \equiv \mathcal{L}'(x') - \mathcal{L}(x) = \mathcal{L}'(x) - \mathcal{L}(x) + \delta x_\mu \partial^\mu \mathcal{L} + \text{higher-order terms.} \tag{B.2}$$

Let $\delta_o \mathcal{L} \equiv \mathcal{L}'(x) - \mathcal{L}(x)$. Then,

$$\delta \mathcal{L} = \delta_o \mathcal{L} + \delta x_\mu \partial^\mu \mathcal{L}. \tag{B.3}$$

In addition to the change in $\mathcal{L}$, the change in $d^4 x$ must be considered. This is just given by the Jacobian:

$$\delta(d^4 x) = d^4 x \partial^\mu \delta x_\mu. \tag{B.4}$$

Thus,

$$\delta S = \int_{\tau_1}^{\tau_2} d^4 x [\delta_o \mathcal{L} + \delta x_\mu \partial^\mu \mathcal{L} + \mathcal{L} \partial^\mu \delta x_\mu]. \tag{B.5}$$

Note that by adding the second and third terms in this integral one obtains

$$\delta x_\mu \partial^\mu \mathcal{L} + \mathcal{L} \partial^\mu \delta x_\mu = \partial^\mu (\mathcal{L} \delta x_\mu). \tag{B.6}$$

Next suppose that $\mathcal{L}$ is a function of the wavefunction $\phi(x)$ and its derivatives. Imagine here that the variations of ϕ and of x are independent:

$$\delta_o \mathcal{L} = \frac{\partial \mathcal{L}}{\partial \phi} \delta_o \phi + \frac{\partial \mathcal{L}}{\partial (\partial^\mu \phi)} \delta_o \partial^\mu \phi, \tag{B.7}$$

where δ_o for ϕ is defined in an analogous manner to that for $\mathcal{L}$, i.e., just the functional change. Note that $\delta_o \partial^\mu \phi = \partial^\mu \delta_o \phi$.

The critical point is to now replace $\partial \mathcal{L}/\partial \phi$ by using the equations of motion:

$$\frac{\partial \mathcal{L}}{\partial \phi} - \partial^\mu \frac{\partial \mathcal{L}}{\partial (\partial^\mu \phi)} = 0, \tag{B.8}$$

$$\delta_o \mathcal{L} = \partial^\mu \left[\frac{\partial \mathcal{L}}{\partial (\partial^\mu \phi)} \right] \delta_o \phi + \frac{\partial \mathcal{L}}{\partial (\partial^\mu \phi)} \partial^\mu \delta_o \phi = \partial^\mu \left[\frac{\partial \mathcal{L}}{\partial (\partial^\mu \phi)} \delta_o \phi \right], \tag{B.9}$$

$$\delta S = \int_{\tau_1}^{\tau_2} d^4x \, \partial^\mu \left[\mathcal{L} \delta^\rho_\mu \delta x_\rho + \frac{\partial \mathcal{L}}{\partial (\partial^\mu \phi)} \partial^\mu \delta_o \phi \right]. \tag{B.10}$$

It is seen that there is almost a conserved current here if $\delta S = 0$. To finish the proof imagine that there are several global (i.e., x-independent) parameters, ω_a, of the transformation and that in terms of these parameters:

$$\delta x_\mu = \frac{\delta x_\mu}{\delta \omega_a} \delta \omega_a; \quad \delta \phi = \frac{\delta \phi}{\delta \omega_a} \delta \omega_a. \tag{B.11}$$

The functional change in ϕ was used for the variation, but the results are now recast using the total variation including the variation due to a change in x_μ. Recall that $\delta \phi = \delta_o \phi + \partial \phi / \partial x_\mu \delta x_\mu$. Thus, finally,

$$\delta S = \int_{\tau_1}^{\tau_2} d^4x \, \partial^\mu \left[\left(\mathcal{L} \delta^\rho_\mu - \frac{\partial \mathcal{L}}{\partial (\partial^\mu \phi)} \partial_\rho \phi \right) \frac{\delta x_\rho}{\delta \omega_a} + \frac{\partial \mathcal{L}}{\partial (\partial^\mu \phi)} \frac{\delta \phi}{\delta \omega_a} \right] \delta \omega_a. \tag{B.12}$$

If the action is invariant under the transformation for all $\delta \omega_a$ and for all τ_1, τ_2, then the current density

$$j^a_\mu = \left[\left(\mathcal{L} \delta^\rho_\mu - \frac{\partial \mathcal{L}}{\partial (\partial^\mu \phi)} \partial_\rho \phi \right) \frac{\delta x_\rho}{\delta \omega_a} + \frac{\partial \mathcal{L}}{\partial (\partial^\mu \phi)} \frac{\delta \phi}{\delta \omega_a} \right] \tag{B.13}$$

is conserved, i.e.,

$$\partial^\mu j^a_\mu = 0. \tag{B.14}$$

Thus, it is seen that symmetries imply conservation laws, a very fundamental result.

APPENDIX C
Clebsch–Gordan Coefficients

The question of how to combine angular momenta or spins together is examined here. That is, what mixture of states will give a final state that is an eigenstate with definite L^2 and L_z? Consider first a single particle. Recall:

$$L_x = -i(y\frac{\partial}{\partial z} - z\frac{\partial}{\partial y}); \quad L_y = -i(z\frac{\partial}{\partial x} - x\frac{\partial}{\partial z}); \quad L_z = -i(x\frac{\partial}{\partial y} - y\frac{\partial}{\partial x}).$$

The commutator of L_x and L_y is:

$$[L_x, L_y] \equiv L_x L_y - L_y L_x = -\left[\left(y\frac{\partial}{\partial z} - z\frac{\partial}{\partial y}\right)\left(z\frac{\partial}{\partial x} - x\frac{\partial}{\partial z}\right)\right.$$
$$\left. - \left(z\frac{\partial}{\partial x} - x\frac{\partial}{\partial z}\right)\left(y\frac{\partial}{\partial z} - z\frac{\partial}{\partial y}\right)\right],$$

$$[L_x, L_y] = \left(-y\frac{\partial}{\partial x} + x\frac{\partial}{\partial y}\right) = iL_z. \tag{C.1}$$

Similarly,

$$[L_y, L_z] = iL_x, \quad [L_z, L_x] = iL_y. \tag{C.2}$$

These commutation relations are the important relations needed to obtain the desired results. Operators with the preceding commutation relations are said to be generators of the group $SU(2)$, which is the group of the two-dimensional unitary matrices with determinant $= +1$. The results to be derived here will hold for any operators with the same commutation relations. Thus, they hold for J, S, and for the concept of isotopic spin.

Consider the total angular momentum, $L^2 \equiv L_x^2 + L_y^2 + L_z^2$:

$$[L^2, L_z] = [L_x L_x, L_z] + [L_y L_y, L_z] + [L_z L_z, L_z],$$

$$[L_x L_x, L_z] = L_x[L_x, L_z] + [L_x, L_z]L_x = -i[L_x L_y + L_y L_x].$$

Similarly, $[L_y L_y, L_z] = i[L_y L_x + L_x L_y]$. Thus, L^2 commutes with the operator for each component of angular momentum:

$$[L^2, L_z] = 0 = [L^2, L_x] = [L^2, L_y]. \tag{C.3}$$

Define

$$L^+ \equiv \frac{1}{\sqrt{2}}(L_x + iL_y) = \text{raising operator}, \tag{C.4}$$

$$L^- \equiv \frac{1}{\sqrt{2}}(L_x - iL_y) = \text{lowering operator}. \tag{C.5}$$

Note that $[L^2, L^+] = [L^2, L^-] = 0$. Thus a raising or lowering operator does not change the value of L^2. It is known from elementary quantum mechanics that the eigenvalues of L^2 are $l(l+1)$. If a wavefunction Ψ is an eigenfunction of L^2, then $L^+\Psi$ and $L^-\Psi$ are also eigenfunctions of L^2 with the same eigenvalue as the original wavefunction.

Consider

$$[L^+, L_z] = \frac{1}{\sqrt{2}}[(L_x + iL_y), L_z] = \frac{1}{\sqrt{2}}(-iL_y - L_x) = -L^+. \tag{C.6}$$

Suppose Ψ is an eigenfunction of both L^2 and L_z. For L_z, suppose the eigenvalue is m. Then,

$$L_z(L^+\Psi) = L^+(L_z\Psi) - [L^+, L_z]\Psi = (m+1)L^+\Psi. \tag{C.7}$$

Thus, $L^+\Psi$ is either an eigenvalue of L_z with eigenvalue $m+1$ or is zero. Similarly $L^-\Psi$ is an eigenfunction of L_z with eigenvalue $m-1$ or zero. This justifies the nomenclature of L^+ and L^- as raising and lowering operators.

L^+ gives a state with eigenvalue of L_z one higher unless Ψ has the highest possible m for the given l, in which case it gives zero. Similarly L^- gives the next lower state unless Ψ has the lowest possible m for the given l, in which case it gives zero. To find the normalizations, let

$$L^+\Psi(l, m) = C_+\Psi(l, m+1); \quad L^-\Psi(l, m+1) = C_-\Psi(l, m).$$

But

$$\langle \Psi(l, m+1) \mid L^+\Psi(l, m) \rangle = \langle \Psi(l, m) \mid L^-\Psi(l, m+1) \rangle^*,$$

since $(L^+)^\dagger = L^-$. This relation is clear once it is noted that L_x, L_y, and L_z are effectively real; the derivatives in the definitions operating on the energy momentum factor $e^{ip\cdot x}$, bring down an additional factor of i. It is perhaps useful here to realize that L_z $(= -i(x\frac{\partial}{\partial y} - y\frac{\partial}{\partial x})) = (xp_y - yp_x)$. Thus $C_+ = C_-^*$ if the Ψ are normalized to one. Next:

C. Clebsch-Gordan Coefficients

$$L^+L^- = \frac{1}{2}(L_x^2 + L_y^2 - i[L_x, L_y]) = \frac{1}{2}(L^2 - L_z^2 + L_z),$$

$$L^+L^-\Psi(l, m+1) = \frac{1}{2}(l(l+1) - m^2 - m)\Psi(l, m+1) = |C_+^2|\Psi(l, m+1),$$

$$|C_+| = \frac{1}{\sqrt{2}}\sqrt{l(l+1) - m(m+1)}. \tag{C.8}$$

Start with a wavefunction Ψ that is an eigenvalue of L^2 and L_z. Using the preceding relations, a connected set of wavefunctions that are eigenvalues of L^2 with all possible L_z eigenvalues can be obtained.

For two particles define $L^\pm = L_1^\pm + L_2^\pm$. These operators can easily be seen to be raising and lowering operators. (Specifically the lowering operator reduces m by one and keeps the eigenvalue of the operator L^2 constant.)

Consider a state $(l_1, m_1 = l_1), (l_2, m_2 = l_2)$ and apply the lowering operator to it. A state of definite total $(l, m) = (l_1 + l_2, l_1 + l_2 - 1)$ must result. By normalizing the state to one, the appropriate mixture is obtained. To find the state with $l = l_1 + l_2 - 1$ construct a state with $m_1 + m_2 = l_1 + l_2 - 1$, which is orthogonal to the state with the same m and $l = l_1 + l_2$. This is a state of pure $l = l_1 + l_2 - 1$ and the highest m. The operations above are then repeated. In this manner, one can eventually construct all of the wavefunctions.

As might be guessed the formulas can become very complicated. Fortunately there are tables of the coefficients of these mixtures. These coefficients are called Clebsch–Gordan coefficients. Figure C.1, taken from the Particle Data Book[17] tables, shows a useful set of these coefficients and the Y_m^l which go with them. The phases of Y_m^l and of Clebsch–Gordan coefficients vary from book to book. In this text, the conventions of the Particle Data Group, of Wigner[227] and of Condon and Shortley[228] are used. P_m^l is the Legendre polynomial.

To illustrate the uses of this table consider the box labeled 2×1. This means the addition of an $L = 2$ state and an $L = 1$ state. Consider the column labeled $\binom{2}{+1}$. This is the column for the wavefunction in the $L = 2, M = +1$ state. Realizing that square roots of the coefficients are to

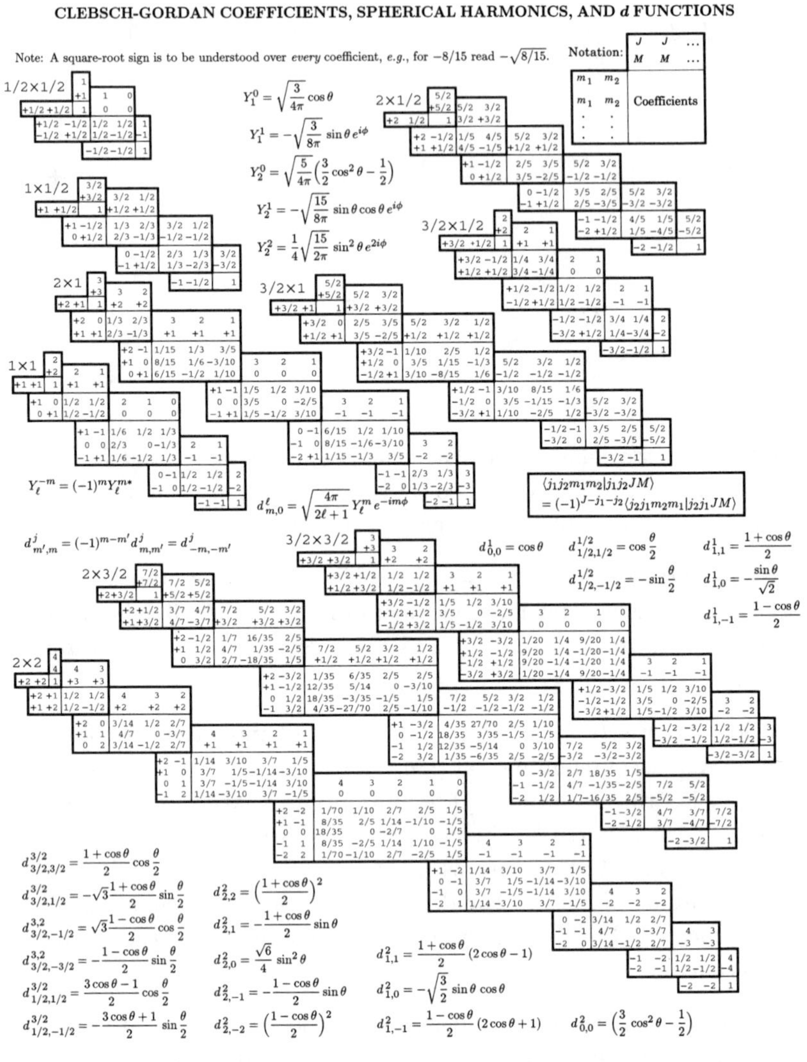

C. Clebsch-Gordan Coefficients

CLEBSCH-GORDAN COEFFICIENTS, SPHERICAL HARMONICS, AND d FUNCTIONS

Note: A square-root sign is to be understood over *every* coefficient, e.g., for $-8/15$ read $-\sqrt{8/15}$.

$$Y_1^0 = \sqrt{\frac{3}{4\pi}}\cos\theta$$

$$Y_1^1 = -\sqrt{\frac{3}{8\pi}}\sin\theta\, e^{i\phi}$$

$$Y_2^0 = \sqrt{\frac{5}{4\pi}}\left(\frac{3}{2}\cos^2\theta - \frac{1}{2}\right)$$

$$Y_2^1 = -\sqrt{\frac{15}{8\pi}}\sin\theta\cos\theta\, e^{i\phi}$$

$$Y_2^2 = \frac{1}{4}\sqrt{\frac{15}{2\pi}}\sin^2\theta\, e^{2i\phi}$$

$$Y_\ell^{-m} = (-1)^m Y_\ell^{m*}$$

$$d_{m,0}^\ell = \sqrt{\frac{4\pi}{2\ell+1}}\, Y_\ell^m\, e^{-im\phi}$$

$$\langle j_1 j_2 m_1 m_2 | j_1 j_2 JM\rangle = (-1)^{J-j_1-j_2}\langle j_2 j_1 m_2 m_1 | j_2 j_1 JM\rangle$$

$$d_{m',m}^j = (-1)^{m-m'} d_{m,m'}^j = d_{-m,-m'}^j$$

$$d_{0,0}^1 = \cos\theta \qquad d_{1/2,1/2}^{1/2} = \cos\frac{\theta}{2} \qquad d_{1,1}^1 = \frac{1+\cos\theta}{2}$$

$$d_{1/2,-1/2}^{1/2} = -\sin\frac{\theta}{2} \qquad d_{1,0}^1 = -\frac{\sin\theta}{\sqrt{2}}$$

$$d_{1,-1}^1 = \frac{1-\cos\theta}{2}$$

$$d_{3/2,3/2}^{3/2} = \frac{1+\cos\theta}{2}\cos\frac{\theta}{2}$$

$$d_{3/2,1/2}^{3/2} = -\sqrt{3}\,\frac{1+\cos\theta}{2}\sin\frac{\theta}{2}$$

$$d_{3/2,-1/2}^{3/2} = \sqrt{3}\,\frac{1-\cos\theta}{2}\cos\frac{\theta}{2}$$

$$d_{3/2,-3/2}^{3/2} = -\frac{1-\cos\theta}{2}\sin\frac{\theta}{2}$$

$$d_{1/2,1/2}^{3/2} = \frac{3\cos\theta-1}{2}\cos\frac{\theta}{2}$$

$$d_{1/2,-1/2}^{3/2} = -\frac{3\cos\theta+1}{2}\sin\frac{\theta}{2}$$

$$d_{2,2}^2 = \left(\frac{1+\cos\theta}{2}\right)^2$$

$$d_{2,1}^2 = -\frac{1+\cos\theta}{2}\sin\theta$$

$$d_{2,0}^2 = \frac{\sqrt{6}}{4}\sin^2\theta$$

$$d_{2,-1}^2 = -\frac{1-\cos\theta}{2}\sin\theta$$

$$d_{2,-2}^2 = \left(\frac{1-\cos\theta}{2}\right)^2$$

$$d_{1,1}^2 = \frac{1+\cos\theta}{2}(2\cos\theta - 1)$$

$$d_{1,0}^2 = -\sqrt{\frac{3}{2}}\sin\theta\cos\theta$$

$$d_{1,-1}^2 = \frac{1-\cos\theta}{2}(2\cos\theta + 1)$$

$$d_{0,0}^2 = \left(\frac{3}{2}\cos^2\theta - \frac{1}{2}\right)$$

Figure C.1 Clebsch–Gordan coefficients.[17]

be taken one has

$$|L=2,\ M=1\rangle = \sqrt{\frac{1}{3}}|m_1=2,\ m_2=-1\rangle + \sqrt{\frac{1}{6}}|m_1=+1, m_2=0\rangle$$

$$-\sqrt{\frac{1}{2}}|m_1=0,\ m_2=+1\rangle,$$

where $M = m_1 + m_2$.

Both the L, M and l_1, m_1, l_2, m_2 representations of the wavefunctions are orthogonal, normalized sets. This is then an orthogonal matrix transformation, i.e., a real unitary matrix transformation ($O\,O^T = O\,O^{transpose} = 1$) and the same table can be used for the inverse transformations. For the same 2×1 box consider the row marked $(+1,\ 0)$

$$|m_1=+1,\ m_2=0\rangle = \sqrt{\frac{8}{15}}|L=3,\ M=1\rangle + \sqrt{\frac{1}{6}}|L=2,\ M=1\rangle$$

$$-\sqrt{\frac{3}{10}}|L=1,\ M=1\rangle.$$

Note that in each case the sum of the squares of the coefficients equals one, which is necessary since one is transforming between two representations of normalized states.

APPENDIX D
Generators for $SU(3)$

$SU(3)$ is the symmetry group of three-dimensional unitary matrices with determinant one. It is needed for describing color, and useful for quark models. The generators of the infinitesimal rotations can be taken to be the eight independent, traceless λ matrices (Gell-Mann matrices) given below:

$$\lambda_1 = \begin{pmatrix} 0 & 1 & 0 \\ 1 & 0 & 0 \\ 0 & 0 & 0 \end{pmatrix}, \quad \lambda_2 = \begin{pmatrix} 0 & -i & 0 \\ i & 0 & 0 \\ 0 & 0 & 0 \end{pmatrix},$$

$$\lambda_3 = \begin{pmatrix} 1 & 0 & 0 \\ 0 & -1 & 0 \\ 0 & 0 & 0 \end{pmatrix}, \quad \lambda_4 = \begin{pmatrix} 0 & 0 & 1 \\ 0 & 0 & 0 \\ 1 & 0 & 0 \end{pmatrix},$$

$$\lambda_5 = \begin{pmatrix} 0 & 0 & -i \\ 0 & 0 & 0 \\ i & 0 & 0 \end{pmatrix}, \quad \lambda_6 = \begin{pmatrix} 0 & 0 & 0 \\ 0 & 0 & 1 \\ 0 & 1 & 0 \end{pmatrix},$$

$$\lambda_7 = \begin{pmatrix} 0 & 0 & 0 \\ 0 & 0 & -i \\ 0 & i & 0 \end{pmatrix}, \quad \lambda_8 = \frac{1}{\sqrt{3}} \begin{pmatrix} 1 & 0 & 0 \\ 0 & 1 & 0 \\ 0 & 0 & -2 \end{pmatrix}.$$

(D.1)

The matrices $\lambda_i/2$ satisfy the commutation relations that characterize the group just as $\tau_i/2$ do for $SU(2)$:

$$\left[\frac{\lambda_i}{2}, \frac{\lambda_j}{2}\right] = i \sum_k f_{ijk} \frac{\lambda_k}{2}. \tag{D.2}$$

The f_{ijk} are known as the structure constants of the group. They are anti-symmetric under the interchange of any two indices. Only a few of the values of f_{ijk} are nonvanishing. Table D.1 lists the nonzero ones whose

Table D.1 Nonzero elements of f_{ijk}.

(ijk)	f_{ijk}
123	1
147	$\frac{1}{2}$
156	$-\frac{1}{2}$
246	$\frac{1}{2}$
257	$\frac{1}{2}$
345	$\frac{1}{2}$
367	$-\frac{1}{2}$
458	$\frac{\sqrt{3}}{2}$
678	$\frac{\sqrt{3}}{2}$

indices are in ascending order. The other nonzero ones are obtained using the anti-symmetry under interchange of any pair of indices.

APPENDIX E
Feynman Rules and Calculation of Matrix Elements

In this text the metric (E, p_x, p_y, p_z) is used. $\not{p} \equiv p_\nu \gamma^\nu = E\gamma^0 - p_x\gamma^1 - p_y\gamma^2 - p_z\gamma^3$. $\not{\partial} \equiv \partial_\mu \gamma^\mu = \partial/\partial t \gamma^0 - \partial/\partial x \gamma^1 - \partial/\partial y \gamma^2 - \partial/\partial z \gamma^3$.

For the present discussion the unitary gauge and Coulomb gauge will be used. The propagators for various spin particles can then be written as

$$\text{Spin 0} \quad \frac{i}{q^2 - m^2 + i\epsilon}, \tag{E.1}$$

$$\text{Spin 1/2} \quad \frac{i(\not{p} + m)}{p^2 - m^2 + i\epsilon}, \tag{E.2}$$

$$\text{Photon (Massless Spin 1)} \quad \frac{-ig^{\mu\nu}}{q^2 + i\epsilon}, \tag{E.3}$$

$$\text{Massive Spin 1} \quad \frac{i(-g^{\mu\nu} + q^\mu q^\nu/m^2)}{q^2 - m^2 + i\epsilon}. \tag{E.4}$$

Here the $i\epsilon$ in the denominator (where ϵ is an infinitesimal) is needed to maintain causality and to determine which sheet to use when integrals in the complex plane are taken. (They are not required for calculations in this text.)

For an initial or final spin 1/2 particle, an appropriate u, v, $\bar{u}$, or $\bar{v}$ must be inserted. The spinors are given explicitly by

$$u_r = \sqrt{E+m} \begin{pmatrix} \chi_r^u \\ \frac{\sigma \cdot p}{E+m} \chi_r^u \end{pmatrix} \text{ where } \chi_1^u = \begin{pmatrix} 1 \\ 0 \end{pmatrix} \text{ and } \chi_2^u = \begin{pmatrix} 0 \\ 1 \end{pmatrix}, \tag{E.5}$$

$$v_r = \sqrt{E+m} \begin{pmatrix} \frac{\sigma \cdot p}{E+m} \chi_r^v \\ \chi_r^v \end{pmatrix} \text{ where } \chi_1^v = \begin{pmatrix} 0 \\ -1 \end{pmatrix} \text{ and } \chi_2^v = \begin{pmatrix} 1 \\ 0 \end{pmatrix}, \tag{E.6}$$

where $r = 1$ for spin up and $r = 2$ for spin down. u and v have dimensions of $E^{1/2}$.

E. Feynman Rules and Calculation of Matrix Elements

Table E.1 Use of u and v in currents.

	fermion	anti-fermion
initial state	u on right	$\bar{v}$ on left
final state	$\bar{u}$ on left	v on right

For a spin 1 particle, the polarization vector ϵ must be included for initial-state particles and ϵ^* for final-state particles. ϵ is a four-vector that is defined to have unit size ($\epsilon \cdot \epsilon = 1$), and to have $q \cdot \epsilon = 0$. Here the four-dimensional scalar product is to be understood, and q is the momentum of the particle. With this definition, there are three independent polarization vectors. For massive vector bosons, if the particle momentum is in the three direction, one possible set of three polarizations is: $(0,1,0,0)$, $(0,0,1,0)$, $(|q|/m, 0, 0, E/m)$. For this choice it is easily verified that the sum over the three polarization vectors gives

$$\sum(\epsilon^{*\mu}\epsilon^{\nu}) = -g^{\mu\nu} + q^{\mu}q^{\nu}/m^2. \tag{E.7}$$

This is true using any set of three orthogonal polarization vectors. Note that the longitudinal polarization involves an extra power of energy. Because of this, the longitudinal bosons are most susceptible to high-energy behavior problems.

For bosons, momenta are taken as positive if the bosons are entering a vertex and minus the physical momenta if they are leaving the vertex. However, for the preceding polarization vectors, the physical momenta are used. For fermions, momenta are in the direction of the fermion arrow. Thus, if the arrow is in the negative direction of time, the momenta are minus the physical momenta.

Next the vertex functions will be examined. Basically, they can be obtained from the interaction terms in the Lagrangian, but a number of modifications must be made and conventions observed.

* The signs of boson charges are taken as given in the Lagrangian if the particles are entering the vertex, and the anti-particles taken if they are leaving the vertex.
* If the derivative (∂_μ) of a particle appears in the Lagrangian, then iq_μ should appear in the matrix element where q is the momentum of the particle with the above sign convention.
* An overall factor of "i" should be added. This basically comes from the perturbation expansion of $e^{i\mathcal{L}}$.

Table E.2 Some electroweak vertex factors.[229]

Vertex	Vertex factor
$\bar{e}e\gamma$	$-ie\gamma^\mu$
$\bar{\nu}\nu Z_\mu$	$(ig/2\cos\theta_W)\gamma^\mu(1-\gamma^5)/2$
$\bar{l}\nu W_\mu^-$	$(ig/\sqrt{2})\gamma^\mu(1-\gamma^5)/2$
$\bar{l}lH$	$-igm_l/(2M_W)$
$Z^\mu(p_1)(W^+)^\nu(p_2)(W^-)^\lambda(p_3)$	$-ig\cos\theta_W[(p_1-p_2)^\lambda g^{\mu\nu}$ $+(p_2-p_3)^\mu g^{\nu\lambda}+(p_3-p_1)^\nu g^{\lambda\mu}]$
$(W^+)^\mu(W^-)^\lambda(W^+)^\nu(W^-)^\rho$	$ig^2 S^{\mu\nu\lambda\rho}$
$(W^+)^\mu(W^-)^\nu HH$	$(ig^2/2)g^{\mu\nu}$
$Z^\mu Z^\nu(W^-)^\lambda(W^+)^\rho$	$-ig^2\cos^2\theta_W S^{\mu\nu\lambda\rho}$
$H(W^+)^\mu(W^-)^\nu$	$igM_W g^{\mu\nu}$
$HZ^\mu Z^\nu$	$igM_Z g^{\mu\nu}/\cos\theta_W$

* If there are n identical particles at the vertex, a factor of $n!$ is added. Thus, the ZZW^+W^- vertex has a factor of 2 and the $ZZHH$ vertex has a factor of four.
* If there are identical particles, the vertex function must be written in a symmetrized form for them. Thus, $W_\nu^- W_\nu^- W_\beta^+ W_\beta^+ - W_\nu^- W_\beta^+ W_\beta^- W_\nu^+$ becomes $(1/2)(2W_\nu^- W_\nu^- W_\beta^+ W_\beta^+ - W_\nu^- W_\beta^- W_\beta^+ W_\nu^+ - W_\beta^- W_\nu^- W_\beta^+ W_\nu^+)$.

The resulting vertex factors for a number of vertices in the electroweak sector of the theory are given in Table E.2. $S^{\mu\nu\lambda\rho} = 2g^{\mu\nu}g^{\lambda\rho} - g^{\mu\lambda}g^{\nu\rho} - g^{\mu\rho}g^{\nu\lambda}$.

Note that three-particle vertices are multiplied by a coupling constant and four-body vertices by a coupling constant squared. Closed loops are to be integrated over $[d^4p/(2\pi)^4])$ and there is a minus sign for each closed fermion loop. There is a factor (-1) between graphs differing from each other only by an interchange of two identical external fermion lines. There also is a symmetry factor S^{-1}, where S is the number of ways that one can reorient a given diagram without changing the topology. Finally, the net result of writing down the Feynman diagrams is equal to iM.

E. Feynman Rules and Calculation of Matrix Elements

Suppose a matrix element includes fermions. Then averaging over initial and summing over final states, one can show (see Section 7.5) that a matrix element of the form $M = \bar{u}Ou$ becomes

$$|M|^2_{\text{ave+sum}} = (1/2)\text{Tr}[O(\not{p}_i + m_i)\gamma^0 O^\dagger \gamma^0(\not{p}_f + m_f)]. \tag{E.8}$$

Some useful identities can be established for traces of gamma matrices. From the relation $\gamma^\mu\gamma^\nu + \gamma^\nu\gamma^\mu = 2Ig^{\mu\nu}$, it is seen that $\not{p}\not{q} + \not{q}\not{p} = 2p\cdot q$. If $p = q$, then $\not{p}\not{p} = p^2$. A trace involving γ^5 and less than four gamma matrices is zero. A trace not involving γ^5, but having an odd number of gamma matrices is zero. $\text{Tr}(\not{a}\not{b}) = 4a\cdot b$; $\text{Tr}(\not{a}\not{b}\not{c}\not{d}) = 4[(a\cdot b)(c\cdot d) - (a\cdot c)(b\cdot d) + (a\cdot d)(b\cdot c)]$; and $\text{Tr}(\gamma^5\gamma^\mu\gamma^\lambda\gamma^\sigma\gamma^\rho) = 4i\epsilon^{\mu\lambda\sigma\rho}$, where $\epsilon^{\mu\lambda\sigma\rho}$ is the completely anti-symmetric tensor with $\epsilon^{1230} = 1$.

If $M = \bar{u}_l(k')\gamma^\alpha u_l(k)$ (the electromagnetic interaction), then the above trace is

$$(L^{em})^{\alpha\beta} = 4[k^\alpha k'^\beta + k'^\alpha k^\beta - g^{\alpha\beta}(k\cdot k' - mm')]. \tag{E.9}$$

If $M = \bar{u}_l(k')\gamma^\alpha[(1-\gamma^5)/2]u_\nu(k)$ (the charged weak interaction), then the above trace is

$$(L^{weak\ CC})^{\alpha\beta} = 2[k^\alpha k'^\beta + k'^\alpha k^\beta - g^{\alpha\beta}(k\cdot k') + i\epsilon^{\alpha\beta\eta\xi}k'_\eta k_\xi]. \tag{E.10}$$

For both the electromagnetic and the weak trace, the same result is obtained if a u is replaced by v. The $m_l m_\nu$ term does not occur in the weak trace even if the mass of the neutrino is not zero.

Given the square of the matrix element, the transition probability is (Section 5.4)

$$d\omega_{\beta\alpha} = \frac{(2\pi)^4 \delta^4(p_\alpha - p_\beta)}{\Pi_\alpha(2E_\alpha)\Pi_\beta(2E_\beta)}|M_{\alpha\beta}|^2 \Pi_\beta \frac{d^3p_\beta}{(2\pi)^3}, \tag{E.11}$$

where α is the set of initial state particles and β is the set of final state particles. In terms of the relativistic phase space, $d\omega_{\beta\alpha}$ is defined as

$$d\omega_{\beta\alpha} = (2\pi)^4|M_{\beta\alpha}|^2/\Pi_\alpha(2E_\alpha)) \times \text{relativistic phase space}, \tag{E.12}$$

where the relativistic phase space is

$$\text{relativistic phase space} \equiv P.S. \equiv \Pi_{\beta_i} \frac{d^3p_{\beta_i}}{(2\pi)^3 2E_{\beta_i}} \delta^4(p_{\beta_1} + p_{\beta_2} + \cdots + p_{\beta_n} - p). \tag{E.13}$$

For a decaying particle, when integrated over the final-state momenta β,

$$\omega_{\beta\alpha} \equiv 1/\tau_\alpha \equiv \Gamma_\alpha, \tag{E.14}$$

where τ_α is the lifetime and Γ_α is the decay width of the particle.

For a scattering problem, the cross section is found from the transition probability:

$$d\sigma_{\beta\alpha} = d\omega_{\beta\alpha}\frac{E_a E_b}{F_{ab}}, \tag{E.15}$$

where

$$\frac{F_{ab}}{E_a E_b} = \frac{\sqrt{(p_a \cdot p_b)^2 - (m_a m_b)^2}}{E_a E_b}. \tag{E.16}$$

APPENDIX F
$W^+ + W^- \to W^+ + W^-$ and a General Theorem for Restrictions to the Standard Model Lagrangian

F.1 $W^+ + W^- \to W^+ + W^-$ SCATTERING

The reactions considered in Sections 11.11 and 11.12 did not require Higgs couplings to eliminate high-energy-behavior problems. However, consider the reaction $W^+ + W^- \to W^+ + W^-$. Consider first the set of diagrams shown in Figure F.1 (a) and (b) involving s-channel Z and γ exchange for Figure F.1 (a) and t-channel exchange for Figure F.1 (b). The matrix elements can be obtained using the methods described above giving:

$$M_a = (-i)\epsilon_\mu^\star(p_3)\epsilon_\nu^\star(p_4)\epsilon_\alpha(p_1)\epsilon_\beta(p_2)$$
$$[g^{\alpha\beta}(p_1 - p_2)^\lambda + g^{\beta\lambda}(p_2 + q)^\alpha + g^{\alpha\lambda}(-q - p_1)^\beta]$$
$$[g^{\mu\nu}(-p_3 + p_4)^\kappa + g^{\nu\kappa}(-p_4 - q)^\mu + g^{\mu\kappa}(q + p_3)^\nu]$$
$$\left[(ie)^2 \frac{-ig_{\lambda\kappa}}{q^2} + (ig\cos\theta_W)^2 \frac{-i(g_{\lambda\kappa} + q_\lambda q_\kappa/M_Z^2)}{q^2 - M_Z^2}\right], \quad \text{(F.1)}$$

$$M_b = (-i)\epsilon_\mu^\star(p_3)\epsilon_\nu^\star(p_4)\epsilon_\alpha(p_1)\epsilon_\beta(p_2)$$
$$[g^{\alpha\mu}(p_1 + p_3)^\lambda + g^{\mu\lambda}(-p_3 + k)^\alpha + g^{\lambda\alpha}(-k - p_1)^\mu]$$
$$[g^{\nu\beta}(-p_4 - p_2)^\kappa + g^{\beta\kappa}(p_2 - k)^\nu + g^{\kappa\nu}(k + p_4)^\beta]$$
$$\left[(ie)^2 \frac{-ig_{\lambda\kappa}}{k^2} + (ig\cos\theta_W)^2 \frac{-i(g_{\lambda\kappa} + k_\lambda k_\kappa/M_Z^2)}{k^2 - M_Z^2}\right]. \quad \text{(F.2)}$$

Consider longitudinally polarized initial- and final-state W's, since these are the most singular states. The center-of-mass system will be assumed. E will be the energy of any of the W's in that system and θ will be the scattering angle between the initial and final W^+. The matrix elements will then be expanded keeping the constant and higher power of energy terms.

F. $W^+ + W^- \to W^+ + W^-$ and a General Theorem

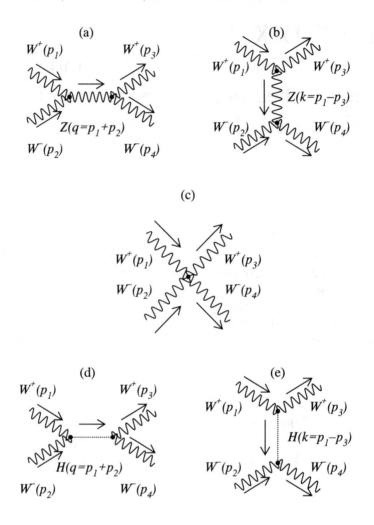

Figure F.1 $W^+ + W^- \to W^+ + W^-$ Scattering. (a) s-channel diagram with Z, γ exchange; (b) t-channel diagram with Z, γ exchange; (c) four W coupling diagram; (d) s-channel diagram with H exchange; (e) t-channel diagram with H exchange.

The evaluation is straightforward, but a bit tedious. Terms in $1/\gamma^2$ come from the difference between p and E in both numerator and propagator, and from the M_Z term in the propagator denominator. The elements are most easily evaluated using one of the several computer programs now available for this purpose. If $\sqrt{s}$ is the total center-of-mass energy ($=2E$), then the matrix elements can be put in the form

$$M = A_2 s^2 + A_1 s + A_0 + O(1/s).$$

For the above elements the result is:

$$A_2^a = \frac{-g^2 \cos\theta}{4M_W^4},$$

$$A_1^a = \frac{-g^2 \cos\theta}{4M_W^2},$$

$$A_0^a = \frac{g^2 \cos\theta}{4\cos^2\theta_W}(5 + 6\cos 2\theta_W), \qquad (F.3)$$

$$A_2^b = \frac{g^2}{32M_W^4}(5 - 4\cos\theta - \cos 2\theta),$$

$$A_1^b = \frac{3g^2}{8M_W^2}(-1 + 5\cos\theta),$$

$$A_0^b = \frac{g^2}{32\sin^2(\theta/2)\cos^2\theta_W}[12 + 4\cos\theta - 8\cos^2\theta$$
$$- 32\cos\theta\cos^2\theta_W + 40\cos^2\theta\cos^2\theta_W]. \qquad (F.4)$$

Not even the highest-order (A_2) term vanishes for the sum of the two diagrams. $M \propto E^4$ and strongly violates the unitarity condition. Something else must be added. In the Standard Model there is a $4W$ vertex. Perhaps that will help. Consider the diagram shown in Figure F.1 (c).

$$M_c = (-i)\epsilon_\mu^*(p_3)\epsilon_\nu^*(p_4)\epsilon_\alpha(p_1)\epsilon_\beta(p_2)(ig^2)(2g^{\mu\beta}g^{\alpha\nu} - g^{\mu\alpha}g^{\nu\beta} - g^{\mu\nu}g^{\alpha\beta}). \qquad (F.5)$$

For M_c one obtains

$$A_2^c = \frac{g^2}{32M_W^4}(-5 + 12\cos\theta + \cos 2\theta),$$

$$A_1^c = \frac{g^2}{2M_W^2}(1 - 3\cos\theta),$$

$$A_0^c = 0. \qquad (F.6)$$

This diagram has certainly helped. $A_2^a + A_2^b + A_2^c = 0$. The worst behaved terms have cancelled. However, the A_1 terms do not cancel and $M \propto E^2$, which still violates the unitarity condition of Chapter 11. Something more is needed. In the discussion in Chapter 11, spin 1/2 and spin 1 particles were introduced and their couplings fixed. It was indicated there that

F. $W^+ + W^- \to W^+ + W^-$ and a General Theorem

spins higher than 1 would not be used because of other serious divergence problems with those fields. This leaves spin 0. It is necessary to introduce the Higgs particles. Consider the diagrams in Figure F.1 (d) and (e). The matrix elements for these diagrams can be obtained with the preceding prescription:

$$M_d = (-i)\epsilon_\mu^*(p_3)\epsilon_\nu^*(p_4)\epsilon_\alpha(p_1)\epsilon_\beta(p_2)(igM_W)^2 g^{\alpha\beta}\frac{i}{q^2-M_H^2}g^{\mu\nu}, \qquad \text{(F.7)}$$

$$M_e = (-i)\epsilon_\mu^*(p_3)\epsilon_\nu^*(p_4)\epsilon_\alpha(p_1)\epsilon_\beta(p_2)(igM_W)^2 g^{\alpha\mu}\frac{i}{k^2-M_H^2}g^{\beta\nu}. \qquad \text{(F.8)}$$

After evaluation, one obtains

$$\begin{aligned} A_2^d &= 0, \\ A_1^d &= \frac{-g^2}{4M_W^2}, \\ A_0^d &= g^2\left(1 - \frac{M_H^2}{4M_W^2}\right), \end{aligned} \qquad \text{(F.9)}$$

$$\begin{aligned} A_2^e &= 0, \\ A_1^e &= \frac{-g^2}{8M_W^2}(1-\cos\theta), \\ A_0^e &= \frac{-g^2}{2}\left(1 + \frac{M_H^2}{2M_W^2} + \cos\theta\right). \end{aligned} \qquad \text{(F.10)}$$

Summing up all five of the matrix elements up, one obtains

$$\begin{aligned} A_2 &= 0, \\ A_1 &= 0, \\ A_0 &= \frac{-g^2 M_H^2}{2M_W^2} + \frac{g^2(7+\cos 2\theta)}{16\cos^2\theta_W \sin^2(\theta/2)}. \end{aligned} \qquad \text{(F.11)}$$

Both the A_2 and A_1 terms now are zero and the matrix element no longer has bad high-energy behavior. In order to accomplish this cancellation and that in Section 11.2, it was necessary to have the Standard Model particles and their couplings.

One can further note that if the mass of the Higgs particle is too high, then the cancellation, which occurs only when $q^2 \gg M_H^2$, fails and unitarity is violated before the cancellation mechanism can take effect. This can be viewed as putting an upper limit on the Higgs mass. This and similar calculations indicate the Higgs mass should be lower than about 1 TeV.

F.2 THE GENERAL THEOREM

In this section an outline of the proof of the general theorem will be given. Proofs of many of the statements will be omitted, but they are mostly generalizations of the calculations shown for the Standard Model in the preceding section and in Section 11.12. For this general treatment of the restrictions that can be put on the form of the Lagrangian, the treatment of Llewellyn Smith[230] will be followed for the most part, with additions from the discussion of Cornwall et al.[231]

Represent all fundamental fermions by a vector:

$$\psi = \begin{pmatrix} \nu_e \\ e \\ \nu_\mu \\ \mu \\ \cdot \\ \cdot \end{pmatrix}. \tag{F.12}$$

Suppose there are g vector mesons represented by real fields W_μ^i ($i = 1, \cdots, g$). For the moment suppose all the vector mesons are massive. The most general Lagrangian for the interaction part of the $f\bar{f} \to WW$ interaction can then be written as

$$\mathcal{L}_1 = \bar{\psi}_\alpha \gamma^\mu \left[(L^i)^{\alpha\beta} \left(\frac{1-\gamma^5}{2} \right) + (R^i)^{\alpha\beta} \left(\frac{1+\gamma^5}{2} \right) \right] \psi_\beta W_\mu^i, \tag{F.13}$$

where L and R are Hermitian matrices. As usual a sum over repeated indices is implied.

A self-interaction must exist, since it is known, for example, that there is a $W^+W^-\gamma$ vertex. The most general form satisfying the preceding assumptions is

$$\mathcal{L}_2 = \frac{1}{2} D_{ij,k} W_\mu^k (W_\alpha^j \partial^\mu (W^i)^\alpha - W_\alpha^i \partial^\mu (W^j)^\alpha) + \frac{1}{2} D_{ij,k}^+ W_\mu^k (W_\alpha^j \partial^\mu (W^i)^\alpha$$

F. $W^+ + W^- \to W^+ + W^-$ and a General Theorem

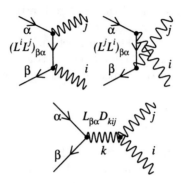

Figure F.2 Relevant Feynman diagrams for $\bar{f}f \to WW$.

$$+ W_\alpha^i \partial^\mu (W^j)^\alpha) + \frac{1}{2} G_{ij,k} \epsilon^{\alpha\beta\gamma\delta} W_\alpha^i W_\beta^j \partial_\gamma W_\delta^k + \frac{1}{4} C_{ij,kl} W_\mu^i (W^j)^\mu W_\nu^k (W^l)^\nu$$
$$+ \frac{1}{4} S_{ijkl} \epsilon^{\alpha\beta\gamma\delta} W_\alpha^i W_\beta^j W_\gamma^k W_\delta^l. \quad \text{(F.14)}$$

Here the couplings are assumed real and appropriately symmetrized and anti-symmetrized.

Consider $WW \to WW$. It can be shown that necessary and sufficient conditions to remove most of the bad terms from $\mathcal{L}_2$ are

$$D^+ = S = G = 0; \quad D_{ij,k} = -D_{ik,j} (\equiv D_{ikj});$$
$$D_{kac} D_{kbd} - D_{kab} D_{kcd} - D_{kad} D_{kbc} = 0;$$
$$2 C_{ij,kl} = D_{ilp} D_{kjp} + D_{ikp} D_{ljp}. \quad \text{(F.15)}$$

However, there are still terms in the matrix element that are proportional to E^2 when all four W's are longitudinal and proportional to E when only three are longitudinal. The relation involving the three pairs of D_{abc} may be recognized as a Jacobi identity, and the above already specifies a Yang–Mills theory, i.e., a theory with a gauge symmetry.

Consider next $\mathcal{L}_1$, the $\bar{f}f \to WW$ interaction. Since the $3W$ vertex has a Yang–Mills form, it can be shown that the leading E^2 pieces here cancel only if the coupling constants represent a "Lie algebra," i.e.,

$$[L_i, L_j] = i D_{ijk} L_k; \quad [R_i, R_j] = i D_{ijk} R_k. \quad \text{(F.16)}$$

The relevant Feynman diagrams and the magnitudes of the couplings involved in this cancellation are shown in Figure F.2 for left-handed leptons. Non-leading ($\propto E$) terms remain unless all fermions are massless or all fermions in a given irreducible multiplet are degenerate and parity is conserved. Nature has not chosen this path. Therefore, to cancel out the remaining bad terms still more particle exchanges must be added. Since general interactions for particles of spin 1/2 and 1 have been included and since spins of greater than 1 are excluded in the assumptions, spin 0 particles must be introduced. Thus to avoid bad high-energy behavior, Higgs particles must be introduced into the theory. Introduce N real scalar particle fields, $\phi^i (i = 1, \cdots, N)$. The most general $f - \phi - W$ interaction is

$$\mathcal{L}_3 = \bar{\psi} \left[C^b_{\alpha\beta} \left(\frac{1+\gamma^5}{2} \right) + C^{+b}_{\alpha\beta} \left(\frac{1-\gamma^5}{2} \right) \right] \psi_\beta \phi^b$$
$$+ \frac{1}{2} W^i_\mu (W^j)^\mu \phi^b K^b_{ij} + \frac{1}{2} T^i_{ba} W^i_\mu (\phi^a \partial^\mu \phi^b - \phi^b \partial^\mu \phi^a)$$
$$+ \frac{1}{4} M^{ij}_{ab} W^i_\mu (W^j)^\mu \phi^a \phi^b + H^i_{ab} W^i_\mu (\phi^a \partial^\mu \phi^b + \phi^b \partial^\mu \phi^a). \quad \text{(F.17)}$$

Again the couplings in the preceding equation are assumed to be real and to be appropriately symmetrized and anti-symmetrized. H^i_{ba} is effectively equivalent to a contact term with inverse mass dimensions. Terms with inverse mass dimensions have been assumed to be absent. It will be assumed that the H^i_{ab} term is absent here although it is difficult to prove this is a necessary condition. It is a sufficient condition to ensure good behavior for $\phi\phi \to \phi\phi$ and $\phi\phi \to \phi W$.

Given this assumption the necessary and sufficient further conditions to guarantee the good behavior of all of the reactions $f\bar{f} \to WW$, $WW \to WW$, $\phi\phi \to WW$, $\phi W \to WW$, and $f\bar{f} \to \phi W$ can then be derived. Let M_i be the vector meson masses and m_j be the fermion masses. Define A, B, and X by

$$2M_c B^a_{ci} = -K^i_{ac}; \quad 2M_t M_c A^d_{tc} = D_{tcd}(M^2_d - M^2_t - M^2_c);$$
$$M^c X^a_{ij} = m_i R^a_{ij} - L^a_{ij} m_j \quad \text{(for } a \leq j\text{);}$$
$$X^a_{ij} = -i C^{a-g}_{ij} \quad \text{(for } a = g+1, \cdots, N\text{).} \quad \text{(F.18)}$$

Further define real anti-symmetric matrices, P^a, for which the upper left block is A^a, the lower right block T, and the off-diagonal blocks B^a for the upper and $-B^a$ for the lower block. The conditions to cancel bad high-energy behavior can then be written as

$$[P^a, P^b] = D_{abc} P^c; \quad L^i X^a - X^a L^i = i P^i_{ca} X^c; \quad M^{ij}_{ab} = -\{P^i, P^j\}_{g+a, g+b}, \quad \text{(F.19)}$$

where $\{A, B\} = AB + BA$ is the anti-commutator of AB. It is clear from the commutation relation in the first of the above equations that there is a "hidden" representation of the Lie algebra present of dimension $N + g$, the number of scalar Higgs particles plus the number of vector mesons.

If there is one massless vector meson (the photon), it can be chosen as $W_\mu^1 = A_\mu^\gamma$. The only changes to the above can be shown to be

$$D_{1ab} = 0 \text{ unless } M_a = M_b;\ T_{ab}^1 = 0 \text{ unless } M_a = M_b;$$
$$R_{ab}^1 = L_{ab}^1 = \delta_{ab}\lambda^a;\ K^b ij = 0;\ P_{1b}^a = P_{b1}^a = X_{ij}^1 = 0. \quad \text{(F.20)}$$

The preceding equations show that the "hidden" Lie algebra representation has dimension one smaller than above, i.e., $N + g - 1$.

References

1. T.D. Lee, *Particle Physics and Introduction to Field Theory*, Revised and Updated First Edition, Harwood Academic Publishers, Chur, Switzerland (1988).

2. R.N. Cahn and G. Goldhaber, *The Experimental Foundations of Particle Physics*, Cambridge University Press, Cambridge, England (1991).

3. *Mathematical Methods for Physicists, Third Edition*, G. Arfken, Academic Press, Inc., Boston (1985).

4. S.P. Ahlen, *Rev. Mod. Phys.* **52**, 121 (1980).

5. T. Stanev et al., *Phys. Rev* **D25**, 1291 (1982);
L. Landau and I. Pomeranchuk, *Dok. Akad. Nauk SSSR* **92**, 535 (1953); **92**, 735 (1953);
A.B. Migdal, *Phys. Rev.* **103**, 1811 (1956); *Zh. Eksp. Teor. Fiz.*, **32**, 633 (1957) (*Sov. Phys.-JETP* **5**, 527 (1957)).

6. F. Sauli, *Principles of Operation of Multiwire Proportional and Drift Chambers*, CERN yellow report, CERN 77-09, (1977).

7. Thomas Ferbel, ed. *Experimental Techniques in High Energy Physics*, Addison Wesley, Menlo Park, CA (1987).

8. C.W. Fabjan and H.G. Fischer, *Rep. Prog. Phys.* , Addison Wesley, Menlo Park, CA (1987).

9. C.M.G. Lattes et al., *Nature* **160**, 453 (1947).

10. L.R. Davis et al., *Phys. Rev. Lett.* **35**, 1402 (1975).

11. J. McDonough et al., *Phys. Rev.* **38D**, 2121 (1988).

12. G.D. Rochester and C.C. Butler, *Nature* **160**, 855 (1947);
L. Leprince-Ringuet and M. L'heritier, *J. Phys. Radium* **7**, 66,69 (1946).

13. M.M. Block et al., *Phys. Rev. Lett.* **3**, 291 (1959).

14. T.D. Lee et al., *Phys. Rev.* **106**, 340 (1957).

15. J.H. Christenson et al., *Phys. Rev.* **126**, 1202 (1962).

16. T.T. Wu and C.N. Yang, *Phys. Rev. Lett.* **13**, 380 (1964).

17. Particle Data Group, L. Montanet et al., *Phys. Rev.* **D50**, 1177 (1994). These tables, which appear here and in *Phys. Lett.* every few years are the standard and indispensable collection of particle data in the field.

18. B. Winstein, Raporteur talk, *The Vancouver Meeting, Particles and Fields '91, Vancouver, Canada 18-22 August 1991* World Scientific, Singapore, Vol. 1, p 209;
    L.K. Gibbons et al., *Phys. Rev. Lett.* **70**, 1203 (1993).

19. H. Burkhardt et al., *Phys. Lett.* **B206**, 169 (1988).

20. M. Woods et al., *Phys. Rev. Lett.* **60**, 1695 (1988).

21. B. Schwarzschild, *Physics Today* **41**, 17 (October 1988).

22. L. Wolfenstein, *Phys. Lett.* **13**, 562 (1964).

23. J.J. Sakurai, *Advanced Quantum Mechanics*, Addison Wesley, Menlo Park, Calif. (1967).

24. S. Gasiorowicz, *Elementary Particle Physics*, John Wiley and Sons, New York (1967).

25. R.H. Dalitz, *Phil. Mag.* **44**, 1068 (1953); *Phys. Rev.* **94**, 1046 (1954).

26. D.H. Perkins, *Introduction to High Energy Physics*, 3rd edition, Addison Wesley, Menlo Park, Calif. (1987).

27. G. Källen, *Elementary Particle Physics*, Addison Wesley, Menlo Park, Calif. (1964).

28. E.D. Commins and P.H. Bucksbaum, *Weak Interactions of Leptons and Quarks*, Cambridge University Press, Cambridge (1983).

29. Harry J. Lipkin, *Lie Groups for Pedestrians*, North-Holland, Amsterdam (1966).

30. V. Barnes et al., it Phys. Rev. Lett. **12**, 204 (1964).

31. F.E. Close, *An Introduction to Quarks and Partons*, Fourth Printing, Academic Press, New York (1981).

32. R.P. Feynman, M. Kislinger, and F. Ravndahl, *Phys. Rev.* **D3**, 2706 (1971).

33. M.J. Alguard et al., *Phys. Rev. Lett.* **37**, 1261 (1976); SLAC preprint, SLAC-PUB-1790 (1976).

34. G. Hanson et al., *Phys. Rev. Lett.* **35**, 1609 (1975).
    G. Hanson et al., *Phys. Rev.* **26D**, 991 (1982).

35. European Muon Collaboration, V. Papavassiliou, reporting in *Proceedings of the International Europhysics Conference on High Energy Physics, Uppsala, Sweden, June 25 - July 1, 1987* 441 (1987).

36. M. Gell-Mann, *Phys. Rev.* **125**, 1067 (1962).
37. S. Okubo, *Prog. Theor. Phys.* **27**, 946 (1962).
38. G. Karl, *Phys. Rev.*, **D45**, 247 (1992).
39. J.J. Aubert et al., *Phys. Rev. Lett.* **33**, 1404 (1974).
40. J.E. Augustin et al., *Phys. Rev. Lett.* **33**, 1406 (1974).
41. M.L. Perl et al., *Phys. Rev. Lett.* **35**, 1489 (1975);
    M.L. Perl et al., *Phys. Lett.* **39**, 1240 (1977).
42. Beijing Electron Synchrotron Group. Result reported by R. Patterson at *XXVII International Conference on High Energy Physics*, Glasgow, July 21-27, (1994).
43. S.W. Herb et al., *Phys. Rev. Lett.* **35**, 1489 (1975);
    W.R. Innes et al., *Phys. Rev. Lett.* **39**, 1240 (1977).
44. C. Berger et al., *Phys. Lett* **B76**, 243 (1978);
    C.W. Darden et al., *Phys. Lett* **B76**, 246(1978); **B78**, 364 (1978);
    J. Beulein et al.*Phys. Lett* **B78**, 360 (1978).
45. D. Andrews et al., *Phys. Rev. Lett.* **44**, 1108 (1980);
    T. Bohringer et al., *Phys. Rev. Lett.* **44**, 1111 (1980);
    D. Andrews et al., *Phys. Rev. Lett.* **45**, 219 (1980).
46. F. Abe et al., CDF group, *Phys. Rev.*, **50D**, 2966 (1994); *Phys. Rev. Lett.*. **73**, 225 (1994).
47. CDF group, F. Abe et al., *Phys. Rev. Lett.*, **74**, 2626 (1995).
48. D0 group, S. Abachi et al., it Phys. Rev. Lett., **74**, 2632 (1995).
49. LEP Electroweak Working Group, *A Combination of Preliminary LEP Electroweak Results for the 1995 Summer Conferences*, CERN/-LEPEWWG/95-02, (1995).
50. O. Adriani et al., *Phys. Lett.* **B292**, 463 (1992).
51. A. Litke et al., *Phys. Rev. Lett.* **30**, 1189 (1973);
    J.-E. Augustin et al., *Phys. Rev. Lett.* **34**, 764 (1975);
    J. Siegrist et al., *Phys. Rev.* **26D**, 969 (1982).
52. A. Browman et al., *Phys. Rev. Lett.* **33**, 1400 (1974).
53. Taken from F. Halzen and A.D. Martin, *Quarks and Leptons: An Introductory Course in Modern Particle Physics*, John Wiley & Sons, New York (1984).
54. P. Jordan and E. Wigner, *Zeit. für Physik* **47**, 631 (1928).
55. E. Fermi, *Zeit. für Physik* **88**, 161 (1934).
56. T.D. Lee and C.N. Yang, *Phys. Rev.* **104**, 254(1956).
57. C.S. Wu, *Phys. Rev.* **105**, 1413 (1957).

58. Frauenfelder et al., sl Phys. Rev. **106**, 368 (1957).

59. Cavanagh et al., *Phil. Mag.* **21**, 1105 (1957).

60. L.A. Page and M. Heinberg, *Phys. Rev* **106**, 1220 (1957).

61. M.L. Goldhaber et al., *Phys. Rev.* **109**, 1015 (1958).

62. B.W. Ridley, *Nucl. Phys.* **25**, 483 (1961).

63. J.S. Allen et al., *Phys. Rev.* **116**, 134 (1959).

64. V.A. Lyubimov et al., *Phys. Lett.* **B94**, 266 (1980);
    S.D. Boris et al., *Phys. Rev. Lett.* **58**, 2019 (1987).

65. J.D. Bjorken and S. Drell, *Relativistic Quantum Fields*, McGraw-Hill, New York, N.Y., 1965.

66. D.A. Bryman et al., *Phys. Rev.* **D33**, 1211 (1986).

67. Heintze et al., *Phys. Lett.* **B70**, 482 (1977);
    K.S. Heard et al., *Phys. Lett.* **B55**, 324 (1975);
    A.R. Clark et al., *Phys. Rev. Lett.* **29**, 1274 (1972).

68. J.L. Brown et al., *Phys. Rev. Lett.* **8**, 450 (1962);
    G. Jensen et al., *Phys. Rev.* **136B**, 1431 (1964).

69. R.D. Bolton et al., *Phys. Rev.* **D38**, 2077 (1988).

70. G. Danby et al., *Phys. Rev. Lett.* **9**, 36 (1962).

71. M. Schwartz, *Phys. Rev. Lett.* **4**, 306 (1960);
    B. Pontecorvo, *Sov. Phys. Jetp* **10**, 1236 (1960).

72. T. Tomoda et al., *Rept. on Prog. in Phys.* **54**, 53 (1991).

73. S.E. Derenzo, *Phys. Rev.* **181**, 1854 (1969).

74. H. Albrecht et al., *Phys. Lett.* **B246**, 278 (1990).

75. J. Dorenbosch et al., *Zeit. für Phys.* **C41**, 567 (1989); Erratum *Zeit. für Phys.* **C51**, 142 (1991).

76. S.E. Willis et al., *Phy. Rev. Lett.* **44**, 522 (1980).

77. M. Jonker et al., *Phys. Lett.* **B93**, 203 (1980).

78. E.R. Cohen and B.N. Taylor, *Rev. Mod. Phys.* **59**, 1121 (1987);
    R.S. van Dyck, Jr. et al., in *Atomic Physics* **9**, edited by R.S. van Dyck, Jr., and E.N. Fortson (World Scientific, Singapore), 53 (1986).

79. J. Bailey et al., *Nucl. Phys.* **B150**, 1 (1979);
    A. Rich and J.C. Wesley, *Phys. Rev. Lett.* **24**, 1320 (1970);
    H.R. Crane et al., *Phys. Rev.* **121**, 1 (1961);
    H.R. Crane et al., *Phys. Rev.* **94**, 7 (1954).

80. Ya.B. Zeldovich and S.S. Gershtein, *ZhETF* **29**, 698 (1955) [*Sov. Phys. JETP* **2**, 576 (1956)];
    R.P. Feynman and M. Gell-Mann, *Phys. Rev.* **109**, 193 (1958).

81. M.L. Goldberger and S.B. Treiman, *Phys. Rev.* **111**, 354 (1958); L. Wolfenstein, *Nuovo Cimento* **8**, 882 (1958).

82. W.K. McFarlane et al., *Phys. Rev.* **D32**, 547 (1985).

83. M. Gell-Mann, *Phys. Rev.* **111**, 162 (1958).

84. Y.K. Lee et al., *Phys. Rev. Lett.* **10**, 253 (1963).

85. C.S. Wu, *Rev. Mod. Phys.* **36**, 618 (1964).

86. S.L. Adler, *Phys. Rev. Lett.* **14**, 1051 (1965).

87. W.I. Weisberger, *Phys. Rev. Lett.* **14**, 1047 (1965); *Phys. Rev.* **143**, 1302 (1966).

88. B.P. Roe, *Phys. Rev. Lett.* **21**, 1666 (1968); *Phys. Rev. Lett.* **23**, 692 (1969).

89. C.A. Piketty and L. Stodolsky, *Nucl. Phys.* **B15**, 571 (1970).

90. J. Bell et al., *Phys. Rev. Lett.* **40**, 1226 (1978).

91. B.R. Holstein and S.B. Treiman, *Phys. Rev.* **D13**, 3059 (1976);
B.R. Holstein, *Phys. Rev.* **C29**, 623 (1984);
H.J. Lipkin, *Phys. Lett.* **B34**, 202 (1971);
A. Halprin et al., *Phys. Rev.* **D14**, 2343 (1976);
P. Langacker, *Phys. Rev.* **D15**, 2386 (1977);
L.J. Carson et al., *Phys. Rev.* **D33**, 1356 (1986).

92. N. Cabibbo, *Phys. Rev. Lett.* **10**, 531 (1963).

93. C. Baltay et al., *Phys. Rev.* **D4**, 670 (1971).

94. T. Akagi et al., *Phys. Rev. Lett.* **67**, 2618 (1991);
A.P. Heinson, *Phys. Rev.* **D44**, R1 (1991).

95. F.J. Hasert et al., *Phys. Lett.* **46B**, 138 (1973); *Nucl. Phys.* **B73**, 1 (1974).

96. J.G.H. deGroot et al., *Zeit. für Physik* **C1**, 143 (1979).

97. S.L. Glashow et al., *Phys. Rev.* **D2**, 1285 (1970).

98. V. Efrimenko et al., *Phys. Lett.* **B88**, 181 (1979).

99. A. Bean et al., *Phys. Rev.* **D35**, 3533 (1987).

100. H. Albrecht et al., *Phys. Lett.* **B255**, 297 (1991).

101. B.P. Roe, *Proceedings of the Topical Conference on Neutrino Physics at Accelerators, Oxford*, p.285 (1978).

102. C.Y. Prescott et al., *Phys. Lett.* **B77**, 347 (1979); SLAC-PUB-2319, May (1979); *Proceedings of the IX International Symposium on Lepton and Photon Interactions at High Energy, Batavia, IL*, p.271 (1979).

103. M. Kobayashi and K. Maskawa, *Prog. Theor. Phys.* **49**, 282 (1972).

104. K. Berkelman and S. Stone, *Ann. Rev. Nucl. and Part. Sci.* **41**, edited by J.D. Jackson, H.E. Gove, and V. Lüth, Annual Reviews, Palo Alto, Calif., p1 (1991).

105. Reported by V. Lüth and M. Swartz at the XVI International Symposium on Lepton-Photon Interactions, Cornell University, Ithaca, New York (1993).

106. P. Ball et al., *Phys. Rev.* **D48**, 2110 (1993).

107. W. Louis et al., *Phys. Rev. Lett.* **56**, 1027 (1986).

108. J.C. Anjos et al., *Phys. Rev. Lett.* **60**, 1239 (1988).

109. M. Artuso et al., *Phys. Rev. Lett.* **62**, 2233 (1989).

110. H. Albrecht et al., *Phys. Lett.* **B192**, 245 (1987).

111. A.I. Vainshtein et al., *Zh. Eksp. Teor. Fiz.* **22**, 123 (1975) (*JETP Lett.* **22**, 55 (1975)).

112. R. Ammar et al., *Phys. Rev. Lett.* **71**, 674 (1993).

113. G. Arnison et al., *Phys. Lett.* **B122**, 103 (1983); P. Bagnaia et al., *Phys. Lett.* **B129**, 130 (1983).

114. R. Davis, in *Proceedings of the 21st International Cosmic Ray Conference, Adelaide, Australia, 1989*, edited by R.J. Protheroe, Graphics Services, Northfield, South Australia, p143 (1990).

115. K. Hirata et al., *Phys. Rev. Lett.* **65**, 1297 (1990); *Phys. Rev.* **D44**, 2241 (1991).

116. GALLEX collaboration, P. Anselmann et al., *Phys. Lett.* **B285**, 376 (1992); *Phys. Lett.* **285**, 390 (1992).

117. T. Bowlier, presented at *Neutrino '92, Granada, Spain* (unpublished); SAGE collaboration, A.I. Abuzov et al., *Phys. Rev. Lett.* **67**, 3332 (1991).

118. S.P. Mikheyev and A. Yu Smirnov, *Yad. Fiz.* **42**, 1411 (1985) [*Sov. J. Nucl. Phys.* **42**, 9913 (1985).

119. L. Wolfenstein, *Phys. Rev.* **D17**, 2369 (1978); *Phys. Rev.* **D20**, 2634 (1979).

120. T.K. Kuo and J. Pantaleone, *Rev. Mod. Phys.* **61**, 937 (1988); S.P. Mikheyev and A. Yu Smirnov, *Usp. Fiz. Nauk.* **153**, 3 (1987) [*Sov. Phys. Usp.* **30**, 759 (1987)].

121. C.H. Llewellyn-Smith, *Phys. Rept.* **3**, 264 (1972).

122. M.N. Rosenbluth, *Phys. Rev.* **79**, 615 (1950).

123. E.B. Dally et al., *Phys. Rev. Lett.* **39**, 1176 (1977).

124. E.B. Dally et al., *Phys. Rev. Lett.* **45**, 232 (1980).

125. J.T. Friedman and H.W. Kendall, *Ann. Rev. Nucl. Sci.* **22**, 203 (1972).

126. V.V. Ammosov et al., *Nucl. Phys.* **B203**, 1 (1982).

127. W.G. Scott, *Proceedings Topical Conference on Neutrino Physics at Accelerators, Oxford July 4-8, 1978*, edited by A.G. Michette and P.B. Renton, p12 (1978).

128. W.C. Leung et al., *Phys. Lett.* **B317**, 655 (1993);
P.Z. Quintas et al., *Phys. Rev. Lett* **71**, 1307 (1993).

129. L. Hand, *Phys. Rev.* **129**, 1834 (1963).

130. F.J. Gilman, *Phys. Rept.* **46**, 95 (1972).

131. D.D. Reeder, *Proceedings Topical Conference on Neutrino Physics at Accelerators*, Oxford July 4-8, 1978, edited by A.G. Michette and P.B. Renton, p1.

132. D.H. Perkins, *Contemp. Phys.* **16**, 173 (1975).

133. J.P. Berge et al., *Phys. Lett.* **B81**, 89 (1979); **B106**, 151 (1981).

134. B. Baru et al., *Phys. Lett.* **B163**, 282 (1985);
A.C. Benevuti et al.*Phys. Lett.* **B189**, 483 (1987);
J. Ashman et al., *Phys. Lett.* **B202**, 603 (1988);
R.G. Arnold et al., *Phys. Rev. Lett.* **52**, 727 (1984).

135. F. Halzen and A.D. Martin, *Quarks and Leptons, An Introductory Course in Modern Particle Physics*, John Wiley and Sons, New York (1984).

136. Reported by D. Gross at the XVI International Symposium on Lepton-Photon Interactions, Cornell University, Ithaca, New York (1993).

137. Quoted by F. Close at the Singapore 1990 HEP Conference.

138. M.B. Gay Ducati et al., *Phys. Rev.* **D48**, 2324 (1993).

139. C.H. Llewellyn-Smith, *Oxford Preprint* **54/79** (1979).

140. M.Y. Han and Y. Nambu, *Phys. Rev.* **139**, B1005 (1965).

141. G.S. Abrams et al., SLAC-PUB 2421, Oct., T/E (1979).

142. M.S. Chanowitz, *Phys. Rev. Lett.* **44**, 59 (1980);
Lawrence Berkeley Laboratory preprint LBL-9639 (1979);
Phys. Rev. Lett. **35**, 977 (1975).

143. L.B. Okun et al., ITEP, Moscow preprint ITEP-79 (1979).

144. B. Iijima and R.L. Jaffe, *Phys. Rev.* **D24**, 177 (1981).

145. D.B. Lichtenberg, *Oxford Theory Preprint*, **26/80** (1980).

146. S. Bethke et al., *Phys. Lett.* **B213**, 235 (1988);
 W. Bartel et al., *Zeit. für Phys.* **C33**, 23 (1986).

147. S. Bethke, *Nucl. Phys. (Proc. Suppl.)* **B29**, 198 (1995).

148. S. Bethke and J. Pilcher, *Ann. Rev. Nucl. Part. Sci. 1992* **42**, 251 (1992).

149. B. Adeva et al., *Phys. Lett.* **B248**, 464 (1990).

150. B. Adeva et al., *Zeit. für Phys.* **C55**, 39 (1992)

151. B. Adeva et al., *Phys. Lett.* **B248**, 227 (1990).

152. ALEPH collaboration, D. Decamp et al., *Phys. Lett.* **B284**, 151 (1992).

153. A. Chodos et al., *Phys. Rev.* **D9**, (1974).

154. W.A. Bardeen et al., *Phys. Rev.* **D11**, 1094 (1975).

155. P. Langacker, *Proceedings of the XXIV International Conference on High Energy Physics, Munich* (1988).

156. C.Y. Prescott et al., *Phys. Lett.* **84B**, 524 (1979).

157. H. Fritzsch and P. Minkowski, *Flavordynamics of Quarks and Leptons, Phys. Rept.* (1980).

158. C.N. Yang and F. Mills, *Phys. Rev.* **96**, 191 (1954);
 O. Klein, in *New Theories in Physics*, International Institute of Co-operation, League of Nations, pp. 77-93 (1938).

159. P. Renton, *Electroweak Interactions, An Introduction to the Physics of Quarks and Leptons*, Cambridge University Press, Cambridge, England (1990).

160. T.W.B. Kibble, *Phys. Rev.* **155**, 1554 (1967).

161. R. Gastmans, *Weak and Electromagnetic Interactions at High Energies*, ed. M. Levy et al., Plenum, New York and London (1975).

162. J.M. Irvine, U. Manchester preprint M/C TH 81/07 (1981).

163. A.M. Hillas and T.E. Cranshaw, *Nature* **184**, 892 (1959);
 reported by H. Primakoff, *Proceedings of the Neutrinos 1978 Conference, Purdue, Indiana*, p.995 (1978).

164. P. Langacker, *Phys. Rept* **72C**, 185 (1981).

165. S. Seidel, et al., *Phys. Rev. Lett.* **61**, 2522 (1988).

166. K.S. Hirata et al., *Phys. Lett.* **B220**, 308 (1989).

167. A.A. Penzias and R.W. Wilson, *Astrophysics J.* **142**, 149 (1965).

168. O. Adriana et al., *Phys. Rept.* **236**, 1 (1993).

169. H. Albrecht et al., *Phys. Lett.* **B192**, 245 (1987).

170. M. Artuso et al., *Phys. Rev. Lett.* **62**, 2233 (1987).

171. B. Adeva et al., *Phys. Lett.* **B252**, 703 (1990).

172. LEP Electroweak Heavy Flavours Working Group and the SLD Heavy Flavour Group, *Combined LEP and SLD Electroweak Heavy Flavour Results for Summer 1995 Conferences Internal Note*, LEPHF/95-02 (1995).

173. ALEPH group, D. Buskulic et al., *Phys. Lett.* **B322**, 441 (1994).

174. O. Adriani et al., *Phys. Lett.* **B301**, 136 (1993); *Phys. Lett.* **B292**, 472 (1992).

175. P.D. Acton et al., *Zeit. für Phys.* **C54**, 193 (1992);
P. Abreu et al., *Zeit. für Phys.* **C53**, 555 (1992);
M.Z. Akrawy et al., *Phys. Lett.* **B246**, 285 (1990).

176. B. Adeva et al., *Phys. Lett.* **B288**, 404 (1992).

177. J-F. Grivas, to appear in *Procedings of the International Europhysics Conference on High Energy Physics*, (1995).

178. K. Riles, presented at Workshop on Physics at Current Accelerators and the Supercollider, Argonne National Laboratory (June 1993).

179. I. Ia. Pomeranchuk, *Soviet Phys. JETP* **34** (7), 499 (1958).

180. S. Weinberg, *Phys. Rev.* **124**, 2049 (1961).

181. M. Froissart, *Phys. Rev.* **123**, 1053 (1961).

182. T. Regge, *Nuovo Cimento* **14**, 951 (1959).

183. G.F. Chew et al., *Phys. Rev.* **126**, 1202 (1962);
S.C. Frautschi et al., *Phys. Rev.* **126**, 2204 (1962);
R. Blankenbecler and M.L. Goldberger, *Phys. Rev.* **126**, 766 (1962);
V.N. Gribov, *Soviet Phys. JETP* **14**, 478, 1395 (1962).

184. M.M. Block et al., *High Energy Behavior of σ_{tot}, ρ, and B — Asymptotic Amplitude Analysis and QCD-Inspired Analysis*, University of Wisconsin preprint MAD/PH/767 (June 1993).

185. L. Lipatov, *Proceedings of the Vth Blois Workshop on Elastic and Diffractive Scattering* (June 1993).

186. F. Halzen, *Proceedings of the Vth Blois Workshop on Elastic and Diffractive Scattering* (June 1993).

187. C. Augier et al., *A Precise Measurement of the Real Part of the Elastic Scattering Amplitude at the $S\bar{p}pS$*, CERN/PPE93-115 (1993).

188. C. Augier et al., *Predictions on the total cross section and real part at LHC and SSC*, CERN-PPE/93-121 (1993).

189. M. Albrow, *Proceedings of the Vth Blois Workshop on Elastic and Diffractive Scattering* (June 1993).

190. M. Jacob, *Hadron Physics in the 200-2000 GeV Energy Range*, Lawrence Berkeley Laboratory, UCR-15-77 (1977).

191. F. Abe et al., *Phys. Rev.* **D41**, 2330 (1990).

192. N. Isgur and M.B. Wise, *Phys. Rev. Lett.* **66**, 1130 (1991); *Phys. Lett.* **B237**, 527 (1990); *Phys. Lett.* **B232**, 113 (1989);
N. Isgur et al., *Phys. Rev.* **D39**, 799 (1989);
B. Grinstein et al., *Phys. Rev. Lett.* **56**, 298 (1986).

193. M. Neubert, *Phys. Repts.* **245**, 259 (1994).

194. B. Grinstein, *Ann. Rev. Nucl. Part. Sci.* **42**, 101 (1992).

195. M. Wirbel, B. Stech, and M. Bauer, *Zeit. für. Phys.* **C29**, 637 (1985).

196. P.A.M. Dirac, *Proc. Roy. Soc.* **A133**, 60 (1931).

197. E.J. Weinberg, *Monopole '83*, edited by J.L. Stone, Plenum Press, New York and London, p1. (1983).

198. Y. Aharonov and D. Bohm, *Phys. Rev.* **115**, 485 (1959).

199. G. T'Hooft, *Nucl. Phys.* **B79**, 276 (1974).

200. A. Polyakov, *JETP Lett.* **20**, 194 (1976).

201. V.A. Rubakov, *JETP Lett.* **33**, 645 (1981), *Inst. Nucl. Res. Report No. P0211*, Moscow (1981), *Nucl. Phys.* **203**, 311 (1982);
V.A. Rubakov and M.S. Serebryakov, *Nucl. Phys.* **B218**, 240 (1983).

202. C.G. Callan, *Phys. Rev.* **D25**, 2141 (1982); *Phys. Rev.* **D26**, 2058 (1982), *Nucl. Phys.* **B212**, 365 (1983).

203. M.S. Turner et al., *Phys. Rev.* **D26**, 1296 (1982).

204. E.N. Parker, *Ap. J.* **160**, 383 (1970).

205. F.C. Adams et al., *Phys. Rev. Lett.* **70**, 2511 (1993).

206. E.W. Kolb and M.S. Turner, *The Early Universe*, Addison-Wesley, Redwood City, California (1990).

207. E. Salpeter et al., *Phys. Rev. Lett.* **49**, 1114 (1982);
J. Arons and R. Blandford, *Phys. Rev. Lett.* **50**, 544 (1983).

208. K. Freese, *Ap. J.* **286**, 216 (1984).

209. K. Freese et al., *Phys. Rev. Lett.* **51**, 1625 (1983);
E.W. Kolb et al., *Phys. Rev. Lett.* **49**, 1373 (1982);
S. Dimopoulos et al., *Phys. Lett.* **B119**, 320 (1982).

210. J.A. Friedman et al., *Ap. J.* **335**, 884 (1988).

211. E.N. Parker, *Astrophys. J.* **160**, 383 (1970).

212. P.J.E. Peebles, *Principles of Physical Cosmology*, Princeton University Press, Princeton, New Jersey, ch 18 (1993).

213. V. Tremble, *Annual Reviews of Astronomy and Astrophysics* **25**, edited by G. Burbidge, D. Layzer, and J.G. Phillips, Annual Review, Palo Alto, Calif. p425 (1987).

214. M. Punch et al., *Nature* **358**, 477 (1992).

215. U. Amaldi et al., *Phys. Lett.* **B260**, 447 (1991).

216. C.B. Dover, *Proceedings, 4th Conference on Intersections Between Particle and Nuclear Physics*, Tuscon, p141 (1991);
J. Madson and P. Haensl, editors, *Proceedings, Strange Quark Matter in Physics and Astrophysics, International Workshop, Aarhus, Denmark, Nucl. Phys. B, Proc. Suppl.* **24B** (1991).

217. E. Witten, *Phys. Rev.* **D30**, 272 (1984).

218. E. Farhi and R.L. Jaffe, *Phys. Rev.* **D30**, 2379 (1984).

219. E. Fischbach and C. Talmadge, *Nature* **356**, 207 (1992).

220. K.F. Smith et al., *Phys. Lett.* **B234**, 191 (1990).

221. J.P. Jacobs et al., *Phys. Rev. Lett.* **71**, 3782 (1993).

222. G. Aardoma et al., *Phys. Lett.* **B194**, 321 (1987);
G.T. Earle et al., *Proceedings of the Second Conference on the Intersections Between Particle Physics*, edited by D.F. Geeseman, American Institute of Physics, Conference Proceedings **150**, New York, p1094 (1986).

223. J.H. Taylor and J.M. Cordes, *Ap. J.* (in press 1993).

224. K. Greisen, *Phys. Rev. Lett.* **16**, 748 (1966).

225. C.A. Meegan et al., *Nature* **355**, 143 (1992);
E. Fenimore et al., *Nature* **357**, 140 (1992).

226. K. Nishijima, *Fundamental Particles*, W.A. Benjamin (1963).

227. E.P. Wigner, *Group Theory*, Academic Press, New York (1959).

228. E.U. Condon and G.H. Shortley, *The Theory of Atomic Spectra*, Cambridge University Press, New York (1953).

229. Ta-Pei Cheng and Ling-Fong Li, *Gauge Theory of Elementary Particle Physics*, Clarendon Press, Oxford (1984, reprinted with corrections 1988).

230. C.H. Llewellyn Smith, *Phys. Lett.* **B46**, 233 (1973).

231. J.M. Cornwall et al., *Phys. Rev.* **D10**, 1145 (1974).

Index

A
absorption cross section 99
absorption factor 97
accelerators
  circular 38
  colliding-beam 38
  fixed target 38
  linear 38
active galactic nuclei 351
Aharonov–Bohm experiment 342, 343
Altarelli–Parisi equation 243
analyticity 314
angular momentum 60, 61, 365–369
anti-commutation relations 152
anti-screening 237
asymptotic energies 319
asymptotic freedom 228, 229, 234–242, 325
asymptotic region 325
axions 352

B
B-decays 195, 306
$B - \overline{B}$ mixing 186, 306, 307
b quarks 136, 305, 308
background integral 317
bag models 239
bare electric charge 16
bare mass 16

baryons 4, 5
  baryon number 4, 73
  baryon number of universe 288
  decuplet 113–115, 125
  magnetic moments 128
  octet 113–117, 123
  singlet 113
beta (β) decay 155
  double beta decay 169, 353
  Fermi terms 157
  Gamow–Teller terms 157
Bhabha scattering 302
bilinear operator 163
bilinear terms 153
Bjorken x and y 212
black disk 323, 326
black holes, primordial 353
Bose–Einstein symmetry 75
bosons 6
Breit–Wigner resonances 100-103
bremsstrahlung 27–31
Brookhaven National Laboratory 39
brown muck 328–333, 339
bubble chamber 49

C
c quark (*see also* J/Ψ *particle*)
  spin 134–136
Cabbibo angle 178, 179
Cabbibo hypothesis 178–179

Cabbibo–Kobayashi–Maskawa (CKM) matrix 184–186, 308, 339, 341
Callen–Gross relation 219, 222
calorimeters 51–54
  electromagnetic 52
  hadronic 53
  sandwich 53
  spaghetti 53
cannibal quarks 225
Casimir operators 110
CERN (Centre Européenne pour la Recherche Nucléaire) 41
charge conjugation (C) 73–76, 82–85, 120
charge conservation 70
charge coupled devices 51
charge equality, electron and proton 285
charge independence 80
charge quantization 285, 343
charge radius of kaon 204
charge radius of pion 204
charm particle threshold 212
charmed particles 181
charmonium 130, 134
Cherenkov counter 50, 51
  differential 50
  ring imaging 51
Cherenkov radiation 33–35
chiral representation 162
chirality operator 162
classical Hamiltonian 358
classical Lagrangian 358
Clebsch–Gordan coefficients 67, 365–367
CLEO 40
CLIC 43
closed loop Feynman diagrams 14
Coleman–Glashow relation 127
color 5, 114, 139–142

color magnetism 329
commutator 360
completeness relations 149
Compton scattering 33
confinement 3
conservation laws 57, 58
  angular momentum 60
  charge 59
  current 58, 59, 201, 205, 333, 363
  energy 61
  leptons 3
  momentum 60
  quarks 3
  vector current (CVC) 174–177, 249
contraction 7
contravariant vector 7
conversion length 33
Cornell Electron Synchrotron 39
cosmic microwave background 288
cosmic ray events 326
cosmic ray photons 356
cosmic strings 346, 353
cosmological constant 261, 281
Coulomb gauge 271, 372
coupling constant evolution 282, 283
coupling constants, variation with time 280
covariant derivative 72, 256
covariant vector 7
CP 76, 155
  violation 90–92, 187, 353, 354
  violation, B decay 186, 357
CPT 76, 90, 190, 288
creation operator 151, 152
critical energy 31, 52
cross section 3, 105, 376
crossed channel 318
crossing symmetry 314

398 Index

current–current interaction
(*see weak interactions, current-current*)

D

D mesons 130
Dalitz plot 107, 108, 154
dark matter 350, 355
  cold 351
  hot 351
dead time 45, 49
decay constants, pseudoscalar 337
decay constants, vector 337
decay width 105
decays of heavy baryons 337
$\Delta I = 1/2$ rule 179, 180, 187
$\Delta S = \Delta Q$ rule 179
delta ray 24
density effect 22
destruction operator 151, 152
DESY (Deutsche Elektronen Synchrotron) 43
detailed balance 63, 64, 66
deuteron 66–68, 78, 79
diffractive collisions 321
dimensions
  Dirac spinors 149
  matrix elements 106, 171
  phase space 106
  transition rates 106
Dirac quantization condition 341
Dirac string 342, 343
Dirac theory 147
Dirac theory free particle wavefunction 148
direct pair production 33
Drell–Yan scattering 123
drift chamber 46, 47
drift velocity 45, 46
dyon 341

E

effective mass 7
eikonal procedure 323
elastic scattering 98, 196, 323
  electron–muon 197, 198
  electron–proton 200–202
electric dipole moments 352, 355
electromagnetic current 58, 59, 246–248
electromagnetic interaction 3, 76, 164, 197, 200, 204, 206, 242
electromagnetic shower 33
electromagnetic-weak force 247
EMC effect 224
electron g-factor 173
energy-loss 21, 23, 25
eta (η) particle 76
exchange potentials 318
exotic meson states 120
extended technicolor 261

F

$(F_2)_{em}/(F_2)_{wkCC}$ 222
Fermi–Dirac statistics (*see Pauli principle*)
Fermilab 38
fermions 5
Feynman diagrams 11, 16, 270, 270, 372
Feynman rules 373
Feynman x and y 319
fifth force 354
fine-tuning 281, 289
flash ADC's 48
flash chamber 48
flavor 2, 110, 237
fly's eye experiment 356
form factors 203–208, 225
  geometrical interpretation 203
  proton 203

weak interaction 332, 337
Yukawa 203
forward inelastic $e--p$ scattering 117
four-jet events 238
four-vector 8
free quarks 143
Froissart bound 315, 316
Furry's Theorem 74

G

g_V, g_A, precision values 302, 305
G-parity 84, 85
gamma matrices 147, 162
gamma ray bursts 356
gauge theory 256–258, 262, 270, 280
  gauge field 256, 261
  gauge heirarchy problem 289
  gauge invariance 71–73, 260
  gauge particles 283
  global gauge invariance 71–73
  local gauge invariance 71–73
  non-abelian 257
  QED 256, 257
  spontaneously broken 258–265
Geiger–Müller counters 45, 48
Gell-Mann matrices 110–112, 371, 372
Gell-Mann–Nishijima formula 86
Gell-Mann–Okubo (GMO) mass relation 126–128
general relativity 355
generalized coordinates 358
generalized momenta 358
generations 2, 110, 137–139, 141, 299, 302
generators 110
ghost particles 207, 270
Glashow–Illiopolis–Maiani (GIM) mechanism 181–183, 279

global gauge invariance (see gauge theory)
glueballs 229
gluons 3, 5, 119, 224, 228–245
  and asymptotic freedom 234, 242
  color 229, 233
  coupling 228
  evidence for 228
  mass 229
  spin 228
gluon ladder diagrams 323
Goldberger–Trieman relation 174
Goldstone boson 260, 261
Goldstone theorem 260
grand unification 281
Grand Unified Theory (GUT) 281
gravitational lensing 353
gravitational radiation 355
gravitons 3
gravity 3, 269
great desert 285

H

Han–Nambu model 233, 234
Heaviside–Lorentz units 9, 14
heavy quark effective field theory 328–340
hedgehog solution for monopole 345
Heisenberg picture 57
Heisenberg representation 63, 359, 362
helicity 150
HERA 43
Hermitian operator 57
Higgs couplings 377
Higgs mechanism 255–280
Higgs particle 255, 261, 262, 264, 270, 277, 279, 280, 285, 311, 352, 357, 379–384
  charged 261

composite 256, 261
  mass 265
high-energy behavior
  bad 268–280
  restrictions placed on Lagrangian by 381
hypercharge 83, 86, 87, 246
hyperon nuclei 86, 353
hyperons 86, 178

I
IHEP at Serpukhov, Russia 43
impact parameter 19–23
inelastic scattering 196
  charged current (*see charged current, weak inelastic scattering*)
  electron–proton 204–206
  forward $e-\,-p$ scattering (*see forward inelastic $e-\,-p$ scattering*)
  neutral current (*see neutral currents, inelastic scattering*)
infinite momentum frame 209, 242
infrared slavery 228
interaction representation 359–362
intermediate bosons (*see vector bosons, Z bosons, and W bosons*)
intrinsic parity 62
invariance 57
invariant cross section 105
ionization 21
ionization chamber 45
isotopic spin 77
  and electromagnetism 83
  multinucleons 78
  nucleon–anti-nucleon pairs 82, 83
  pion nucleon system 80, 81
  pions 79, 80
  single nucleons 77, 78

J
J/Ψ particle 129, 130
  spin, parity 132
  width 132
Jacobi identity 382
JADE algorithm 238
jets 119, 239

K
K decays 87–92, 168, 169, 178
K°
  K°–$\overline{K}^\circ$ oscillations 89–92
  mass difference 89, 180
  regeneration 90
K capture 153
KEK at Tsukubo, Japan 43
Klein–Gordon equation 147
KLN theorem 268
knock-on electron 24
Koba–Nielsen–Oleson (KNO) scaling 327
Kobayashi–Maskawa matrix (*see Cabibbo–Kobayashi–Maskawa matrix*)
Kurie Plots 159–161

L
Lambda (Λ) hyperon decay 180
Landau gauge 271
Landau tail 25
large hadron collider (LHC) 43
LEAR 41
LEP 41
leptons 3
  doublets 246
  form factors 310
  mass 264

new heavy 352
lepton number 73
  additive 172
  conservation 354
  multiplicative 172
leptoquarks 282, 284
Lie algebra 279, 382–384
lifetime, decaying particle 375
linear potentials 319
local field theory 76
local gauge invariance (see gauge invariance)
logarithmic divergences 16
Lorentz gauge 71
Lorentz transformation 8
low-Q^2 processes 314
lowering operator 366
LPM effect 30
luminosity 55

M
M matrix 104, 269
MACHO's 353
magnetic moments 10
  baryon (see baryon, magnetic moment)
  Bohr magneton 10
  muon (see muon, magnetic moment)
  nuclear magneton 10
  proton anomalous moment (see proton anomalous magnetic moment)
Majorana theory 62, 190, 353
Mandelstam variables 10, 11, 13
mass divergence 268
mass relations 125–127
  electromagnetic dependence 127
  p–n mass difference 127
  spin dependence 127

matrix elements 163
  calculation of 372
Maxwell gauge 71
Maxwell's equations 341
meson nonets 120
mesonic atom 67
mesonic x-rays 67
metric tensor 7
Michel ρ parameter 172
Mikheyev–Smirnov–Wolfenstein (MSW) effect 191
minimal electromagnetic interaction 257
Molière radius 52
monopoles 341–349, 355
  color 344
  cosmic ray flux 347
  Dirac 341
  extended gauge theories 343
  galactic magnetic fields 347
  mass 345
  Parker bound 347, 349
  proton decay catalysis 346
  t'Hooft Polyakov 345
  vector potential 342
Möller flux factor 105
multi-wire proportional chamber 46
multiple scattering 25, 26
multiplicity distribution 327
muon
  decay 171–174
  g-factor 173
  magnetic moment 173
  μ–e universality 169, 171
muon pair production in e^+e^- annihilation 199, 200

N
Nachtmann–Reiter angle 238
natural units 9

neutral currents, weak 180, 181, 246, 248, 252
  charm-changing 183
  inelastic scattering 252
  kaon mixing 87, 92
  μ–e universality 183
  neutrino interactions 251, 252
  strangeness-changing 183
neutrinos
  charm production by 223
  electron scattering by 251, 272
  left and right handed 157
  mass limits 159–161, 354
  oscillations (*see also Mikheyev-Smirnov-Wolfenstein (MSW) effect*) 189, 191, 354
  several 169, 170
  total cross section 219
  two component 161
neutron electric dipole moment 77
NLC (Next Linear Collider) 39
Noether's theorem 58, 363, 364
non-leptonic weak interactions (*see weak interactions, non-leptonic*)
nuclear β-decay 153

O

octet 228
odderons 323
optical theorem 99

P

pair production 32, 33
parity (P) 61, 62, 76, 77, 150
  of anti-fermion 150
  pion (*see pions, parity*)
  violation 76, 77, 153–155, 208
  violation in K decays 154

partial conservation of the axial vector current (PCAC) 174, 177, 249
partial wave 99
particle accelerators 37–44
particle detectors 44–54
particle–anti-particle pairs 75
particle–anti-particle oscillation 89, 90, 186
partons 209, 225
Pauli principle 68, 75, 79, 152
  anti-commutation 152
  generalized 79
Pauli spin matrices 77, 148
penguin diagrams; applications to $\Delta I = 1/2$ and to CP-violation 187
PEP 39
peripheral collisions 322
perturbation theory 11, 359, 362
phase shifts 91, 97
phase space 103, 106, 107
  three body 107
  two body 106
photoelectric absorption 33
photomultiplier tube 49
photon charge conjugation 74
photons (γ) 3
  mass 73, 257
pions (π) 64
  charge conjugation 74
  charged π decay 165–167
  G-parity 84
  intrinsic parity 67–70
  π^0 decay 142
  spin 65–67
pion factory 40
Planck mass 283
polarization vector 272, 373
polarized $e--d$ scattering 250
Pomeranchuk theorem 315

Pomeranchuk trajectory (Pomeron) 318
positronium 70, 75
preons 351, 352
projection operator, polarization 150
projection operator, positive frequency 149
propagator 12–14, 19, 271, 373
 for a massive vector particle 270
 for various spins 271, 373
proportional counter 46
proportional wire chamber 45
proton anomalous magnetic moment 202
proton decay 286
proton wavefunction 116
pseudo-rapidity 319
pseudoscalar 155
pseudoscalar meson nonet 120
pseudovector 155

Q

Q value 88, 89, 160
quadrupoles 55
quantum chromodynamics 228, 243
quarks 2–6, 110–146, 208, 225, 242, 329–341
 baryon states 113, 120
 charge 142, 233, 234
 doublets 246
 flow diagrams 122
 form factor 310
 fraction of proton momentum carried by 224
 heavy 2, 130, 328–340, 302, 305–309
 mass 2, 264
 meson states 120
 proton spin carried by 119, 120, 225
 quark model 110–146, 329–341
 sea 119, 209, 222, 223, 225
 spin 119, 222
 strange sea 185, 211
 valence 4, 119, 223, 224
quark–gluon plasma 353, 355
quartic terms in Lagrangian 258
quasars and AGN's 351
quencher gas 45

R

R value in $e+e-$ collisions 140
radiation length 26, 30
radiative corrections 302
raising operator 366
rank 110
rapidity 8, 319
rapidity plateau 321, 327
reaction cross section 99
Regge theory 315–319
 Regge poles 317
 Regge recurrences 319
 Regge trajectories 318
relativistic phase space 106, 107, 375
relativity 6–9
renormalizable field theory 261
renormalization terms 15
resistive plate chamber 48
resonances 81, 100–103, 108
RHIC 39
right-left symmetric theory 351
rishons 351
rotation invariance 60
rotation operator 61
Rutherford Scattering 19, 25

S

S matrix 103, 105, 362

s–channel process 13, 314
Sargeant's rule 160
scalar product 7
scale invariance 211
scattering amplitude 97
scattering angle 21
Schrödinger representation 359
scintillation counter 49
screening 234, 237
sea quarks 222, 223
second class currents 177
second quantization 151
semi-simple group 279
shrinkage and anti-shrinkage 319
silicon strip detectors 51
simple group 285
SLC 39
$SO(10)$ 289
solar neutrinos 350, 354, 355
Sommerfeld–Watson transform 316
space inversion 61
spacelike momentum transfer 14
spark chamber 45
SPEAR 39
sphalerons 288
spinors 14, 148, 372
spontaneous symmetry breaking 255, 258–265, 279, 280
SPS 41
Standard Model 188, 246–281, 293, 343, 345, 351, 357, 381
  Lagrangian for 265, 266
standard solar models 350
Stanford Linear Accelerator Center (SLAC) 39
sterile neutrino 191
straggling 24, 31
strangeness 85, 92
strangeness-changing weak decays 178, 179
straws 47
streamer chamber 48
string theory 269
strong interactions 3, 76, 228, 242, 243
  Lagrangian for 266
structure constants of a group 112, 258, 266, 277, 370
structure functions 211
  evolution 242, 243
$SU(2)$ 77, 365
$SU(2) \otimes U(1)$ 246
$SU(3)$ 110–114, 120, 122, 229–233, 370
$SU(5)$ 282, 346
$SU(6)$ 114, 128
$SU(N)$ 283
sum rules
  Adler 225
  Gottfried 225
  Gross – Llewellyn-Smith 224
  proton spin by quarks 225
Superconducting Supercollider 40
supergravity theory 290, 352
superstring theory 291
supersymmetry (SUSY) 290, 352
  low-energy 289
  particles 352
superweak force 3, 92, 353
symmetries 57, 58
symmetry of Lagrangian 363
symmetry operations, invariance under 76, 77
synchrotron radiation 54

T

T matrix 104
t–channel process 13, 314
t quark 137, 206, 302
  mass 136, 137, 298, 302
τ particle 132

decay 172
  lepton decay branching ratios 142
technicolor 261, 352
theory of everything 281
three jet events 238
time expansion chamber (TEC) 48
time invariance 61
time projection chamber 47
time reversal (T) 63, 76
timelike momentum transfer 14
total cross section 99, 123, 222, 323
traces 163, 164
  charged weak interaction 375
  electromagnetic interaction 375
  identities 275, 375
transition probability 362, 375
transition radiation 25
transition radiation detector 51
transition rate 104, 105
translation invariance 60
translation operator 60
transverse mass 320
tree level calculation 207, 269, 279
triangle anomalies 142, 252
TRISTAN 43

U

u and v in currents 154, 374
U-spin 112, 126
unification-energy 242, 283, 285
unitarity bound 261, 269
  tree unitarity 279
unitary gauge 207, 270, 372
unitary operators 57, 63
UNK 43
upsilon (Υ) particle 134

V

V−A (*see weak interaction, V−A*)
V-spin 112
vacuum polarization 15, 234
vacuum state 258, 261, 263, 345
  chiral symmetric 326
vector bosons 188, 189 (*see also Z and W bosons*)
  massive 261
  massless 258, 261
vector meson nonet 122
Veltman parameter 294
vertex functions 373, 374
vertex renormalization 15

W

W bosons (*see also vector bosons*) 3
  mass 249, 264
  W–W interactions 311
  W–W production from ν–$\bar{\nu}$ 274–277
$WWWW$ vertex 379
Ward–Takahashi identities 201
wave function renormalization 15
wave optics 97, 103
weak hypercharge 246
weak interaction 3, 76, 147, 164, 192, 225, 242
  charged current 174, 177, 206, 208, 225, 247, 248
  charged current neutrino interactions 206–208, 211, 216–224
  current–current 170, 171, 174, 177
  non-leptonic 179
  neutral current (*see neutral current*)
  neutral current neutrino interactions 252

V−A 159, 172, 177, 178, 183
weak isospin 246, 281
weak magnetism 175, 176
Weinberg angle 248–252, 285, 286, 293, 298, 308
Weinberg model of Higgs boson 263
Weinberg–Salam Model (*see Standard Model*)
Weizsäcker–Williams method 27–30
Weyl representation 162

X
X, Y particles 284

Y
Yang–Mills theory 277, 278, 382

Yukawa distribution 203

Z
Z bosons 3
  mass 250, 264
Z decays 292, 312
  asymmetries 304
  Born approximation 294
  exotic 311
  into b quarks 305, 306
  into u, d 310
  invisible modes 299
  non-photonic corrections 295
  radiative 310
  radiative corrections 295
  studies 293–313
Zweig's rule 122, 124